MACRO- TO MICROSCALE
HEAT TRANSFER

Series in Chemical and Mechanical Engineering

G. F. Hewitt and C. L. Tien, *Editors*

Carey, Liquid–Vapor Phase-Change Phenomena: An Introduction to the Thermophysics
  of Vaporization and Condensation Processes in Heat Transfer Equipment
Diwekar, Batch Distillation: Simulation, Optimal Design and Control
Tzou, Macro- to Microscale Heat Transfer: The Lagging Behavior

FORTHCOMING TITLES

Tong and Tang, Boiling Heat Transfer and Two-Phase Flow, Second Edition

MACRO- TO MICROSCALE HEAT TRANSFER
The Lagging Behavior

D. Y. Tzou

Department of Mechanical and Aerospace Engineering
University of Missouri—Columbia

USA	Publishing Office:	Taylor & Francis
		1101 Vermont Avenue, N.W., Suite 200
		Washington, DC 20005-3521
		Tel: (202) 289-2174
		Fax: (202) 289-3665
	Distribution Center:	Taylor & Francis
		1900 Frost Road, Suite 101
		Bristol, PA 19007-1598
		Tel: (215) 785-5800
		Fax: (215) 785-5515
UK		Taylor & Francis Ltd
		1 Gunpowder Square
		London EC4A 3DE
		Tel: 171 583 0490
		Fax: 171 583 0581

MACRO- TO MICROSCALE HEAT TRANSFER: The Lagging Behavior

1 2 3 4 5 6 7 8 9 0 B R B R 9 8 7 6

This book was set in Times Roman by D. Y. Tzou. The editors were Christine Williams and Carol Edwards. Cover design by Michelle Fleitz. Printing and binding by Braun-Brumfield, Inc.

A CIP catalog record for this book is available from the British Library.
⊚ The paper in this publication meets the requirements of the ANSI Standard Z39.48-1984 (Permanence of Paper)

Library of Congress Cataloging-in-Publication Data

Tzou, D. Y.
    Macro- to microscale heat transfer: the lagging behavior / D.Y. Tzou.
        p. cm.—(Series in chemical and mechanical engineering)

    1. Heat—Transmission. I. Title. II. Series.
QC320. T96   1996
536'.2—dc20
ISBN 1-56032-435-X (case)

96-30320
CIP

CONTENTS

Transient response has been a long-time concern in almost every discipline of engineering science. In the classical theory of heat diffusion, for example, the major concern lies in the transient behavior of temperature resulting from the time-rate of change of internal energy in transition between thermodynamic states. Such transient behavior varies sensitively with the time-dependent boundary conditions or body heating, but the result is always diffusion, no matter how short the response time.

Additional physical mechanisms may need to be incorporated to describe the process of heat transport as the transient time shortens. The wave theory in heat conduction, for instance, uses the relaxation behavior to describe pulsed heat transport at short times. As the transient time becomes comparable to the mean free time of energy carriers, the microscale effect in space needs to be further accommodated because the thermal penetration depth developed in this time frame may cover only several tens of angstroms. Consequently, the thermal penetration depth may become comparable to the characteristic length in heat transport. Existing efforts to model microscale events include the mechanisms of phonon-electron interaction in metal films; phonon scattering in dielectric crystals, insulators and semiconductors; and fracton transport through fractal networks. The approaches used for describing microscale heat transport may have very different physical bases from those in macroscopic phenomenological approaches, exemplified by the semiclassical Boltzmann transport equation, equation of phonon radiative transport, and fractal geometry describing the motion of random walkers in percolating structures. From a broader point of view, therefore, the transient response in heat conduction involves more than obtaining the time-dependent solution satisfying a certain energy equation. It also includes new physical mechanisms, which may involve new philosophical bases in fundamental descriptions that arise in association with shortening of the response time.

This book is concerned with the lagging behavior in heat transport that describes the short-time response from macroscale to microscale in a self-consistent framework. The development includes three major phases: (1) description of the phase-lag concept, with emphases on the perfect reduction of the dual-phase-lag model to the existing microscopic models in limiting cases (Chapter 2) and the second-law admissibility (Chapter 3); (2) phenomenological comparison with the experimental results, including temperature pulses propagating in superfluid liquid helium (Chapter 4), ultrafast pulse-laser heating on metal films (Chapter 5), nonhomogeneous lagging response in porous media (Chapter 6), and thermal lagging in amorphous materials (Chapter 7); and (3) direct consequences and possible outcomes of thermal lagging in the short-time transient, including the effect of material defects (Chapter 8) and thermomechanical coupling (Chapter 9). Some high-order effects in thermal lagging are discussed in Chapter 10.

Thermal lagging is a special response in time. It provides a phenomeno-logical approach that describes the noninstantaneous response between the heat flux vector and the temperature gradient in the process of heat transport. The relaxation behavior resides in the phase lag, or time delay, of the heat flux vector. It describes the fast-transient effect of thermal inertia. The finite time required for the energy exchange/thermal activation in microscale, on the other hand, resides in the phase lag of the temperature gradient. It describes the microstructural interaction effect in space, in terms of the resulting delayed response in time. The two phase lags interweave in the history of heat transport, resulting in the flux-precedence type of heat flow and the gradient-precedence type of heat flow. The dual-phase-lag model thus developed still employs the continuum concepts such as temperature gradient and heat flux vector across a representative volume. However, the resulting lagging response seems to be descriptive for several unique features observed in the experimental and analytical results of microscale heat transport. The phase-lag concept would be found effective in making the transition from macroscale to microscale.

The use of mathematics, sometimes seeming intensive, is unavoidable because the phase-lag concept results in new types of energy equations in heat transport. For the sake of a rigorous treatment, the lagging temperature or heat flux vector has been derived in detail, whenever possible, to extract the physical meaning behind the mathematical solutions. I believe that this is necessary because the structure of thermal lagging, unlike heat diffusion, is somewhat new to the field. Experienced researchers, however, may want to neglect these details and proceed directly to the physics of thermal lagging.

The phase-lag concept results from my research supported by the National Science Foundation (NSF) of the United States during 1990 and 1995, by the Chemical and Thermal System Program and the Mechanics and Materials Program. The support from NSF was vital to materializing the phase-lag concept. Invaluable discussions with Professor Necati M. Özisik at North Carolina State University and Professor Chang-Lin Tien at the University of California at Berkeley had a direct impact on my research. Their pioneer works in the wave theory of heat conduction and microscale heat transport have intrinsic influences on the outgrowth of the dual-phase-lag model. The patience and professional assistance from Mrs. Lisa Ehmer Carolyn Ormes, and Christine E. Williams at Taylor & Francis were essential for the final appearance of this book. Finally, I am deeply indebted to my son Andy for his patience and understanding during the composition of this book. It took away several years from his childhood. This book is dedicated to him.

D. Y. Tzou
Columbia, Missouri

NOMENCLATURE

A • dimensionless coefficient (2, 4, 7, 8)
• positive coefficient, m/W K (3)
• parameter in the Laplace transform solution, 1/m (6)
• amplitude of the near-tip temperature, K (8)
• dimensionless radius (10)

A_i • $i = 1$ to 6. Positive coefficients; $[A_1, A_2, A_4] =$ W/m K^3; $[A_3] =$ m/W K; $[A_5]$ = 1/Pa K s; $[A_6] =$ W/m^3 K (3)
• $i = 1, 2, 3$. Coefficients in Laplace transform solutions, K s (5), dimensionless (9); A_e, A_l: ratio of thermal diffusivity (5)

A_{ij} $i, j = 1, 2, 3$. Coefficients in Laplace transform solutions, K s (5)

a • parameter in the normalized autocorrelation function, dimensionless (5, 9)
• radius of the circular or spherical cavity, m (8)

a_i $i = 1, 2, 3$. Generalized coefficients in the boundary conditions, dimensionless (8)

B • $\tau_T/(2\tau_q)$ (2); τ_T/τ_q (4, 8, 10)
• positive coefficient, 1/Pa K s (3)
• coefficient in Laplace transform solutions, 1/m (5)
• Coefficients of eigenfunctions, dimensionless (8)

B_i • $i = 1$ to 4. Positive coefficients. $[B_1, B_2] =$ J s/m^3 K; $[B_3] = 1/K^2$; $[B_4] =$ 1/K s (3)
• $i = 1, 2$. Coefficients in Laplace transform solutions, 1/m (5); $i = 1, 2, 3$, dimensionless (9)

b parameter in the Laplace transform solution, dimensionless (9, 10)

b_i $i = 1, 2, 3$. Generalized coefficients in the boundary conditions, dimensionless (8)

C thermal wave speed, m/s

$C_{(e, l)}$ volumetric heat capacity of electron gas (e) and metal lattice (l), J/m^3 K (1, 5, 8, 10)

C_i
- $i = 1,2$. Dimensionless coefficients in Fourier transform solutions (2).
- coefficients in Laplace transform solutions, dimensionless (4, 5, 10); $[C_1] = $ m K/W, $[C_2] = $ m K s/W when used with dimensions (5)
- $i = 1$ to 4. $[C_1] = $ m^4 s/kg W K; $[C_2] = $ J/kg K Pa^2; $[C_3] = $ m^2 J/kg K^3; $[C_4] = $ J/kg K (3)
- $i \equiv p, v, \kappa$. Volumetric heat capacity, J/kg K; $i = 1, 2, 3$. Coefficients involving Poisson ratio, dimensionless (9)

C_{ij} $i, j = 1, 2$. Coefficients in Laplace transform solutions, $[C_{11}, C_{21}] = $ m K/W; $[C_{12}, C_{22}] = $ m s K/W (5)

C_p volumetric heat capacity, J/m^3 K

C_T speed of T wave, m/s (10)

C_v speed of CV wave, m/s (10)

$C^{(i)}$ $i = 1, 2$. The ith wave speed in thermomechanical coupling, m/s (9)

c
- mean phonon speed, m/s (2)
- $v/2\alpha$, 1/m (8)
- parameter in the Laplace transform solution, dimensionless (9, 10)

D
- dimensionless coefficients in Laplace transform (2)
- dimensionless radius (4)
- coefficient in Laplace transform solutions, K s (6)
- fractal and fracton dimensions, dimensionless (7)
- dimensionless function (10)

D_i $i = 1, 2, 3$. Dimensionless

coefficients (5, 8, 9)
- $i = 1$ to ∞. Fourier coefficients, dimensionless (10)

d dimensionality of heat source or conducting media, dimensionless (7)

d_i
- $i = 1, 2, 3$. Distance traveled by phonons or electrons, nm (1)
- $i = 1$ to 4. Coefficients in the asymptotic expansion, dimensionless (9)

E
- phonon/electron energy, J (1, 10)
- conjugate tensor to the Cauchy strain tensor, W/m^3 K (3)
- averaged error threshold, dimensionless (6)
- Young's modulus in elasticity, Pa (9, 10)

E_i $i = 1, 2$. Dimensionless coefficients in Laplace transform solutions (5)

e
- Cauchy strain tensor, mm/mm (3, 9, 10)
- volumetric or one-dimensional strain, dimensionless (9, 10)

F dimensionless coefficients (5, 8)

F_i $i = 1$ to 5. Dimensionless Coefficients (9, 10)

f
- temperature rise relative to its maximum value, dimensionless (6)
- time function of diffusive temperature, $1/\sqrt{s}$ (7)
- transformation function or eigenfunction, dimensionless (8)

f_i $i = 1, 2$. Nonhomogeneous functions, dimensionless (9)

G
- electron-phonon coupling factor, W/m^3 K (1, 2, 3, 5, 8,

10)
• solid-gas energy coupling factor, W/m^3 K (6)
• dimensionless heat intensity (7)

g • heat intensity per unit area, J/m^2 (7)
• reciprocal of the laser penetration depth, 1/m (9)
• spatial distribution of the oscillating heat source (10)

g_i $i = 1, 2, 3$. Transformation function, dimensionless (8)

G_i conjugate vector to the temperature gradient, W/m^2 K^2 (3)

H • dimensionless coefficient (5)
• angular distribution of the near-tip temperature, dimensionless (8)
• unit step function (9, 10)
• complex amplitude of the temperature wave, K (10)

H_i $i = 1, 2$. Coefficient in Laplace transform solutions, s (5)

h • Planck constant, J s (1, 5, 10)
• unit step function (2, 4, 5)
• specific enthalpy per unit mass, J/kg (9)

I power intensity of laser beam, W/m^2 (5)

I_n modified Bessel function of the first kind of order n (4, 8, 9)

i number of terms in a series (2)

J • entropy flux vector, W/m^2 K (3)
• energy intensity of laser pulse, J/m^2 (5, 9)

J_n Bessel function of the first kind of order n (4)

j • number of terms in a series

(2)
• mass flux density, kg/m^2 s (4)

K • bulk modulus in elasticity, Pa (9)
• thermal conductivity of the electron gas, W/m K (10)

k thermal conductivity, W/m K

L • thin-film thickness, μm (1, 5)
• length of the one-dimensional solid, μm (2, 4, 10)
• effective mean free path in phonon collision, μm (1)

l • dimensionless length of the one-dimensional solid (2, 4)
• half-length of the sand container, m (6)

M • number of terms in the Riemann-sum approximation or Taylor series expansion (2)
• number of data points in the experiment (6)
• v/C, thermal Mach number (8)
• atomic mass, kg (10)

m • time exponent of surface temperature; slope in the logarithmic temperature-versus-time curve, dimensionless
• effective mass of electrons, kg (10)

N number of terms in the series truncation (2, 6, 7)

n • number density per unit volume, $1/m^3$ (1, 5, 10)
• unit normal of the differential surface area, dimensionless (9)

n_c critical model number for the occurrence of the thermal resonance (10)

P	transient matrix element (10)
p	• Laplace transform parameter, dimensionless (2, 4, 5, 7, 8, 9, 10); 1/s when used with dimensions (5, 6)
	• transformation function (8)
Q	• volumetric heat source, W/m^3 (2, 10)
	• angular distribution of the heat flux vector (8)
	• dimensionless laser absorption rate (9)
	• dimensionless heat flux, $q/(C_p T_0 C_L)$ (10)
Q_i	conjugate vector to the heat flux vector, 1/m K (3)
q	heat flux, W/m^2
R	• reflectivity, dimensionless (5)
	• mean distance traveled by random walkers (7)
	• rigidity propagator in heat transport, W/m^3 (10)
R_c	C_e/C_l (10)
Re	real part of a function
r	position, μm
r_i	$i = 1, 2$. Dimensionless coefficients in Fourier transform (2)
S	• energy-absorption rate, W/m^3 (5, 9)
	• surface area, m^2 (9)
	• volumetric heat source, W/m^3 (10)
S_{ij}	$i, j = 1, 2, 3$. Conjugate tensor to the Cauchy stress tensor, 1/K s (3)
s	• entropy per unit mass, J/kg K (3, 4, 9)
	• eigenvalues (r dependency) of the near-tip heat flux vector, dimensionless (8)
T	absolute temperature, K
t	physical time, s
t_i	$i = 1, 2, 3$. Travel times of

	phonons or electrons in successive collisions (1)
U	dimensionless displacement (9)
U_i	$i = 1, 2$. Coefficients in Laplace transform solutions, W/m K (5)
u	• velocity vector, m/s (3)
	• displacement, m (9, 10)
u_i	$i = 1, 2$. First and second sound speeds in liquid helium (4)
V	volume, m^3 (9)
V_i	$i = 1, 2$. Coefficients in Laplace transform solutions, K s (5)
v	• specific volume, m^3/kg (3, 9)
	• flow velocity, m/s (4)
	• crack velocity, m/s (8)
v_s	phonon velocity or speed of sound, m/s (1, 5, 10)
W_i	$i = 1, 2$. Coefficients in Laplace transform solutions, K s/m (5)
w	• displacement vector, m (3)
	• transformation variable, dimensionless (9)
x	one-dimensional space variable, m
x_i	$i = 1, 2, 3$. Cartesian coordinates, m (8)
y	transformed or integral variables, dimensionless (2, 7, 9)
Z	τ_T/τ_q (7)
z	• transformed or integral variable, dimensionless (2, 4, 8)
	• ratio of phase lag (τ) to diffusion time (l^2/α), dimensionless (6)
	• dimensionless phase lags (9, 10)

Greek Symbols

α thermal diffusivity, m^2/s

β dimensionless time

Δ
- Dirac-delta function (4)
- average volume of the unit cell, m^3 (10)

δ
- dimensionless space (1, 2, 4, 5, 8, 9, 10)
- Kronecker delta (3)
- penetration depth, nm (5)
- delta function (7)

ε specific internal energy per unit mass, J/kg (3, 9)

ε_i i = 1, 2, and 3. Radii of circles around the branch points, dimensionless (2)

Φ amplitude function, K (8)

ϕ
- heat-flux potential, W/m (2)
- azimuthal angle, rad (8)
- Lamé displacement potential, Pa m^2 (10)

ϕ_n n = 1 to ∞. Spatial eigenfunctions of the undamped T wave (10)

Γ
- time amplitude of temperature, K/m$^\lambda$ (8); K (10)
- Gamma function when noted, dimensionless (8)

γ
- real axis in the Bromwich contour, dimensionless (2); 1/s when used with dimensions (6)
- transformation function, dimensionless (8)

η
- dimensionless heat flux vector (2, 4, 7, 8, 10)
- thermomechanical coupling factor, dimensionless (9, 10)

κ
- Boltzmann constant, J/K (1, 5, 10)
- coefficient of thermal

expansion, 1/K (strain) or Pa/K or J/m^3 K (stress)

Λ time amplitude of the near-tip heat flux vector, W/m$^{\lambda+2}$ (8); defined constant (10)

λ
- intrinsic length scale, m (1)
- positive coefficient, Pa or J/m^3 (3)
- characteristic length, m (7)
- eigenvalues (r dependency) of the near-tip temperature, dimensionless (8)
- Lamé modulus in elasticity, Pa (9, 10)

λ_T, λ_q lengths of nonlocal response in the temperature gradient and the heat flux vector, m (10)

μ
- direction cosine (1)
- coefficient of viscosity, Pa s (3)
- shear modulus in elasticity, Pa (9, 10)

ν
- vibration frequency of metal lattice, 1/s (1)
- Poisson ratio, dimensionless (9, 10)

Ω oscillating frequency, 1/s (10)

ω
- frequency in the Fourier transform domain
- angular velocity of the running crack, m/s (8)

ω_n n = 1 to ∞. Frequency of the undamped T wave, 1/s (10)

Θ dimensionless temperature (8, 10)

θ
- dimensionless temperature (1, 2, 4, 5, 7, 9, 10)
- azimuthal angle, rad (8, 10)

ρ
- integral variables, dimensionless (2)
- mass density, kg/m^3 (3, 4,

9, 10)

Σ • entropy production rate per unit volume, W/m^3 K (3)
• dimensionless stress (9)

σ Cauchy stress tensor, Pa (3, 9, 10)

τ • mean free time or relaxation time, s (1, 5, 8)
• half-period of wave oscillations, s (2)
• phase lags, s (1, 2, 3, 4, 5, 6, 7, 8, 9)
• time delay between the heating and probing laser, s (5)

ξ dimensionless space variable (7, 8, 10)

ξ_i • $i \equiv D, W$. Correlation length, m (7)
• $i = 1, 2$. Material coordinates convecting with the crack tip (8)

ζ transformation variable, dimensionless (8)

ψ modal parameter in the auto-correlation function of laser pulses, 1/s (5)

∇ gradient operator, 1/m

Subscripts and Superscripts

0 • initial value at $t = 0$ (2, 4, 6, 7, 8, 9, 10)
• equilibrium conditions (3, 5)
• dimensionless quantity (9, 10)

a atom (1, 2, 5, 10)
b boundary (4, 5, 9, 10)
D • diffusion (1, 7, 8)
• Debye temperature (10)
E • equivalent quantity (5, 6)
• elastic dilatation (9)
e electron
F quantities calculated at the Fermi surface

g gaseous phase (6)
l • metal lattice (1, 2, 5, 10)
• fractal (7)
L longitudinal waves (10)
L^{-1} inverse Laplace transform
M • mechanical field (3)
• thermal Mach wave (8)
max maximum value
N normal process of phonon collision (2)
n • normal viscous fluid component (4)
• fracton (7)
• $n = 1$ to ∞. Wave mode (10)
p • isobaric (2, 4, 9)
• parallel assembly (5)
• the full-width-at-half-maximum pulse (5)
q heat flux vector
R umklapp process of phonon collision (2)
r r component
s • boundary quantities (2, 4); Fourier transform (2)
• superfluid (4)
• pulse quantities (5, 6)
• solid phase (6)
(s) steady state (8)
T • temperature gradient (2, 4)
• thermal field or temperature (3, 6, 7, 9, 10)
T_i $i = 0, 1$. Boundary temperatures (1)
(t) transient state (8)
v constant volume (9)
W • wave (1)
• quantities at the wall (2, 4, 5, 10)
$\bar{X}$ vector X
$\overline{X}$ • Laplace transform of X
• complex conjugate (10)
$\dot{X}$ time derivative of X, $\partial X/\partial t$
X^* • shifted, equivalent, or

apparent quantities of X (2, 9)

• dimensionless frequencies normalized with respect to τ_q (10)

X^+ approaching from the side greater than X (8)

X' • deviatoric component (3)

• derivative of X with respect to its argument (8, 9)

$X_{,i}$ $\partial X/\partial x_i$, spatial derivatives (3)

$X_{(i)}$ $i = 1, 2$. Material properties of X in the ith layer

(X) $X = $ I, II. Quantities in the subsystem (X)

X_i the ith components of a vector (3)

ε strain (9)

θ θ component

σ stress (9)

ONE

HEAT TRANSPORT BY PHONONS AND ELECTRONS

From a microscopic point of view, the process of heat transport is governed by phonon-electron interaction in metallic films and by phonon scattering in dielectric films, insulators, and semiconductors. Conventional theories established on the macroscopic level, such as heat diffusion assuming Fourier's law, are not expected to be informative for microscale conditions because they describe macroscopic behavior averaged over many grains. This holds even more true should the transient behavior at extremely short times, say, on the order of picoseconds to femtoseconds, become major concerns. A typical example is the ultrafast laser heating in thermal processing of materials. The quasi-equilibrium concept implemented in Fourier's law further breaks down in this case, along with the termination of macroscopic behavior in heat transport.

This chapter provides a brief summary of existing microscale heat transfer models, including the microscopic two-step model (phonon-electron interaction model), phonon-scattering model, phonon radiative transfer model, and the thermal wave model. The first three models emphasize microscale effects in space, while the fourth, the thermal wave model, describes microscale effects in time. Rather than a detailed review, however, emphasis is placed on the special behavior depicted by these models that might reveal possible lagging behavior in heat transport. For developing a more complete picture of these models, the readers may refer to the original papers cited in each section. The review articles by Tien and Chen (1994) and Duncan and Peterson (1994) for a broad perspective in microscale heat transfer and those by Joseph and Preziosi (1989, 1990), Tzou (1992a), and Özisik and Tzou (1994) for the wave theory in heat conduction would be helpful in developing an overall understanding of the models' development.

1.1 CHALLENGES IN MICROSCALE HEAT CONDUCTION

Regardless of the type of conducting medium, heat transport requires sufficient collisions among energy carriers. In metals, such energy carriers include electrons and phonons. In dielectric crystals, insulators, and semiconductors, on the other hand, phonons are the primary energy carriers. The phonon gas can be viewed as a group of "mass particles" that characterize the energy state of a metal lattice. For a metal lattice vibrating at a frequency v at a certain temperature T, the energy state of the metal lattice, and hence the energy state of the phonon, is

$$E = hv \tag{1.1}$$

with h being the Planck constant. The lattice frequency is of the order of tens of terahertz (10^{13} 1/s) at room temperature. It is conceivable that the lattice frequency increases with the temperature of the metal lattice. Energy transport from one lattice to the other can thus be thought of as the consequence of a series of phonon collisions in time history, as illustrated in Figure 1.1.

Bearing energy hv at time t_1, phonon 1 collides with phonon 2 at t_2 and with phonon 3 at t_3. In the course of such successive collisions, energy is transferred from phonon 1 to phonons 2 and 3, causing a successive change of vibrating frequency of phonon 1. For the ease of illustration, the mean free path (d, in space) is defined as the algebraic mean of the distances traveled by phonon 1 between the two successive collisions with phonons 2 and 3:

$$d = \frac{d_1 + d_2 + d_3}{3}. \tag{1.2a}$$

The mean free time τ, in a similar fashion, can be defined as the algebraic mean of the times traveled by phonon 1 between the two successive collisions with phonons 2 and 3:

$$\tau = \frac{\left(t_2 - t_1\right) + \left(t_3 - t_2\right) + \left(t - t_3\right)}{3} = \frac{t - t_1}{3}. \tag{1.2b}$$

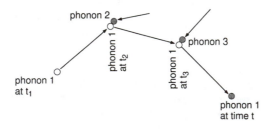

Figure 1.1   Energy transport through phonon collision. The mean free path for phonon 1 in successive collisions is $(d_1 + d_2 + d_3)/3$. The mean free "time" for phonon 1 in successive collisions is $(t - t_1)/3$.

Two collisions for phonon 1 are used in this example for the ease of illustration. In order to have a meaningful statistical ensemble space, of course, a "sufficient" number of collisions must be collected for defining the mean free path and mean free time.

The macroscopic models assume the physical domain for heat transport is so large that it allows hundreds of thousands of phonon collisions to occur before an observation/description is made for the process of heat transport. Phonon collision requires a finite period of time to take place. Hundreds of thousands of such collisions also imply a sufficiently long time for the process of heat transport to occur. The macroscopic models, therefore, not only necessitate a sufficiently large physical domain for conducting heat (much larger than the mean free path), but also a sufficiently long time for heat conduction to take place (much longer than the mean free time). The sufficiently long time for the stabilization of energy transport by phonons should not be confused with that required for the steady state to be developed. The sufficiently long time required in phonon collisions is for a statistically meaningful concept in mean free path and time. The process of heat transport can still be time dependent after phonon transport becomes stabilized. In a phenomenological sense, the mean free time illustrated in Figure 1.1 is parallel to the characteristic time describing the relaxation behavior in the fast-transient process. For metals, the mean free time, or relaxation time, is of the order of picoseconds. The relaxation time is longer for dielectric crystals and insulators, roughly of the order of nanoseconds to picoseconds. As a rough estimate, therefore, any response time being shorter than a nanosecond should receive special attention. The fast-transient effect, such as wave behavior in heat conduction, may dramatically activate and introduce some unexpected effects in heat transport. Such a threshold value of nanoseconds, however, depends on the combined effect of geometric configuration (of the specimen) and thermal loading imposed on the system. It may vary by 1 order of magnitude should the system involve an abrupt change of geometric curvatures (such as in the vicinity of a crack or notch tips) or be subject to discontinuous thermal loading (irradiation of a short-pulse laser, for example).

The mean free path for electrons is of the order of tens of nanometers (10^{-8} m) at room temperature. The mean free path is a strong function of temperature, however. It may increase to the order of millimeters in the liquid helium temperature range, approximately 4 K. The mean free path in phonon collision and phonon scattering (from grain boundaries) is much longer. For type IIa diamond film at room temperature (Majumdar, 1993), for example, the mean free path is of the order of tenths of a micron (10^{-7} m). As a rough estimate, again, the physical device with a characteristic dimension in submicrons deserves special attention. The microstructural interaction effect, such as phonon-electron interaction or phonon scattering, may significantly enhance heat transfer in short times. Enhancement of heat transfer enlarges the heat-affected zone and promotes the temperature level, which may thereby lead to early burnout of micro-devices if not properly prevented.

Because the physical dimension in microscale heat transfer is of the same order of magnitude as the mean free path and, consequently, the response time in heat transport is of the same order of magnitude as the mean free time, the quantities based on the macroscopically averaged concept must be reexamined for their true physical meanings. The temperature gradient that has been taken for granted in macroscale heat transfer, for example, may lose its physical ground for a thin film of thickness of the same order of magnitude as the mean free path. As illustrated in Figure 1.2, it is true that we could still divide the temperature difference, $T_2 - T_1$, by the film thickness l ($\cong d$, the mean free path of phonon interaction/scattering) to obtain a "gradient-like" quantity, but the temperature gradient obtained in this manner loses its usual physical meaning because there are no sufficient energy carriers between the two surfaces of the film and, consequently, the temperature field is *discontinuous* across the film thickness. Should the concept of temperature gradient fail, the conventional way of defining the heat flux vector according to Fourier's law becomes questionable. Ambiguity of the concepts in both the temperature gradient and the heat flux vector is the first challenge that the microscale effect in space has raised against conventional theories in macroscale heat transfer.

A similar situation exists in the response time for temperature. The typical response time in the thin film is of the same order of magnitude as the mean free time, as a result of phonons traveling in the threshold of the mean free path. If the response time of primary concern (for temperature or heat flux vector) is of the same order of magnitude as the mean free time (relaxation time), the individual effects of phonon interaction and phonon scattering must be taken into account in the short-time transient of heat transport. This is the second challenge that the microscale effect in time has raised against conventional theories in macroscale heat transfer. From Figure 1.2, most important, it is evident that the microscale effect in space interferes with the microscale effect in time. They are not separable and must be accommodated *simultaneously* in the same theoretical framework. This is also clear from considering the finite speed of phonon transport in short times. Phonons propagate at the speed of sound, on average, which is of the order of 10^4 to 10^5 m/s at room temperature, depending on the type of solid medium. A response time of the order of picoseconds (10^{-12} s) thus implies a traveling distance (the penetration depth of heat by phonon transport) of the order of submicrons (10^{-8} to 10^{-7} m). Such a penetration depth is on the microscopic level, necessitating a simultaneous consideration of the microscale effect in space.

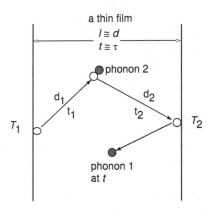

Figure 1.2 Phonon interaction and scattering in a thin film of thickness of the same order of magnitude as the mean free path, illustrating the challenge of the microscale effect in space to the concept of the temperature gradient.

1.2 PHONON-ELECTRON INTERACTION MODEL

1.2.1 Description

Phenomenologically, the phonons illustrated in Figures 1.1 and 1.2 can be replaced by phonons/electrons to depict the phonon-electron interaction for heat transport in metals. Owing to much smaller heat capacity of the electron gas, about 1 to 2 orders of magnitude smaller than that of the metal lattice, however, the heating mechanism involves excitation of the electron gas and heating of the metal lattice through phonon-electron interaction in short times. The phonon-electron interaction model was proposed to describe this two-step process for energy transport in microscale. The early version of the two-step model (phonon-electron interaction model) was proposed by Kaganov et al. (1957) and Anisimov et al. (1974) without a rigorous proof. It remained as a phenomenological model for about four decades, until the recent effort by Qiu and Tien (1993), which places the two-step model on a quantum mechanical and statistical basis. In the absence of an electrical current during short-time heating, the generalized *hyperbolic* constitutive equation for heat transport through the electron gas was derived from the Boltzmann transport equation. As the relaxation time of the electron gas calculated at the Fermi surface vanishes, the hyperbolic two-step model perfectly reduces to the *parabolic* two-step model originally proposed by Kaganov et al. (1957) and Anisimov et al. (1974). For a progressive presentation, we shall review the parabolic two-step model in this chapter and leave the more involved hyperbolic two-step model to Section 10.1, where a special type of T wave shall be introduced to account for the ballistic behavior of heat transport through the electron gas.

For metals, the two-step model describes heating of the electron gas and the metal lattice by a two-step process. Mathematically,

$$C_e \frac{\partial T_e}{\partial t} = \nabla \cdot \left(K \nabla T_e \right) - G \left(T_e - T_l \right), \text{ heating of the electron gas,} \quad (1.3a)$$

$$C_l \frac{\partial T_l}{\partial t} = G \left(T_e - T_l \right), \text{ heating of the metal lattice,} \quad (1.3b)$$

with C denoting the volumetric heat capacity, K the thermal conductivity of the electron gas, and subscripts e and l standing for electron and metal lattice, respectively. The effect of heat conduction through the metal lattice is neglected in equation (1.3b). Such an effect shall be reinstated in Section 5.8. The externally supplied photons, such as those from an intensified laser, first increase the temperature of the electron gas according to equation (1.3a). As clearly shown, *diffusion* is assumed at this stage, rendering a parabolic nature for heat transport through the electron gas. Through phonon-electron interactions, which is the second stage of heat transport, represented by equation (1.3b), the hot electron gas heats up the metal lattice by phonon-electron interaction. The energy exchange between phonons and electrons is characterized by the phonon-electron coupling factor G (Kaganov et al., 1957):

$$G = \frac{\pi^2}{6} \frac{m_e n_e v_s^2}{\tau_e T_e} \quad \text{for} \quad T_e >> T_l \tag{1.4}$$

where m_e represents the electron mass, n_e the number density (concentration) of electrons per unit volume, and v_s the speed of sound,

$$v_s = \frac{\kappa}{2\pi h} \left(6\pi^2 n_a\right)^{-\frac{1}{3}} T_D. \tag{1.5}$$

The quantity h in equation (1.5) stands for the Planck constant, κ the Boltzmann constant, n_a the atomic number density per unit volume, and T_D represents the Debye temperature. The electron temperature (T_e) is much higher than the lattice temperature (T_l) in the early-time response. The condition of $T_e >> T_l$ in equation (1.4) for the applicability of the G expression is thus valid in the fast-transient process of electron-phonon dynamics. Within the limits of Wiedemann-Frenz's law, which states that for metals at moderate temperatures (roughly for $T_l > 0.48T_D$) the ratio of the thermal conductivity to the electrical conductivity is proportional to the temperature and the constant of proportionality is independent of the particular metal (a metal-type-independent constant), the electronic thermal conductivity can be expressed as

$$K = \frac{\pi^2 n_e \kappa^2 \tau_e T_e}{3m_e}, \tag{1.6a}$$

resulting in

$$m_e = \frac{\pi^2 n_e \kappa^2 \tau_e T_e}{3K}. \tag{1.6b}$$

Substituting equation (1.6b) into equation (1.4) for the electron mass gives

$$G = \frac{\pi^4 (n_e v_s \kappa)^2}{18K}. \tag{1.7}$$

The phonon-electron coupling factor, therefore, depends on the thermal conductivity (K) and the number density (v_s) of the electron gas. Through the speed of sound, in addition, the cooling factor further depends on the number density of atoms (n_a) and the Debye temperature (T_D). The coupling factor does not show a strong dependence on temperature (T_e) and it does not seem to be affected by the electronic relaxation time (τ_e).

In order to estimate the value of G according to equation (1.7), the number density of the electron gas, n_e, is a key quantity. Qiu and Tien (1992) assumed one free electron per atom for noble metals (silver (Ag) and gold (Au), for example) and employed the s-band approximation for the valence electrons in transition metals. Owing to the relatively heavy mass of the d-band electrons in the valence electrons, only a fraction of the s-band electrons can be viewed as free electrons. The value of n_e, therefore, is chosen as a fraction of the valence electrons. The phonon-electron coupling factor thus calculated, and the experimentally measured values are listed in Table 1.1 for comparison. Except for copper (Cu) and lead (Pb), which may exhibit certain ambiguous transition mechanisms, the s-band approximation seems to agree well with the experimental results. As a general trend, a higher free electron number density (n_e) and a higher Debye temperature (T_D) would result in larger values of G and smaller values of the relaxation time (τ).

From a mathematical point of view, equations (1.3a) and (1.3b) provide two equations for two unknowns, the electron-gas temperature (T_e) and the metal-lattice temperature (T_l). They can be solved in a coupled manner, or they can be combined to give a *single* energy equation describing heat transport through phonon-electron interaction in microscale. The combined energy equation, from an alternate point of view, can be derived from the phase-lag concept in the temporal response, which is a central topic in Chapter 2. Such a coincidence strongly supports the dual-phase-lag model.

Complexity of solutions for equations (1.3a) and (1.3b) lies in the temperature-dependent heat capacity of the electron gas, i.e., $C_e \equiv C_e(T_e)$. For an electron-gas temperature lower than the Fermi temperature, which is of the order of 10^4 K, the electron heat capacity is proportional to the electron temperature. Such a temperature dependence makes equations (1.3a) and (1.3b) nonlinear. For a gold film subjected to femtosecond laser heating, Qiu and Tien (1992) employed the Crank-Nicholson scheme of finite difference to obtain the solutions. With regard to the comparison with the experimental result, the normalized temperature change in the electron gas is identical to the normalized reflectivity change on the film surfaces:

$$\frac{\Delta R}{(\Delta R)_{max}} = \frac{\Delta T_e}{(\Delta T_e)_{max}} \tag{1.8}$$

Table 1.1 Phonon-electron coupling factor G for some noble and transition metals (reproduced from Qiu and Tien (1992))

Metal	Calculated, $\times 10^{16}$ W/m^3 K	Measured, $\times 10^{16}$ W/m^3 K
Cu	14	4.8 ± 0.7 (Brorson et al., 1990)
		10 (Elsayed-Ali et al., 1987)
Ag	3.1	2.8 (Groeneveld et al., 1990)
Au	2.6	2.8 ± 0.5 (Brorson et al., 1990)
Cr	45 $(n_e/n_a = 0.5)$	42 ± 5 (Brorson et al., 1990)
W	27 $(n_e/n_a = 1.0)$	26 ± 3 (Brorson et al., 1990)
V	648 $(n_e/n_a = 2.0)$	523 ± 37 (Brorson et al., 1990)
Nb	138 $(n_e/n_a = 2.0)$	387 ± 36 (Brorson et al., 1990)
Ti	202 $(n_e/n_a = 1.0)$	185 ± 16 (Brorson et al., 1990)
Pb	62	12.4 ± 1.4 (Brorson et al., 1990)

where R denotes the reflectivity and the subscript "max" refers to the maximum value occurring in the transient process. Both ratios in equation (1.8), therefore, are less than 1. The left side of equation (1.8) can be measured by the front-surface-pump and back-surface-probe technique (Brorson et al., 1987; Elsayed-Ali, 1991; Qiu et al., 1994). The right side of equation (1.8), on the other hand, can be calculated by solving equations (1.3a) and (1.3b) for the electron temperature and normalizing with respect to the maximum value in the transient response at various times. For a gold film subjected to irradiation of a 96 femtosecond (fs, 10^{-15} s) laser with an energy flux of 1 mJ/cm^2, the results of reflectivity change at the front surface of the film are reproduced in Figure 1.3 from the work by Qiu and Tien (1992). The time delay marked on the horizontal axis is the time difference between the pump (heating) and the probe (detecting) lasers, which is equivalent to the physical time in the transient response. For both thicknesses of the films, 0.05 and 0.1 μm, the microscopic two-step model accommodating the phonon-electron interaction effect nicely captures the heating ($0 \leq t \leq 0.096$ picosecond (ps)) and thermalization ($0.096 \leq t \leq 3$ ps) processes of the electron temperature. The temperature level, as expected, increases as the thickness of the film decreases. The classical theory of diffusion, which assumes an immediate equilibrium between phonons (lattice) and electrons and is called the one-step heating model by Qiu and Tien (1992), fails to describe the fast energy transport process. Particularly in the thermalization stage, it *overestimates* the transient temperature by several times. The transient temperature at the front surface does not seem to depend on the film thickness according to the diffusion model. The transient temperature remains almost at the same level as the film thickness increases from 0.05 to 0.1 μm.

Figure 1.4 shows the transient reflectivity change at the rear surface of the film. The heating and thermalization processes remain similar, with the response times, however, increasing. The time at which the electron temperature reaches its

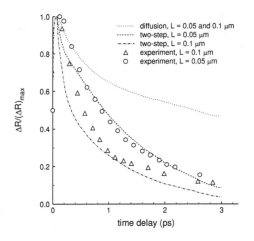

Figure 1.3 Transient reflectivity change at the front surface of gold films (thickness 0.05 and 0.1 μm) subject to laser irradiation (pulse width 96 fs, energy flux 1 mJ/cm^2). Reproduced from the work by Qiu and Tien (1992) with the experimental results obtained by Brorson et al. (1987).

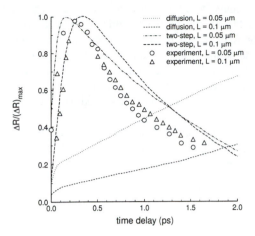

Figure 1.4 Transient reflectivity change at the rear surface of gold films (thickness = 0.05 and 0.1 μm) subject to laser irradiation (pulse width 96 fs, energy flux 1 mJ/cm^2). Reproduced from the work by Qiu and Tien (1992) with the experimental results obtained by Brorson et al. (1987).

maximum value (the instant of time separating the heating and thermalization stages) increases with film thickness, which is a "wave-like" behavior even under the assumption of diffusion for heat transport through the electron gas. Section 5.4 discusses this special behavior in detail. The one-step heating model (diffusion) completely fails to describe the thermalization process at the rear surface of the film. Unlike the situation shown in Figure 1.3 (the front surface), where at least the qualitative trend was preserved in the same domain of response times, the heating stage predicted by the diffusion model shown in Figure 1.4 (the rear surface) lasts beyond the threshold of 2 ps, resulting in a transient response of reflectivity change that significantly differs from the experimental result both quantitatively and qualitatively. From Figures 1.3 and 1.4 it is clear that, for metals, the microscopic phonon-electron interaction is an important effect to be incorporated for an accurate description of microscale heat transport. In addition to the familiar thermal properties such as heat capacity and thermal conductivity, the phonon-electron coupling factor describing the short-time energy exchange between phonons and electrons is a dominating property in the fast-transient process of laser heating. Along with the equivalent thermal *wave* speed in the *parabolic* two-step model, typical values of the phonon-electron coupling factor (G) are listed in Chapter 5, Table 5.1 for copper, gold, silver, and lead. They are of the order of 10^{16}, in units of W/m^3 K, for metals.

1.2.2 Single Energy Equation

Equations (1.3a) and (1.3b) can be combined to give a *single* energy equation governing heat transport through the metal lattice or the electron gas (Tzou, 1995a, b). Given their present forms, solving equations (1.3a) and (1.3b) for T_e and T_l in a simultaneous manner may be more efficient from a numerical point of view. Combining equations (1.3a) and (1.3b) together to give a single energy equation describing the electron temperature or the lattice temperature alone, however, is more indicative for the fundamental behavior in microscale heat transport. With emphasis on the characteristics of lattice and electron temperatures, all of the thermal properties, including heat capacities of the electron gas (C_e) and the metal lattice (C_l) as well as the thermal conductivity (K) are assumed to be temperature-independent (constant).

A single energy equation governing the lattice temperature can be obtained by eliminating the electron temperature, T_e, from equations (1.3a) and (1.3b). From equation (1.3b), the electron temperature can be expressed in terms of the lattice temperature and its time derivative:

$$T_e = T_l + \frac{C_l}{G}\frac{\partial T_l}{\partial t} . \tag{1.9}$$

Substituting equation (1.9) into (1.3a) and using the result of $G(T_e - T_l)$ from equation (1.3b) results in

$$\nabla^2 T_l + \left(\frac{C_l}{G}\right)\frac{\partial}{\partial t}\nabla^2 T_l = \left(\frac{C_l + C_e}{K}\right)\frac{\partial T_l}{\partial t} + \left(\frac{C_l C_e}{KG}\right)\frac{\partial^2 T_l}{\partial t^2} . \tag{1.10}$$

Equation (1.10), governing the lattice temperature alone, introduces a new type of energy equation in conductive heat transfer. It has an usual diffusion term ($\partial T_l/\partial t$), a thermal *wave* term ($\partial^2 T_l/\partial t^2$), and a mixed-derivative term ($\partial[\nabla^2 T_l]/\partial t$) that reflect the combined effect of microscopic phonon-electron interaction and macroscopic diffusion. In the case that the phonon-electron coupling factor approaches infinity ($G \rightarrow \infty$), implying that energy transfer from electrons to phonons is occurring at an *infinite* rate, equation (1.10) reduces to the conventional diffusion equation employing Fourier's law, with the coefficient ($C_e + C_l$)/K appearing as the equivalent thermal diffusivity. A detailed discussion for the physical significance of the coefficients in equation (1.10) is provided later in Section 5.4, where the physical concepts of phase lags are introduced in their entirety.

A single energy equation describing the electron temperature can be obtained in the same manner. From equation (1.3a),

$$T_l = T_e - \frac{k}{G}\nabla^2 T_e + \frac{C_e}{G}\frac{\partial T_e}{\partial t} . \tag{1.11}$$

Substituting equation (1.11) into (1.3b) and using the result of $G(T_e - T_l)$ from equation (1.3a) yields

$$\nabla^2 T_e + \left(\frac{C_l}{G}\right)\frac{\partial}{\partial t}\nabla^2 T_e = \left(\frac{C_l + C_e}{K}\right)\frac{\partial T_e}{\partial t} + \left(\frac{C_l C_e}{KG}\right)\frac{\partial^2 T_e}{\partial t^2} . \tag{1.12}$$

Note that equation (1.12), governing the electron temperature, has *exactly* the same form as equation (1.10), governing the lattice temperature.

1.3 PHONON SCATTERING MODEL

Guyer and Krumhansl (1966) solved the linearized Boltzmann equation for the pure phonon field in terms of the eigenvectors of the normal-process, phonon-collision operator. Major emphasis was placed on heat transport by phonon collision/ scattering, and the contribution from the electron gas in conducting heat was neglected. The other interactions in which momentum is lost from the phonon system were also neglected in their analysis. The formal solution was represented by two equations relating the temperature deviation (rise) and the heat current (flux):

$$C_p \frac{\partial T}{\partial t} + \nabla \bullet \vec{q} = 0 \tag{1.13a}$$

$$\frac{\partial \vec{q}}{\partial t} + \frac{c^2 C_p}{3} \nabla T + \frac{1}{\tau_R} \vec{q} = \frac{\tau_N c^2}{5} \left[\nabla^2 \vec{q} + 2\nabla \left(\nabla \bullet \vec{q} \right) \right] \tag{1.13b}$$

where c is the average speed of phonons (speed of sound), τ_R stands for the relaxation time for the "umklapp" process, in which momentum is lost from the phonon system (the momentum-nonconserving processes), and τ_N is the relaxation time for normal processes in which momentum is conserved in the phonon system. As recognized, the first equation (1.13a) is the usual energy equation in a rigid conductor. The second equation (1.13b), through a quite complicated procedure, is derived from the generalized phonon-thermal-conductivity relation. Equations (1.13a) and (1.13b) provide two equations for two unknowns, temperature T and heat flux vector $\vec{q}$. Equation (1.13b) serves as the constitutive equation in heat conduction, which relates the heat flux vector and the temperature gradient. It is equivalent to Fourier's law in the classical diffusion model and the CV wave equation in the thermal wave model (Cattaneo, 1958; Vernotte, 1958, 1961).

All quantities in equation (1.13b) occur at the same instant of time and are ready for further combination with the energy equation (1.13a). Taking divergence of equation (1.13b) and substituting the result for the divergence of the heat flux vector in terms of the time-rate of change of temperature from equation (1.13a) gives

$$\nabla^2 T + \frac{9\tau_N}{5} \frac{\partial}{\partial t} \left(\nabla^2 T \right) = \frac{3}{\tau_R c^2} \frac{\partial T}{\partial t} + \frac{3}{c^2} \frac{\partial^2 T}{\partial t^2}. \tag{1.14}$$

Equation (1.14) was first derived by Joseph and Preziosi (1989), with emphasis on the interrelation with Jeffrey's type of heat flux equation. It was derived again later by Tzou (1995a) with emphasis on the lagging behavior in microscale.

Regardless of the completely different mechanisms in microscale, it is note-worthy that equation (1.14), describing heat transport by phonon collision, has *exactly* the same form as equation (1.10) or equation (1.12), describing heat transport through phonon-electron interaction. The *universal* form of equations (1.10), (1.12), and (1.14), describing microscale heat transport in various environments, in fact, stimulates the development of the generalized concept of lagging behavior for the fast-transient process of heat transport at small scales. As is shown later in Section 2.3, the generalized form of equation (1.10) or (1.14) can be derived alternately from the phase-lag concept. The complicated coefficients in the microscopic formulations, equation (1.10), (1.12), or (1.14), can be characterized into two *delayed times* to reflect the microscopic effect in heat transport.

In passing, note that equations (1.3a) and (1.3b), describing the phonon-electron interaction (Qiu and Tien, 1993), and equations (1.13a) and (1.13b), describing the phonon scattering/collision (Guyer and Krumhansl, 1966), are derived directly from the solutions of the linearized Boltzmann transport equation. The heat flux vector in heat flow, for example, results from the integration of the distribution function for electrons or phonons in momentum space. Their original works are highly recommended to readers who want to explore microscopic treatments in more detail.

1.4 PHONON RADIATIVE TRANSFER MODEL

The phonon radiative transfer model (PRT) proposed by Majumdar (1993) starts from the same approach as other models, employing the solution of the linearized Boltzmann transport equation. However, the PRT model distinguishes itself from the others by describing the Stefan-Boltzmann radiative heat equation for an acoustically thin medium (in steady states) and the *CV* wave equation for an acoustically thick medium (in transient heat conduction) in limiting cases. The acoustically "thin" or "thick" medium weighs the thickness of the film structure relative to the phonon mean free path. For an acoustically thick medium, thickness of the film is much greater than the mean free path of phonons. For an acoustically thin medium, on the other hand, thickness of the film is much less than the mean free path of phonons. Since phonons propagate at the speed of sound in solids, the mean free path of phonons is equal to the speed of sound multiplied by the relaxation time (the mean free time) in phonon collisions.

In the case of one-dimensional heat transport, Majumdar (1993) derived the equation of PRT from the Boltzmann transport equation employing the relaxation-time approximation:

$$\frac{\partial f_\omega}{\partial t} + v_x \frac{\partial f_\omega}{\partial x} = \left(\frac{\partial f_\omega}{\partial t}\right)_{scattering} \cong \frac{f_\omega^0 - f_\omega}{\tau} \tag{1.15}$$

where f_ω denotes the distribution function of phonons with vibrating frequency ω, v_x is the one-dimensional (assumed to be the x direction without loss in generality) phonon velocity, and τ is the relaxation time (the mean free time in phonon scattering). Disturbance of the distribution function from its equilibrium state, f_ω^0 in equation (1.15), results from phonon scattering for the duration of the relaxation time.   The scattering term on the right side of equation (1.15), to the first-order approximation in the short-time transient, can thus be replaced by the amount of deviation of the distribution function divided by the relaxation time.   This is called the relaxation-time approximation, popularly used for obtaining the solution of the linearized Boltzmann transport equation.

The phonon intensity in heat transport, I_ω, can be obtained by summing up the three phonon polarizations (p) over the distribution function:

$$I_\omega(\theta,\phi,x,t) = \sum_p \bar{v}(\theta,\phi) f_\omega(x,t) h\omega D(\omega) \qquad (1.16)$$

with $\bar{v}(\theta,\phi)$ denoting the velocity vector of phonons in the direction defined by (θ, ϕ) in a spherical coordinate system within a solid angle $d\Omega = \sin\theta\, d\theta d\phi$, as shown in Figure 1.5, h the Planck constant, and $D(\omega)$ the density of states per unit volume in the frequency domain of lattice vibrations.   The projection of the velocity vector onto the x axis is clearly $v_x = v\cos\theta$.   Multiplying equation (1.15) by $v_x h\omega D(\omega)$ and summing up the result over the three phonon polarizations according to equation (1.16) gives

$$\frac{\partial}{\partial t}\sum_p v_x h\omega D(\omega) f_\omega + v_x \frac{\partial}{\partial x}\sum_p v_x h\omega D(\omega) f_\omega = \sum_p v_x h\omega D(\omega)\left[\frac{f_\omega^0 - f_\omega}{\tau}\right].$$

$$(1.17)$$

Denoting

$$v_x = v\mu, \quad \text{with} \quad \mu = \cos\theta, \quad \text{and} \quad I_\omega^0 = \sum_p v_x h\omega D(\omega) f_\omega^0, \qquad (1.18)$$

and dividing the entire equation by the phonon speed, v, equation (1.17) becomes

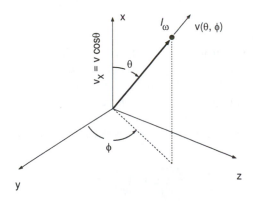

Figure 1.5   Phonon intensity I_ω and the azimuthal angles θ and ϕ defining the velocity vector.

$$\frac{1}{v}\frac{\partial I_\omega}{\partial t} + \mu\frac{\partial I_\omega}{\partial x} = \frac{I_\omega^0 - I_\omega}{\tau v}, \tag{1.19}$$

with $v\tau$ being the mean free path in phonon collisions. The right side of equation (1.19) represents disturbance of an equilibrium state by mutual interactions of phonons. Consequently, it is called the equation of phonon radiative transfer (EPRT) by Majumdar (1993).

Based on the phonon intensity thus obtained, the heat flux (q) and the internal energy (e) at any point in space can be calculated as

$$q = \int_{\Omega=4\pi} \int_0^{\omega_D} \mu I_\omega \, d\omega \, d\Omega, \tag{1.20a}$$

$$e = \int_{\Omega=4\pi} \int_0^{\omega_D} \frac{I_\omega}{v} \, d\omega \, d\Omega, \quad \text{with} \quad d\Omega = \sin\theta \, d\theta \, d\phi \tag{1.20b}$$

with ω_D being the Debye cutoff phonon frequency. For phonon transport with azimuthal symmetry in ϕ, in particular, q and e result in

$$q = 2\pi \int_{-1}^{1} \int_0^{\omega_D} \mu I_\omega \, d\omega \, d\mu \tag{1.21a}$$

$$e = 2\pi \int_{-1}^{1} \int_0^{\omega_D} \frac{I_\omega}{v} \, d\omega \, d\mu. \tag{1.21b}$$

Multiplying equation (1.19) by 2π and integrating the resulting equation over μ and ω in the range $-1 < \mu < 1$ and $0 < \omega < \omega_D$ gives, with the assistance of equations (1.21a) and (1.21b),

$$\frac{\partial e}{\partial t} + \frac{\partial q}{\partial x} = 2\pi \int_{-1}^{1} \int_0^{\omega_D} \left[\frac{I_\omega^0 - I_\omega}{\tau v}\right] d\omega \, d\mu \tag{1.22}$$

Two features are immediately noted. (1) The equilibrium intensity function I_ω^0 on the right side of equation (1.22), according to equation (1.18), is a function of temperature only, i.e., $I_\omega^0 \equiv I_\omega^0(T(x))$. It can thus be extracted out of the integral signs when performing the integrations with respect to ω and μ. (2) The left side of equation (1.22) is zero definite because it is the one-dimensional form of the energy equation in a rigid conductor. Based on these observations, equation (1.22) gives

$$2\int_0^{\omega_D} \left(\frac{I_\omega^0}{\tau v}\right) d\omega = \int_0^{\omega_D} \left(\int_{-1}^{1} \frac{I_\omega}{\tau v} d\mu\right) d\omega, \tag{1.23a}$$

rendering a particular solution for $I_\omega^0(T(x))$:

$$I_\omega^0 = \frac{1}{2}\int_{-1}^{1} I_\omega d\mu \,. \tag{1.23b}$$

Substituting equation (1.23b) into equation (1.19), the EPRT takes the final form of an integro-differential equation to be solved for the phonon intensity function $I_\omega(x, t, \mu)$:

$$\frac{1}{v}\frac{\partial I_\omega}{\partial t} + \mu\frac{\partial I_\omega}{\partial x} = \frac{\dfrac{1}{2}\displaystyle\int_{-1}^{1} I_\omega d\mu - I_\omega}{\tau v}. \tag{1.24}$$

Once the phonon intensity is obtained by solving equation (1.24), the temperature distribution is obtained from the Bose-Einstein distribution function at an equilibrium state:

$$I_\omega^0(T) = \frac{1}{2}\int_{-1}^{1} I_\omega d\mu = \sum_p v_p \frac{\hbar\omega D(\omega)}{\exp\left[\dfrac{\hbar\omega}{\kappa T(x)}\right] - 1} \tag{1.25}$$

As is clearly shown by equation (1.24), the phonon intensity travels in the solid as a wave, with wave speed $v\mu = v\cos\theta$. Owing to the complicated mathematical structure, an analytical method determining the phonon intensity I_ω satisfying equation (1.24) is difficult to obtain. Joshi and Majumdar (1993) solved numerically for heat transport in a one-dimensional medium by using the explicit upstream differencing method, i.e., employing a backward differencing scheme in space for the intensity wave propagating in the positive x direction (the physical domain with $\mu > 0$) and adopting a forward differencing scheme in space for the intensity wave propagating in the negative x direction ($\mu < 0$). Numerical stability is ensured by choosing $\Delta x \geq (v/|\mu|)\,\Delta t$, a standard criterion for ensuring the numerical stability in solving wave equations. Other details were discussed by Joshi and Majumdar (1993) that will not be repeated here.

With the EPRT thus developed, Joshi and Majumdar (1993) compared the temperature profiles with those obtained by the use of the *macroscopic* diffusion and thermal wave models:

$$\frac{\partial T}{\partial t} = \alpha\frac{\partial^2 T}{\partial x^2} \qquad \text{(diffusion equation)} \tag{1.26}$$

$$\tau\frac{\partial^2 T}{\partial t^2} + \frac{\partial T}{\partial t} = \alpha\frac{\partial^2 T}{\partial x^2} \qquad \text{(thermal wave equation)} \tag{1.27}$$

where α is the thermal diffusivity and τ is the relaxation time in phonon collisions. The physical basis of equation (1.27) is explored in more detail in the next section. To use EPRT to obtain the temperature distribution, as mentioned above, equation

(1.24) is first solved for obtaining the phonon intensity function. Equation (1.25) is then used for obtaining the temperature distribution. The boundary conditions considered for the macroscopic models were

$$T = T_1 \quad \text{at} \quad x = 0, \quad T = T_0 \quad \text{at} \quad x = L, \tag{1.28}$$

where $T_1 > T_0$ was assumed without loss of generality. The boundary conditions used in the numerical solution for EPRT, from equations (1.25) and (1.28), were

$$I_\omega = \begin{cases} I_\omega^0(T_1) & \text{at} \quad x = 0, \quad \text{for} \quad \mu > 0, \\ I_\omega^0(T_0) & \text{at} \quad x = L, \quad \text{for} \quad \mu < 0. \end{cases} \tag{1.29}$$

The medium was assumed to be disturbed from an initial temperature T_0 (at $t = 0$) with a zero time-rate of change of temperature ($\partial T/\partial t = 0$ as $t = 0$).

Figure 1.6 compares the temperature profiles obtained by the classical diffusion model (macroscopic in both space and time), the *CV* wave model (macroscopic in space but microscopic in time), and the EPRT (microscopic in both space and time). The dimensionless temperature (θ), dimensionless space (δ), and dimensionless time (β) are defined as

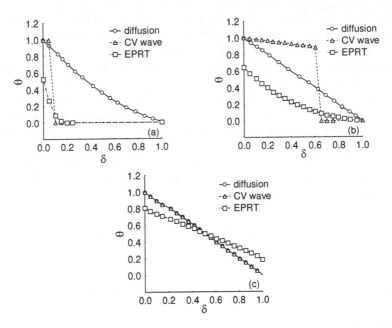

Figure 1.6   Temperature profiles in a one-dimensional solid predicted by the classical diffusion, *CV* wave, and EPRT models. $L = 0.1$ μm at (*a*) $\beta = 0.1$, (*b*) $\beta = 1.0$, and (*c*) steady state. Here $\delta = x/L$ and $\beta = t/(l/v)$. Reproduced from the work by Joshi and Majumdar (1993).

$$\theta = \frac{T - T_0}{T_1 - T_0}, \quad \delta = \frac{x}{L}, \quad \beta = \frac{t}{(l/v)} \tag{1.30}$$

with l being the effective mean free path, $\tau = l/v$. The film thickness was taken as 0.1 μm in Figure 1.6. Sharp wavefronts exist in the temperature profiles predicted by the *CV* wave model in Figure 1.6(a) at $\beta = 0.1$ and Figure 1.6(b) at $\beta = 1.0$. The macroscopic models in space, including both diffusion and *CV* wave models, seem to significantly overestimate the transient temperature. As the transient time lengthens, illustrated by the steady-state distributions shown in Figure 1.6(c), the sharp wavefront in the *CV* wave model vanishes, while the temperature profile collapses onto that predicted by diffusion. It is important to note that EPRT does *not* reduce to the diffusion model employing Fourier's law at steady state. According to Figure 1.6(c), EPRT predicts a lower temperature level at the high-temperature side ($x = 0$) and a higher temperature at the low-temperature side ($x = L$).

The difference between the EPRT and the diffusion (or *CV* wave) model at steady state vanishes as the film thickness exceeds the acoustical limit, i.e., $L \gg v\tau$ ($= l$, the effective mean free path). This is illustrated in Figure 1.7 for $L = 10$ μm, which is 2 orders of magnitude thicker than the value used in Figure 1.6. The temperature profiles predicted by EPRT, diffusion, and *CV* wave models collapse onto each other at steady state, showing a *linearly* decayed distribution of temperature with respect to x, as required by the steady-state energy equation (1.26) or (1.27) under $\partial T/\partial t = \partial^2 T/\partial t^2 = 0$.

For an acoustically thin film with film thickness (L) of the same order of magnitude as the effective mean free path (l), i.e., $L \cong l$, the EPRT developed by Majumdar (1993) results in the following steady-state heat flux:

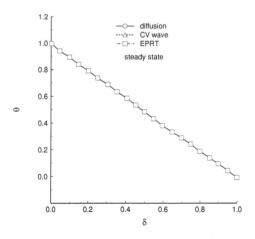

Figure 1.7 Coalescence of the steady-state temperature profiles in an acoustically thick medium, $L = 10$ μm. Reproduced from the work by Joshi and Majumdar (1993).

$$q = \frac{\sigma(T_1^4 - T_0^4)}{\frac{3}{4}\left(\frac{L}{l}\right) + 1}, \tag{1.31}$$

with σ denoting the Stefan-Boltzmann constant in radiative heat transfer. Equation (1.31) describes the transition mode from diffusion to ballistic (radiative) heat transport associated with shrinkage of the film thickness. In the Casimir limit, where $L/l \rightarrow 0$ for an extremely thin film, equation (1.31) reduces to the same form as the Stefan-Boltzmann law in radiative heat transfer. In the transition of the film thickness from the acoustically thick domain to the acoustically thin domain, equation (1.31) may be the most elegant feature in EPRT.

From an analytical point of view, retrieval of the heat radiation behavior in acoustically thin media, equation (1.31), cannot be described by the lagging behavior, although the temperature with lagging may appear significantly lower than that predicted by the classical theory of diffusion, like those shown in Figure 1.6. Including the study of microstructural effects on thermal properties, this can be seen as a salient feature in microscopic approaches.

1.5 RELAXATION BEHAVIOR IN THERMAL WAVES

The classical thermal wave model proposed by Cattaneo (1958) and Vernotte (1958, 1961), in addition to the steady-state response proportional to the temperature gradient (∇T), accounts for the *increase* of the heat flux vector due to the phonon collision in a duration of the mean free time (τ):

$$\vec{q} + \tau \frac{\partial \vec{q}}{\partial t} = -k\nabla T. \tag{1.32}$$

In the framework of wave theory in heat conduction, as mentioned above, the mean free time (τ) is usually termed the relaxation time, which is the effective mean free path (l) divided by the phonon speed (v_s, the speed of sound). Mathematically, $\tau = l/v$. In the absence of the relaxation time, $\tau = 0$, a mathematical idealization from either a *zero* mean free path ($l = 0$) or an *infinite* phonon speed ($v \rightarrow \infty$) in phonon collisions, equation (1.32) reduces to the classical Fourier's law. An infinite speed of heat propagation in phonon transport, therefore, is an intrinsic assumption made in the classical theory of diffusion assuming Fourier's law.

Equation (1.32) must be coupled with the energy equation (1.13a),

$$-\nabla \bullet \vec{q} = C_p \frac{\partial T}{\partial t}, \tag{1.13a'}$$

for resolving two unknowns, the temperature T and the heat flux vector $\vec{q}$, that describe the process of heat flow. Taking the divergence of equation (1.32) results in

$$\nabla \bullet \vec{q} + \tau \frac{\partial}{\partial t}\left(\nabla \bullet \vec{q}\right) = -k\nabla^2 T. \qquad (1.33)$$

All the thermal properties, including thermal conductivity (k), volumetric heat capacity (C_p), and relaxation time (τ) are assumed to be constant (temperature independent) in this treatment. Substituting equation (1.13a) into (1.33) gives a single energy equation governing temperature,

$$\nabla^2 T = \frac{1}{\alpha}\frac{\partial T}{\partial t} + \frac{1}{C^2}\frac{\partial^2 T}{\partial t^2}, \quad \text{with} \quad \alpha = \frac{k}{C_p} \quad \text{and} \quad C = \sqrt{\frac{\alpha}{\tau}}. \qquad (1.34)$$

Equation (1.34) is the *thermal wave* equation hypothesized by Morse and Feshbach (1953) without a rigorous proof. It depicts a temperature disturbance propagating as a wave, with thermal diffusivity appearing as a damping effect in heat propagation. It is hyperbolic in nature, in contrast to the parabolic diffusion equation employing Fourier's law, and is sometimes called the hyperbolic theory in heat conduction or the non-Fourier effect in heat conduction. The quantity C is the thermal wave speed (finite speed of heat propagation). It depends on the ratio of thermal diffusivity to relaxation time. In the absence of relaxation time, $\tau = 0$, the thermal wave speed approaches infinity, and equation (1.34) reduces to the classical diffusion equation.

Chester (1963), from a kinetic point of view, interpreted the thermal wave in terms of a coherent propagation of density disturbances in the phonon gas. It may be in this sense that equation (1.32) inherited a truncated appearance of a more extensive relation in the kinetic theory of ideal gas (Maxwell, 1867). The phonon gas can be viewed as a group of particles, each of which moves with the same speed of sound (v_s) in a homogeneous solid. For heat propagation in an isotropic solid, the thermal wave speed is identical in the three principal directions (x, y, and z in Cartesian coordinates, for example), each of which is denoted by C, implying $C_x = C_y = C_z = C$. The resultant velocity must be identical to the speed of sound at which phonons travel,

$$C_x^2 + C_y^2 + C_z^2 = C^2 + C^2 + C^2 = v_s^2 \quad \text{or} \quad C = \frac{v_s}{\sqrt{3}}. \qquad (1.35)$$

The thermal wave speed, in other words, is about 57.7% of the speed of sound. With the relaxation time denoted by τ, the *critical* value of the circular frequency for the activation of a wave behavior in heat conduction is

$$f_c = \frac{1}{2\pi\tau} = \frac{1}{2\pi}\frac{C^2}{\alpha} = \frac{1}{2\pi\alpha}\frac{v_s^2}{3} = \frac{v_s^2}{6\pi\alpha} \qquad (1.36)$$

where v_s and α are tabulated thermal properties of the material. For type IIa diamond film, for example, $v_s = 1.229 \times 10^4$ m/s and $\alpha = 1.053 \times 10^{-3}$ m^2/s at room temperature. The critical frequency at room temperature, according to equation (1.36), is thus 7.607 gigahertz (10^9 hertz). Any frequency of phonon collisions lower than this threshold ensures a wave behavior in heat propagation. The relaxation time can also be calculated according to the speed of sound and thermal diffusivity:

$$\tau = \frac{3\alpha}{v_s^2}.$$ (1.37)

Under the same values of v_s and α for type IIa diamond film, the relaxation time is about 20.9 ps at room temperature. The thermal wave behavior would become pronounced should the transient time be of the same order of magnitude or shorter than this threshold value. Note that equations (1.35) to (1.37) are derived on the basis of phonon transport in microscale. They do not allow dispersion in the course of phonon collisions, nor do they apply to superior conductors, where heat transport by phonon-electron interaction dominates the process in the short-time transient. The relaxation time still exists in the presence of the phonon-electron interaction, but the expression has an alternate, much more involved form (Kittel, 1986):

$$\tau(T) = \frac{3m_e}{(\pi\kappa)^2 n_e T} K(T)$$ (1.38)

with m_e being the effective mass of electrons, n_e the number density of electrons per unit volume, κ the Boltzmann constant, K the thermal conductivity of the electron gas, and T the absolute temperature in Kelvins. The relaxation time becomes a function of temperature and is proportional to the thermal conductivity in this case. The values of the relaxation time for various semiconductors, superconductors, metals, and insulators (such as glasses) can be found in the work by Vadavarz et al. (1991).

1.5.1 Engineering Assessment of the Relaxation Time

Obviously, the relaxation time τ is the crucial parameter in the wave theory of heat conduction. It results in a finite value for the thermal wave speed, giving rise to a sharp wavefront in the history of heat propagation. While equations (1.37) and (1.38) provide clear physical interpretations for the relaxation time from a microscopic point of view, for a better understanding from an engineering point of view, Tzou (1993a) made an attempt to address its engineering significance from a macroscopic approach. The relaxation time was interpreted in two ways. It is the physical instant of time at which the *intrinsic length scale* in diffusion merges onto that in waves. It is also the *time delay* between the heat flux vector and the temperature gradient in the fast-transient process.

Intrinsic length scale. For media with constant thermal properties, the intrinsic length scale in the classical theory of diffusion can be extracted by re-arranging equation (1.27) in the following form:

$$\frac{\partial T}{\partial (\alpha t)} = \frac{\partial^2 T}{\partial x^2}.$$ (1.39)

For the sake of dimensional consistency, clearly, the quantity (αt) must have a dimension in length squared, rendering

$$\lambda_D = \sqrt{\alpha t} \qquad (1.40)$$

as the *intrinsic* length in diffusion. In fact, equation (1.40) leads to the well-known error-function solution for temperature (Carslaw and Jaeger (1959), for example) when used in defining the similarity transformation for diffusion in a one-dimensional solid.

The highest order differentials in equation (1.27) dominate the fundamental characteristics in thermal wave propagation:

$$\frac{\partial^2 T}{\partial x^2} = \frac{1}{C^2}\frac{\partial^2 T}{\partial t^2} + \text{low order terms}, \quad C = \sqrt{\frac{\alpha}{\tau}}. \qquad (1.41)$$

Equation (1.41) represents a wave equation without damping, which becomes exact for responses in extremely short times $(t \to 0)$. By the same reasoning of dimensional consistency, the intrinsic length scale in the case of thermal waves becomes

$$\lambda_W = Ct. \qquad (1.42)$$

When used in the similarity transformation, equation (1.42) leads to the famous logarithmic-type distribution in the one-dimensional solid.

Equations (1.41) and (1.42) reveal an intrinsic transition of length scales in the history of heat propagation. In extremely short time responses, the length scale is depicted by the wave behavior, as shown by equation (1.42). When the transient time lengthens, the length scale is depicted by the diffusion behavior, which is shown by equation (1.41). It is conceivable that at a certain instant of time in the transient process, two length scales, λ_D and λ_W, become equal. This important instant of time, denoted by t_c, can be found by equating λ_D to λ_W, resulting in

$$\sqrt{\alpha t_c} = Ct_c \quad \text{or} \quad t_c = \frac{\alpha}{C^2} \equiv \tau. \qquad (1.43)$$

The relaxation time described in the wave theory of heat conduction is thus shown to be the physical instant at which the intrinsic length scales in diffusion and waves merge together.

Time delay in the heat flux vector. The relaxation time can also be viewed as the time delay between the heat flux vector and the temperature gradient. Mathematically,

$$\vec{q}(\vec{r}, t+\tau) = -k\nabla T(\vec{r}, t) \quad \text{for} \quad \tau > 0. \qquad (1.44)$$

The temperature gradient, ∇T, is established across a material volume located at $\bar{r}$ at time t. Owing to the finite time required for phonon collision (and hence heat transport) to take place within the material volume, an effective heat current, $\bar{q}$, starts to flow across the material volume at a *later* time $(t + \tau)$. The time delay τ used here, in other words, is to address the finite time required for phonon collision to take place in microscale heat transport. Assuming that the time delay τ is much shorter than the type response time in a transient process, i.e., $\tau \ll t$, the first-order Taylor series expansion can be applied to equation (1.44), rendering

$$\bar{q}(\bar{r},t) + \tau \frac{\partial \bar{q}}{\partial t}(\bar{r},t) \cong -k\nabla T(\bar{r},t) \quad \text{for} \quad \tau \ll t. \qquad (1.45)$$

Clearly, equation (1.45) has exactly the same form as equation (1.32), which is the constitutive equation in the *CV* wave model. Identity of the two equations supports the fact that, at least to the first-order approximation, the relaxation time is indeed equivalent to the time delay of the heat flux vector relative to the temperature gradient in the fast-transient process.

1.5.2 Admissibility With Phonon Radiative Transport Phenomena

Along the same lines as Tavernier's (1962) and Majumdar's (1993) work, equation (1.19) can be manipulated to yield the *CV* equation in the wave theory of heat conduction. Multiplying the direction cosine, μ, on both sides of equation (1.19) and integrating over the range $-1 \leq \mu \leq 1$ and $0 \leq \omega \leq \omega_D$ results in

$$\frac{1}{v}\frac{\partial q}{\partial t} + \int_0^{\omega_D} \int_{-1}^1 \mu^2 \frac{\partial I_\omega}{\partial x} d\mu \, d\omega = -\int_0^{\omega_D} \frac{q_\omega}{v\tau} d\omega \qquad (1.46)$$

Assuming that the relaxation time in phonon collisions is a constant, the term on the right side of equation (1.46) gives

$$-\int_0^{\omega_D} \frac{q_\omega}{v\tau} d\omega = -\frac{1}{v\tau}\int_0^{\omega_D} q_\omega \, d\omega = -\frac{q}{v\tau}. \qquad (1.47)$$

The second term on the left side of equation (1.46), for an acoustically *thick* film where the physical concept of temperature gradient is appropriate, reduces to

$$\int_0^{\omega_D} \int_{-1}^1 \mu^2 \frac{\partial I_\omega}{\partial x} d\mu \, d\omega = \left[\int_0^{\omega_D} \int_{-1}^1 \mu^2 \left(\frac{\partial I_\omega}{\partial T}\right) d\mu \, d\omega\right]\left(\frac{\partial T}{\partial x}\right). \qquad (1.48)$$

Combining equations (1.47) and (1.48) into equation (1.46), the resulting equation can be arranged into the following form:

$$q + \tau \frac{\partial q}{\partial t} = -k \frac{\partial T}{\partial x} \qquad (1.49)$$

with the effective thermal conductivity defined as

$$k = \int_0^{\omega_D} \int_{-1}^{1} v\tau\mu^2 \left(\frac{\partial I_\omega}{\partial T}\right) d\mu \, d\omega \; . \tag{1.50}$$

Equation (1.49) is the *CV* equation in the wave theory of heat conduction. It is now derived on the basis of the phonon radiative transport equation, which is based on the Boltzmann transport equation.

1.6 THIN-FILM THERMAL PROPERTIES

In addition to the heat transfer models describing the microscopic process of heat transport, another important area of research in microscale heat transfer is the effect of microstructures on thermal properties. The heat transfer models and thermal properties reflecting the effect of microstructures are obviously interconnected and not separable.

From a microscopic point of view, the mean free path in phonon and/or electron collisions is the dominant parameter for heat transport in microscale. Tien et al. (1968) demonstrated the size effect on thermal conductivity. Geometric configurations of the microdevices include both thin metallic films and metal wires at cryogenic temperatures. The film thickness and the wire diameter considered in the transverse direction were comparable to the mean free path of energy carriers. Owing to free-edge effects and, consequently, a much shorter mean free path in the transverse direction, the thermal and electrical conductivity is significantly lower than that in the bulk material.

Research along the same lines includes the thermal conductivity measured by Savvides and Goldsmid (1972) for single-crystal silicon at room temperature; the thermal conductivity of copper films (400 to 8000 angstroms) in the temperature range from 100 to 500 K by Nath and Chopra (1974); the transverse lattice thermal conductivity of 20 to 40 nanometer bismuth films by Volklein and Kessler (1986) with the effect of elevated temperature from 80 to 400 K; measurement of the thermal conductivity and the interfacial resistance between a film and a substrate in dielectric optical films by Lambropoulos et al. (1989); measurement of the electrical conductivity (and hence the thermal conductivity according to the Wiedemann-Franz law) for sputtered films of rare-earth transition metals by Anderson (1990); and determination of the effective thermal conductivity of amorphous silicon dioxide layers by Goodson et al. (1993). For a thin-film structure, owing to different geometric constraints imposed on the mean free path of electrons/phonons in the transverse and longitudinal directions, most important, the thermal conductivity becomes a *structural* property rather than an intrinsic thermal property of the material (Flik and Tien, 1990). Owing to the shorter mean free path in the transverse direction of the film, the thermal conductivity in the transverse direction is only a fraction of the value of that in the longitudinal direction. This is especially true when approaching a free surface where the microscale effect in space becomes pronounced; in other words, the thermal conductivity becomes *orthotropic* because of the reduction of the mean free path. In a continuous effort, moreover, Flik et al. (1991) found that the small-scale effect in the transverse direction needs to be taken into account for film thicknesses of less than about 7 times the mean free path ($L <$ $7l$). In the longitudinal direction of the film, on the other hand, such a threshold value decreases to about 4.5 times ($L < 4.5l$) for the microscale effect to activate. The characteristic dimension for which microscale effects in heat conduction become important is, in general, a combined result of the thermal loading condition, thermal properties of the thin film, and geometric configuration of the microstructures (thin films, wires, or even the vicinity of a crack/notch tip).   Such complicated dependencies were also discussed by Tien and Chen (1994).   From these representative works, it is clear that the mean free path of energy carriers, including both phonons in dielectric crystals or semiconductors and electrons in metals, is a dominant parameter for the thermal conductivity in thin-film structures.

In applying the microscopic heat transfer models to evaluate the thermal performance of microdevices, a knowledge of the effective thermal conductivity alone may not be sufficient. As verified in a series of papers by Qiu and Tien (1992, 1993, 1994), for example, the phonon-electron coupling factor controls the energy exchange between phonons and electrons in the short-time transient, in addition to the thermal diffusivity. Such a physical mechanism was summarized in Section 1.2 and will serve as the basis for Chapter 5. To date, the experimental values (by Brorson et al., 1987; Elsayed-Ali et al., 1987; Groeneveld et al., 1990) and the theoretical values (by Qiu and Tien, 1992; Qiu et al., 1994) for the phonon-electron coupling factor have been obtained for several representative metals, including Cu, Ag, Au, Cr, W, V, Nb, Ti, Pb, NbN, and V_3Ga. They are of the order of 10^{16} to 10^{18} W/m^3 K and depend only *weakly* on the temperatures of electrons and phonons; see equation (1.7). The phonon radiative transfer model proposed by Majumdar (1993) is another example illustrating the importance of additional microscopic properties other than the thermal conductivity.   Reflected by equation (1.24), the phonon velocity (v), the relaxation time (τ) in phonon collisions, and the density of states per unit volume in the frequency domain ($D(\omega)$) are all influential quantities in the determination of the phonon intensity and, consequently, the temperature in dielectrics.   The two phase lags in the dual-phase-lag model proposed by Tzou (1995a) dominate the short-time transient in small scales in addition to the thermal conductivity and heat capacity of the conductor. While the latter depicts the transient response in a relatively long time ($t \gg \tau$), the values of the phase lags describe the short-time response in the order of the relaxation time ($t \cong O(\tau)$). The precedence sequence resulting from the time delays between the heat flux vector and the temperature gradient depicts completely different characteristics in microscale heat transport; see Chapter 4 for the gradient-precedence type of heat flow and Chapters 5 to 7 for the flux-precedence type of heat flow. While the microscale heat transfer models have been continuously developed for applications to conductors with different microstructures, the new thermal properties involved in the models should be continuously explored through various analytical and experimental means.

TWO
LAGGING BEHAVIOR

The lagging response, in general, describes the heat flux vector and the temperature gradient occurring at *different* instants of time in the heat transfer process. If the heat flux precedes the temperature gradient in the time history, the heat flux is the *cause* and the temperature gradient is the *effect* of heat flow. If the temperature gradient precedes the heat flux, on the other hand, the temperature gradient becomes the cause and the heat flux becomes the effect. This concept of precedence does not exist in the classical theory of diffusion because the heat flux vector and the temperature gradient are assumed to occur simultaneously. This chapter establishes the theoretical foundation for the lagging response in conductive heat transfer. It results in a new type of energy equation, capturing the classical theories of diffusion (macroscopic in both space and time), thermal waves (macroscopic in space but microscopic in time), phonon scattering model and phonon-electron interaction model (microscopic in both space and time) in the same framework. The resulting model employing the two phase lags in describing the transient process is called the *dual-phase-lag* model. Universality of the dual-phase-lag model facilitates a consistent approach describing the intrinsic transition from one type of behavior (diffusion, for example) to another (the phonon-electron interaction) associated with shortening of the response time.

2.1 PHASE-LAG CONCEPT

In the classical theory of diffusion, the heat flux vector ($\vec{q}$) and the temperature gradient (∇T) across a material volume are assumed to occur at *the same* instant of time. Fourier's law of heat conduction,

$$\vec{q}(\vec{r},t) = -k\nabla T(\vec{r},t),\qquad(2.1)$$

with $\bar{r}$ denoting the position vector of the material volume and t the physical time, dictates such an *immediate* response. It results in an *infinite* speed of heat propagation, implying that a thermal disturbance applied at a certain location in a solid medium can be sensed immediately anywhere else in the medium. Because the heat flux vector and the temperature gradient are simultaneous, there is no difference between the cause and the effect of heat flow. In the wave theory of heat conduction, on the other hand, the heat flux vector and the temperature gradient across a material volume are assumed to occur at different instants of time. In parallel to Fourier's law, the constitutive equation can be written as (Tzou, 1989a, b, 1990a, b, 1992a)

$$\bar{q}(\bar{r},t+\tau) = -k\nabla T(\bar{r},t),\qquad(2.2)$$

where τ is the time delay, called the "relaxation time" in the wave theory of heat conduction. The first-order expansion of τ in equation (2.2) with respect to t bridges all the physical quantities at the same instant of time. It results in the expression

$$\bar{q}(\bar{r},t)+\tau\frac{\partial\bar{q}}{\partial t}(\bar{r},t) \cong -k\nabla T(\bar{r},t),\qquad(2.3)$$

which is the *CV* wave model originated by Cattaneo (1958) and Vernotte (1958, 1961). The *CV* wave model removes the paradox of infinite speed of heat propagation assumed in Fourier's law. The relaxation time, indeed, relates to the thermal wave speed by (Chester, 1963)

$$\tau = \frac{\alpha}{C^2}\qquad(2.4)$$

where α is the thermal diffusivity and C denotes the thermal *wave* speed. In the case of C approaching infinity, the relaxation time decreases to zero ($\tau = 0$), and the *CV* wave model, equation (2.2) or (2.3), reduces to Fourier's law, equation (2.1). The thermal wave model has been one of the major research areas in conductive heat transfer. Detailed reviews can be found in the articles by Joseph and Preziosi (1989, 1990), Tzou (1992a) and Özisik and Tzou (1994).

In order to solve for the two unknowns $\bar{q}$ and T, equation (2.1) (the diffusion model) or (2.3) (the thermal wave model) is combined with the energy equation established at a general time t during the transient process,

$$-\nabla \bullet \bar{q}(\bar{r},t)+Q(\bar{r},t) = C_p\frac{\partial T}{\partial t}(\bar{r},t),\qquad(2.5)$$

with C_p being the *volumetric* heat capacity, $C_p = \rho c_p$, and Q the volumetric heat source. Although allowing for a delayed response between the heat flux vector and the temperature gradient, evidently, the *CV* wave model still assumes an *immediate* response between the temperature gradient and the energy transport. This response occurs right after a temperature gradient is established across a material volume; in other words, the *CV* wave model assumes an instantaneous heat flow. The temperature gradient is always the cause for heat transfer, while the heat flux is always the effect.

The dual-phase-lag model aims to remove the precedence assumption made in the thermal wave model. It allows either the temperature gradient (cause) to precede the heat flux vector (effect) or the heat flux vector (cause) to precede the temperature gradient (effect) in the transient process. Mathematically, this can be represented by (Tzou, 1995a to c)

$$\vec{q}(\vec{r}, t + \tau_q) = -k\nabla T(\vec{r}, t + \tau_T) \qquad (2.6)$$

where τ_T is the phase lag of the temperature gradient and τ_q is the phase lag of the heat flux vector. For the case of $\tau_T > \tau_q$, the temperature gradient established across a material volume is a result of the heat flow, implying that the heat flux vector is the cause and the temperature gradient is the effect. For $\tau_T < \tau_q$, on the other hand, heat flow is induced by the temperature gradient established at an earlier time, implying that the temperature gradient is the cause, while the heat flux vector is the effect. The first-order approximation of equation (2.6), in a form parallel to equation (2.3), reads

$$\vec{q}(\vec{r}, t) + \tau_q \frac{\partial \vec{q}}{\partial t}(\vec{r}, t) \cong -k\left\{\nabla T(\vec{r}, t) + \tau_T \frac{\partial}{\partial t}\left[\nabla T(\vec{r}, t)\right]\right\}. \qquad (2.7)$$

Including the cause-and-effect relationship in the transient process involves more than an addition of the phase lag into the temperature gradient. When equation (2.7) is combined with the energy equation (2.5), the precedence switching results in capricious situations in transient conduction.

In passing, three important characteristics in the dual-phase-lag model are noted:
1. The heat flux vector and temperature gradient shown in equation (2.6) are the *local responses* within the solid medium. They must not be confused with the *global* quantities specified in the boundary conditions. When applying a heat flux to the boundary of a solid medium, namely in a flux-specified boundary value problem, the temperature gradient established at a material point *within* the solid medium can still precede the heat flux vector. Application of a heat flux at the boundary does not warrant the precedence of the heat flux vector to the temperature gradient at all. In fact, whether the heat flux vector precedes the temperature gradient or not depends on the combined effect of thermal loading, geometry of the specimen, and thermal properties of the material.
2. There are actually *three* characteristic times involved in the dual-phase-lag model. The instant of time $(t + \tau_T)$ at which the temperature gradient is established across a material volume, $(t + \tau_q)$ for the onset of heat flow, and t for the occurrence of heat transport (at which conservation of energy is described). The constitutive equation describing the lagging behavior, equation (2.6), differs from the *CV* wave model, equation (2.2), by a shift in the time scale, $\tau = \tau_q - \tau_T$, which is immaterial as far as the constitutive equation itself is

concerned. When combined with the energy equation (2.5), however, such a shift in the time scale becomes nontrivial, resulting in a completely different response from that depicted by the *CV* wave model. A detailed discussion is postponed until section 2.3.

3. The two phase lags τ_T and τ_q, like the thermal diffusivity and thermal conductivity, are *intrinsic* thermal properties of the bulk material. For solids with internal structures such as interstitial gas in porous media or second-phase constituents in composites, however, they become *structural* properties, which also depend on the detailed configurations of the substructures.

2.2 INTERNAL MECHANISMS

The lagging behavior in the transient process is caused by the *finite* time required for the substructural interaction to occur. These interactions may take place on the order of several seconds (such as the delayed response induced by the low-conducting pores in sand media) to nanoseconds (the delayed response caused by the inert behavior of molecules at low temperatures (Tzou, 1995a) to picoseconds or femtoseconds (the delayed response due to phonon scattering or phonon-electron interactions (Tzou, 1995a to c). The delayed times in the lagging response are thus of the same orders of magnitude.

Figure 2.1 describes the delayed response caused by the phonon-electron interactions in metallic structures. The energy state of a metal lattice is discretized into quanta, called phonons in quantum mechanics. When excited by a short-pulse laser, photons from the laser beam first heat up the electron gas at a certain time t shown in Figure 2.1(a). At this moment, no appreciable temperature change in the metal lattice, distant from the heat source, can be detected. Through the phonon-electron interactions, Figure 2.1(b), energy transport from the hot electron gas to phonons follows, giving rise to an appreciable increase of temperature in the metal lattice at a *later* time $t + \tau$. The phonon-electron interaction requires a finite time to take place, usually of the order of picoseconds (Qiu and Tien, 1992, 1993). The

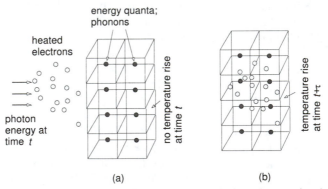

(a)   (b)

Figure 2.1 The finite time required for the phonon-electron interaction in microscale. (a) The first stage of electron-gas heating by photons and (b) the second stage of metal lattice heating by phonon-electron interactions.

resulting phase lag (τ) is thus of the same order of magnitude. For dielectric films, insulators, and semiconductors, where heat transport by electrons is negligible, phonon scattering from grain boundaries dominates microscale heat transport. The same argument can be applied in this case, but the phase lag results from the finite time required in the process of phonon scattering. As shown in Sections 1.2 and 1.3, a detailed understanding of the microscopic phonon scattering and phonon-electron interaction mechanisms requires a profound knowledge in quantum mechanics and statistical thermodynamics. Based on the dual-phase-lag model developed in this chapter, Chapter 5 is dedicated to this type of lagging behavior, with a detailed comparison to the novel experiments by Brorson et al. (1987) and Qiu and Tien (1992, 1993) for ultrafast laser heating on metal films.

The inert behavior of molecules at extremely low temperatures is another cause for the possible lagging response. Heat transport from one location to another relies on molecular collisions. At extremely low temperatures, such as in liquid helium of approximately 1 to 4 K, molecules become inert, and it is conceivable that they need a finite time to establish their activation energy before heat transfer can really take place. Bertman and Sandiford (1970) performed a transient experiment for heat propagation in liquid helium. It will be used as the basis in Chapter 4, where the possible lagging behavior at low temperatures is studied.

The finite time required for heat flow to circulate around the low-conducting aerial closures in porous media is another known cause for lagging behavior. This is illustrated in Figure 2.2 for the case where the heat flux precedes the temperature gradient ($\tau_T > \tau_q$). The material volume under consideration is enclosed by the dashed line. Owing to the presence of pores between the heater and the material volume, the heat flow produced by the heater at a general time t arrives at the material volume at a *later* time $t + \tau_q$. The internal pores within the material volume cause an additional delay in heat transport, prolonging the establishment of the temperature gradient at $t + \tau_T$. This type of delayed response depends on the detailed configuration of the solid particles and the interstitial gas within the material volume. Should a smaller material volume be chosen in the neighborhood of the heater, the configuration of the discrete structure will have a more pronounced effect. Consequently, the lagging response reflected by the two phase lags will display a different pattern. Along with some experimental results, the *structural* properties of the phase lags will be discussed in detail in Chapter 6.

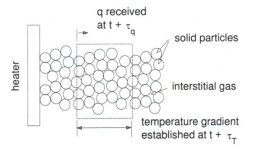

Figure 2.2 The lagging response induced by the interstitial gas in a porous medium.

2.3 TEMPERATURE FORMULATION

Analysis in terms of boundary value problems provides an efficient way to study the fundamental characteristics of lagging behavior. The energy equation (2.5) and the constitutive equation (2.6) depicting the lagging response are to be combined for this purpose. These two equations, generally speaking, exhibit two coupled, *delayed* differential equations for two unknowns, T and $\vec{q}$. No general solution has yet been found for this type of equation with delays. The *linear* version based on the first-order expansions of τ_T and τ_q, equation (2.7), provides an approximation to describe the lagging behavior in the simplest case. It is not as general as equation (2.6), but should reveal *all* the fundamental behavior in the lagging response. Such a linear approximation approaches the exact formulation when τ_T and τ_q are small, a situation existing in the phonon scattering and phonon-electron interaction models where the values of τ_T and τ_q are of the order of picoseconds to femtoseconds. The linear expansion is sufficient in these cases.

At this initial stage of development, however, exploring the physical conditions under which the lagging behavior may become pronounced and searching for the appropriate experimental evidence are much more important than making a full expansion into the nonlinear regime of τ_T and τ_q. Based on the fundamental understanding thus developed, the high-order responses including the nonlinear effect of τ_T and τ_q can always be incorporated for a more refined analysis. Development of the *CV* wave model, in fact, bears this merit. While equation (2.3) serves as a linear approximation of equation (2.2), depicting a general delayed response in short times, it has been used as the basis for the hyperbolic theory of heat conduction. Finding the physical environment in which the *linear* wave theory can be applied is still the most important task for researchers in this area (Özisik and Tzou, 1994).

Equations (2.5) and (2.7) are repeated here for convenience:

$$-\nabla \bullet \vec{q}(\vec{r},t) + Q(\vec{r},t) = C_p \frac{\partial T}{\partial t}(\vec{r},t) \tag{2.5$'$}$$

$$\vec{q}(\vec{r},t) + \tau_q \frac{\partial \vec{q}}{\partial t}(\vec{r},t) \cong -k\left\{ \nabla T(\vec{r},t) + \tau_T \frac{\partial}{\partial t}\left[\nabla T(\vec{r},t) \right] \right\}. \tag{2.7$'$}$$

To reiterate, three characteristic times are involved in the process of heat transport: the instant of time $(t + \tau_T)$ at which the temperature gradient is established across a material volume, time $(t + \tau_q)$ at which heat flows through the material volume, and general time t at which conservation of energy is imposed. The linear expansion of τ_T and τ_q, shown by equation (2.7), bridges all the physical quantities at the *same* instant of time, and equations (2.5) and (2.7) are ready for a further combination.

The temperature representation of the energy equation results from elimination of the heat flux vector from the two equations. Assuming constant thermal properties, the divergence of equation (2.7) gives

$$\nabla \bullet \bar{q} + \tau_q \frac{\partial}{\partial t} \left[\nabla \bullet \bar{q} \right] = -k \nabla^2 T - k \tau_T \frac{\partial}{\partial t} \left[\nabla^2 T \right]. \tag{2.8}$$

Substituting the expression of $\nabla \bullet \bar{q}$, in equation (2.5), into equation (2.8),

$$\nabla \bullet \bar{q} = Q - C_p \frac{\partial T}{\partial t}, \tag{2.9}$$

and introducing the thermal diffusivity $\alpha = k/C_p$ gives

$$\nabla^2 T + \tau_T \frac{\partial}{\partial t} \left[\nabla^2 T \right] + \frac{1}{k} \left[Q + \tau_q \frac{\partial Q}{\partial t} \right]$$

$$= \frac{1}{\alpha} \frac{\partial T}{\partial t} + \frac{\tau_q}{\alpha} \frac{\partial^2 T}{\partial t^2} \quad (T \text{ representation}). \tag{2.10}$$

This equation describes the temperature response with lagging in the linearized framework accommodating the first-order effect of τ_T and τ_q. It captures several representative models in heat transfer as special cases.

In the absence of the two phase lags, $\tau_T = \tau_q = 0$, equation (2.10) reduces to the diffusion equation employing Fourier's law. In the absence of the phase lag of the temperature gradient, $\tau_T = 0$, equation (2.10) reduces to the *CV* wave model. The phase lag of the heat flux vector (τ_q) reduces to the relaxation time defined by Chester (1963), equation (2.4). The phase lag of the heat flux vector introduces an *apparent* heat source, $(\tau_q/k)(\partial Q/\partial t)$, in addition to the real heat source applied to the solid (Frankel et al., 1985; Tzou 1989a, b, 1990a, b). The apparent heating is the physical basis for the thermal resonance phenomena explored by Tzou (1991b, c, 1992d, e) in the wave theory of heat conduction. The two popular models used for describing the macroscopic heat transfer are thus captured in the framework of the dual-phase-lag model under special cases.

Compared with the microscopic models for heat transport, the energy equation employing the dual-phase-lag model, equation (2.10), has *exactly* the same form as the energy equation in the phonon scattering model, equation (1.14), and the energy equation in the phonon-electron interaction model, equation (1.10) or (1.12). Such *perfect* correlations, encouragingly, facilitate a direct determination of the two phase lags in terms of the microscopic thermal properties. Comparing equation (2.10) with equation (1.14) results in

$$\alpha = \frac{\tau_R c^2}{3}, \quad \tau_T = \frac{9 \tau_N}{5}, \quad \tau_q = \tau_R. \tag{2.11}$$

The phase lag of the heat flux vector (τ_q) is equivalent to the relaxation time (τ_R) in the umklapp process describing the momentum lost in phonon scattering. The phase lag of the temperature gradient, τ_T, on the other hand, displays a simple stretch from the relaxation time, τ_N, in the normal process conserving the momentum in phonon collisions.   For heat transport in dielectric films, insulators, and semiconductors, where phonon scattering is the dominant mechanism in microscale, therefore, the dual-phase-lag model is equivalent to the phonon scattering model.   The relaxation times for the umklapp and normal processes, however, have alternate interpretations that could be better envisioned by engineers.   Comparing equation (2.10) with (1.10) or (1.12) for the phonon-electron interaction model, again, a perfect correlation results:

$$\alpha = \frac{K}{C_e + C_l}, \quad \tau_T = \frac{C_l}{G}, \quad \tau_q = \frac{1}{G}\left[\frac{1}{C_e} + \frac{1}{C_l}\right]^{-1}. \tag{2.12}$$

Heat capacities of the metal lattice (C_l) and the electron gas (C_e) and the phonon-electron coupling factor (G) are thus the microscopic properties determining the two phase lags.   For fast-transient heat transport in metals, therefore, the dual-phase-lag model successfully captures the microscopic phonon-electron interaction model in its framework.

The perfect correlations between the dual-phase-lag model and the microscopic phonon scattering and phonon-electron interaction models demonstrates the feasibility of modeling the microstructural interaction effect *in space* by its delayed response *in time*.   Equations (2.11) and (2.12) explicitly indicate the microscopic properties causing such time delays in the heat flux vector and the temperature gradient.

While capturing the existing macroscopic (diffusion and wave) and microscopic (phonon scattering and phonon-electron interaction) models in a consistent approach, the dual-phase-lag model introduces a new type of energy equation in conductive heat transfer.   As shown in equation (2.10), the *mixed-derivative* term, containing the first-order derivative with respect to time and the second-order derivative with respect to space, appears as the highest order differential in the equation, which will dramatically alter the fundamental characteristics of the solution for temperature.   A wave term, the second-order derivative with respect to time, still exists on the right side of the energy equation, but the mixed-derivative term completely destroys the wave structure, and the energy equation is *parabolic* in nature.   It predicts a higher temperature level in the heat-penetration zone than diffusion but does not have a sharp wavefront in heat propagation.   The fundamental solutions of equation (2.10) are illustrated in Sections 2.6 to 2.8.

2.4 HEAT FLUX FORMULATION

The lagging behavior depicted by equation (2.7) makes it difficult to apply the T representation of the energy equation (equation (2.10)) to problems involving flux-specified boundary conditions.   This can be illustrated by a direct integration of equation (2.7) for the heat flux vector:

$$\vec{q}(\vec{r},t) = -\left(\frac{k}{\tau_q}\right)e^{-\frac{t}{\tau_q}}\int_0^t e^{\frac{\eta}{\tau_q}}\left[\nabla T(\vec{r},\eta) + \tau_T\frac{\partial}{\partial\eta}[\nabla T(\vec{r},\eta)]\right]d\eta. \tag{2.13}$$

In the dual-phase-lag model, clearly, the heat flux vector depends not only on the temperature gradient established at the same instant of time, but also on the *entire history* in which the temperature gradient and its time-rate of change are established. This *path dependency* reveals a special behavior with "memory," which is completely different from the pointwise relationship depicted by Fourier's law. Should equation (2.13) be used to specify the heat flux in a boundary condition, the *T* representation, equation (2.10), will involve an integral equation in the boundary conditions. The resulting boundary value problem becomes difficult to handle, implying the need for a direct formulation in terms of the *heat flux*. This situation does not exist in the classical theory of diffusion. Specifying the heat flux at a boundary is equivalent to specifying the gradient of temperature according to Fourier's law. The boundary condition switches from a Dirichlet type to a Neumann type, but no special difficulty would result.

The *q* representation, formulation of the energy equation in terms of heat fluxes, results from elimination of temperature from equations (2.5) and (2.7). Taking the gradient of equation (2.5) gives

$$C_p\frac{\partial}{\partial t}(\nabla T) = -\nabla(\nabla\bullet\vec{q}) + \nabla Q. \tag{2.14}$$

Substituting the result for $\partial(\nabla T)/\partial t$ from equation (2.14) into equation (2.7) gives

$$\vec{q} + \tau_q\frac{\partial\vec{q}}{\partial t} = -k\nabla T + \alpha\tau_T\left[\nabla(\nabla\bullet\vec{q}) - \nabla Q\right]. \tag{2.15}$$

Differentiating equation (2.15) with respect to time and substituting equation (2.14) for the resulting time derivative of the temperature gradient, finally, gives the result

$$\nabla(\nabla\bullet\vec{q}) + \tau_T\frac{\partial}{\partial t}[\nabla(\nabla\bullet\vec{q})] - \left[\nabla Q + \tau_T\frac{\partial}{\partial t}(\nabla Q)\right]$$

$$= \frac{1}{\alpha}\frac{\partial\vec{q}}{\partial t} + \frac{\tau_q}{\alpha}\frac{\partial^2\vec{q}}{\partial t^2} \quad (q\ \text{representation}). \tag{2.16}$$

The *q* representation shown by equation (2.16) has exactly the same structure in time as the *T* representation shown by equation (2.10). The apparent heating, however, switches from the time derivative ($\partial Q/\partial t$ in equation (2.10)) to the gradient (∇Q in equation (2.16)) of the real heat source. Unlike the *T* representation, the *q* representation represents a set of three coupled partial differential equations to be solved for the three components of the heat flux vector.

The relationship between the T representation and the q representation can be better understood by introducing a heat flux potential,

$$\vec{q} = \nabla\phi. \qquad (2.17)$$

Substituting equation (2.17) into (2.16) gives

$$\nabla\left\{\left[\nabla^2\phi + \tau_T \frac{\partial}{\partial t}\left(\nabla^2\phi\right)\right] - \left[Q + \tau_T \frac{\partial Q}{\partial t}\right]\right\} = \nabla\left\{\frac{1}{\alpha}\frac{\partial\phi}{\partial t} + \frac{\tau_q}{\alpha}\frac{\partial^2\phi}{\partial t^2}\right\}. \qquad (2.18)$$

A general solution of equation (2.18) is

$$\nabla^2\phi + \tau_T \frac{\partial}{\partial t}\left[\nabla^2\phi\right] - \left[Q + \tau_T \frac{\partial Q}{\partial t}\right] = \frac{1}{\alpha}\frac{\partial\phi}{\partial t} + \frac{\tau_q}{\alpha}\frac{\partial^2\phi}{\partial t^2} + f(t) \qquad (2.19)$$

with $f(t)$ being an arbitrary function of time. Except for a sign difference in front of the apparent heat source and the arbitrary time function, equation (2.19) governing the heat flux potential is identical to equation (2.10) governing the temperature. It is thus informative to conclude that the heat flux potential has a very similar behavior to temperature because the fundamental characteristics of a differential equation are dictated by the highest order differentials.

In a multidimensional problem, the heat flux potential provides a powerful transformation to solve the coupled differential equations shown by equation (2.16). In fact, the heat flux potential is comparable to the Lamé potential in the theory of elasticity, where Navier's equation describing the conservation of momentum is decoupled in the same fashion. The q representation, especially for one-dimensional problems, is more convenient to use for problems involving a flux-specified boundary condition. It avoids the use of equation (2.13), and hence the complicated conversion between the temperature gradient and the heat flux vector, in solving the boundary value problems.

There are some problems existing in practice that involve both temperature and heat flux in the boundary conditions. The lagging behavior in discrete sand media is a typical example. In this situation, solving the energy equation (equation (2.5)) and the constitutive equation (equation (2.7)) simultaneously is the only way to avoid the complicated conversion between the heat flux vector and the temperature gradient. This procedure is demonstrated in Chapters 4 to 7.

2.5 MATHEMATICAL METHODS

Since the lagging behavior is a special response in time, consideration of one-dimensional problems in *space* is sufficient to illustrate its fundamental characteristics. In addition, from a mathematical point of view, the lagging behavior introduces the *highest* order differentials in the energy equation, reflected by the mixed-derivative and the wave terms in equations (2.10) and (2.16). These terms characterize the fundamental solutions of the energy equation employing the dual-phase-lag model. Consideration of a multidimensional problem will *not* alter the qualitative behavior depicted by the one-dimensional problem, yet it complicates the analysis to a great extent. The heat source term in the energy equation, by the same argument, shall be dropped for the time being because it does not provide additional insight into the fundamental solution.

Heat propagation driven by a suddenly raised temperature at the boundary of a semi-infinite solid may be the simplest example to illustrate the lagging behavior (Tzou et al., 1994). The same example was considered by Baumeister and Hamill (1969, 1971) to reveal the fundamental properties in thermal wave propagation. The lagging behavior is described by equation (2.10). In a one-dimensional situation without body heating, it reduces to

$$\frac{\partial^2 T}{\partial x^2} + \tau_T \frac{\partial^3 T}{\partial x^2 \partial t} = \frac{1}{\alpha}\frac{\partial T}{\partial t} + \frac{\tau_q}{\alpha}\frac{\partial^2 T}{\partial t^2}. \tag{2.20}$$

The suddenly raised temperature at $x = 0$ is described by

$$T(0,t) = T_W \quad \text{for} \quad t > 0, \tag{2.21}$$

while no disturbance can be detected at a distance far from the heated boundary,

$$T(x,t) \rightarrow T_0, \quad x \rightarrow \infty. \tag{2.22}$$

The quantity T_0 in equation (2.22) is the initial temperature of the solid at $t = 0$:

$$T(x,0) = T_0 \quad \text{and} \quad \frac{\partial T}{\partial t}(x,0) = 0 \quad \text{for} \quad x \in [0,\infty) \tag{2.23}$$

where the initial time-rate of change of temperature, $\partial T/\partial t$ at $t = 0$, is assumed to be zero for the time being. This is an important factor postponed to Section 2.7 for more detailed study.

The lagging response of temperature, reflected by equations (2.20) to (2.23), is characterized by three parameters: the thermal diffusivity α, the phase lag of the temperature gradient τ_T, and the phase lag of the heat flux vector τ_q. An analysis based on dimensionless variables assists in further identifying the dominant parameter. Introducing

$$\theta(x,\ t) = \frac{T(x,t) - T_0}{T_W - T_0}, \quad \beta = \frac{t}{2\tau_q}, \quad \delta = \frac{x}{2\sqrt{\alpha\tau_q}}, \tag{2.24}$$

equations (2.20) to (2.23) become

$$\frac{\partial^2 \theta}{\partial \delta^2} + B\frac{\partial^3 \theta}{\partial \delta^2 \partial \beta} = 2\frac{\partial \theta}{\partial \beta} + \frac{\partial^2 \theta}{\partial \beta^2} \quad \text{with} \quad B = \frac{\tau_T}{2\tau_q} \tag{2.25}$$

$$\theta(\delta,0) = 0 \quad \text{and} \quad \frac{\partial\theta}{\partial\beta}(\delta,0) = 0 \quad \text{for} \quad \delta \in [0,\infty) \tag{2.26}$$

$$\theta(0,\beta) = 1 \quad \text{and} \quad \theta(\delta,\beta) \to 0 \quad \text{as} \quad \delta \to \infty. \tag{2.27}$$

The factor "2" in defining the dimensionless variables for β and α is for a precise reduction to the Baumeister and Hamill's (1969) formulation for a *CV* wave in the limit of $\tau_T \to 0$ ($B \to 0$). The dimensionless analysis clearly shows that the lagging response is indeed characterized by a *single* parameter B, the ratio between the two phase lags shown in equation (2.25). In terms of the microscopic properties, according to equations (2.11) and (2.12), the dominant parameter B can be expressed by

$$B = \frac{\tau_T}{2\tau_q} = \begin{cases} \dfrac{9}{5}\dfrac{\tau_N}{\tau_R} & \text{(phonon scattering)} \\[2ex] \dfrac{1}{2}\left[1 + \left(\dfrac{C_l}{C_e}\right)\right] & \text{(phonon - electron interaction).} \end{cases} \tag{2.28}$$

The ratio of τ_N (relaxation time in the normal process) to τ_R (relaxation time in the umklapp process) in the phonon scattering model and the ratio of C_l (volumetric heat capacity of the metal lattice) to C_e (volumetric heat capacity of the electron gas) in the phonon-electron interaction model thus characterize the lagging response.

Reductions to the existing models under various situations become even more clear in the following cases:

1. When the values of τ_T and τ_q are selected according to the perfect correlation shown by equation (2.11), the dual-phase-lag model describes an equivalent phenomenon reflecting the phonon scattering in microscales. Equation (2.25) is the energy equation describing the lagging response, and the coefficient B is shown by the first expression in equation (2.28). This type of response is important for heat transport in dielectric films, insulators, and semiconductors in short times.

2. When the values of τ_T and τ_q are selected according to the perfect correlation shown by equation (2.12), the dual-phase-lag model captures the microstructural effect of phonon-electron interactions. The energy equation is shown by equation (2.25), and the coefficient B is given by the second expression in equation (2.28). This is the dominant mode of heat transport in metallic structures.

3. In the case of $\tau_T = 0$, implying that $B = 0$ according to equation (2.28), equation (2.25) reduces to the dimensionless form of the *CV* wave equation derived by Baumeister and Hamill (1969). The remaining phase lag, τ_q, reduces to the conventional relaxation time (τ in equation (2.4)), and the dual-phase-lag model reduces to the *CV* wave model. Because the classical *CV* wave model only captures the inertia effect in the fast-transient process (in time) while the spatial response (in space) remains *macroscopic*, the additional mixed-derivative term led by τ_T in equation (2.25) reflects the microstructural interaction effect in *space*. When such an effect is activated, the conventional relaxation time (τ)

transits into the phase lag of the heat flux vector (τ_q), implying that the small-scale responses in space and time are interconnected.   They must be accommodated *simultaneously* in the theoretical framework.

4.   In the case of $\tau_T = \tau_q$, not necessarily equal to zero, equation (2.20) can be re-arranged into the following form:

$$\left(\frac{\partial^2 T}{\partial x^2} - \frac{1}{\alpha} \frac{\partial T}{\partial t} \right) + \tau_q \frac{\partial}{\partial t} \left(\frac{\partial^2 T}{\partial x^2} - \frac{1}{\alpha} \frac{\partial T}{\partial t} \right) = 0 \, . \tag{2.29}$$

For a homogeneous initial temperature, it has a general solution

$$\frac{\partial^2 T}{\partial x^2} - \frac{1}{\alpha} \frac{\partial T}{\partial t} = 0 \, , \tag{2.30}$$

which is the classical *diffusion* equation. When the two phase lags are equal, $\tau_T = \tau_q$; in other words, the dual-phase-lag model reduces to the diffusion model employing Fourier's law.   This becomes obvious in view of equation (2.6) because equal phase lags imply a trivial shift in the time scale, while an *instantaneous* response between the heat flux vector and the temperature gradient still exists.   Equation (2.25), the dimensionless form of the energy equation, reduces to

$$\frac{\partial^2 \theta}{\partial \delta^2} = 2 \frac{\partial \theta}{\partial \beta} \tag{2.31}$$

in this case of diffusion.

2.5.1 Method of Laplace Transform

The method of Laplace transform is especially suitable for equations involving special structures in time. Defining the Laplace transform pair as usual,

$$\overline{\theta}(\delta; p) = \int_0^\infty \theta(\delta, \beta) \, e^{-p\beta} d\beta$$

$$L^{-1}\left[\overline{\theta}(\delta; p)\right] \equiv \theta(\delta, \beta) = \frac{1}{2\pi i} \int_{\gamma - i\infty}^{\gamma + i\infty} \overline{\theta}(\delta; p) \, e^{p\beta} dp, \tag{2.32}$$

the Laplace transform solution satisfying equations (2.25) to (2.27) can be obtained in a straightforward manner:

$$\overline{\theta} = \frac{e^{-\sqrt{\frac{p(p+2)}{1+Bp}}\delta}}{p} \, . \tag{2.33}$$

Partial expansion technique. The fundamental behavior of a lagging response can be directly extracted from equation (2.33) by the method of partial expansions, along with the limiting theorems in the Laplace transform. As the Laplace transform parameter p approaches infinity, the arguments in equation (2.33) have the following asymptotic behavior:

$$1 + Bp \rightarrow Bp, \quad p(p+2) \rightarrow p^2 . \tag{2.34}$$

The limiting behavior of $\theta(\delta,\beta)$ as β approaches zero (the extremely short time behavior) is thus

$$\lim_{\beta \to 0} \theta(\delta,\beta) \sim \lim_{p \to \infty} \overline{\theta}(\delta; p) \cong \frac{e^{-\left(\frac{\delta}{\sqrt{B}}\right)\sqrt{p}}}{p} . \tag{2.35}$$

Note that the multiplier p in front of $\overline{\theta}$ has been removed in the limiting theorem to render an inverse solution with a familiarized form. The equal sign is thus replaced by the approximate sign ($\sim$) in assembling the asymptotic behavior. Recognizing that (Abramowitz and Stegun, 1964)

$$L^{-1}\left[\frac{e^{-\left(\frac{\delta}{\sqrt{B}}\right)\sqrt{p}}}{p}\right] = 1 - \text{erf}\left(\frac{\delta}{2\sqrt{B\beta}}\right) = \text{erfc}\left(\frac{\delta}{2\sqrt{B\beta}}\right), \tag{2.36}$$

equation (2.35) results in

$$\lim_{\beta \to 0} \theta(\delta,\beta) \sim 1 - \text{erf}\left(\frac{\delta}{2\sqrt{B\beta}}\right) = \text{erfc}\left(\frac{\delta}{2\sqrt{B\beta}}\right). \tag{2.37}$$

Equation (2.37) shows that the lagging temperature resembles *diffusion* at extremely short times, reflected by the complimentary error function that exists for heat diffusion in a semi-infinite, one-dimensional medium. Although a wave term is present in equation (2.25), the mixed-derivative term on the left side effectively destroys the sharp wavefront and, hence, the entire wave structure. For further justifying the method of partial expansions, the solution for a *CV* wave is retrieved by taking $B = 0$ ($\tau_T = 0$) in equation (2.33):

$$\overline{\theta} = \frac{e^{-\sqrt{p(p+2)}\delta}}{p} \quad (CV \text{ wave model}). \tag{2.38}$$

The asymptotic behavior in this case becomes

$$\lim_{\beta \to 0} \theta(\delta, \beta) \sim \lim_{p \to \infty} \overline{\theta}(\delta; p) \cong \frac{e^{-p\delta}}{p} \tag{2.39}$$

and the Laplace inversion gives a unit-step function (Abramowitz and Stegun, 1964)

$$\lim_{\beta \to 0} \theta(\delta, \beta) \cong L^{-1} \left[\frac{e^{-p\delta}}{p} \right] = h(\beta - \delta) \quad (CV \text{ wave model}). \tag{2.40}$$

This solution describes a *discontinuity* at $\delta = \beta$, i.e., at the thermal wavefront at $x = Ct$ according to equations (2.4) and (2.24), as depicted by the CV wave model. The partial expansion technique developed here is thus capable of capturing the fundamental characteristics of heat diffusion and thermal waves.

The long-time behavior of temperature with lagging can be obtained in the same fashion. The result is

$$\lim_{\beta \to \infty} \theta(\delta, \beta) \cong 1 - \text{erf}\left(\frac{\delta}{\sqrt{2\beta}} \right) = \text{erfc}\left(\frac{\delta}{\sqrt{2\beta}} \right). \tag{2.41}$$

At large times, while retaining the same behavior as in diffusion, the lagging behavior becomes *independent* of the parameter B.

Bromwich contour integrations. The full response of lagging temperature depicted by equations (2.25) to (2.27) relies on the Laplace inversion of equation (2.33). Bromwich contour integration is the standard procedure for obtaining the inverse solution. The procedure is somewhat tedious, but like the early examples in diffusion and thermal wave models, it is a necessary process associated with the birth of a new-type energy equation.

The lagging temperature is obtained by the Laplace inversion:

$$\theta(\delta, \beta) = \frac{1}{2\pi i} \int_{\gamma - i\infty}^{\gamma + i\infty} \overline{\theta}(\delta; p) e^{p\beta} dp, \tag{2.42}$$

with $\overline{\theta}(\delta; p)$ given by equation (2.33). There exist three branch points, at $p = 0, -2$, and $-1/B$. Mathematically, the branch point at $(-1/B)$ can be located on either side of -2. For fast-transient laser heating on metals (Chapter 5), however, the ratio of τ_T to τ_q (and hence the parameter B) is of the order of several tens (Tzou, 1995a to c). This can also be seen from the correlation to the phonon-electron interaction model in equation (2.28). At room temperature (Qiu and Tien, 1992, 1993), the heat capacity of the metal lattice (C_l) is about 2 orders of magnitude larger than that of the electron gas (C_e). The value of B is consequently of the order of several tens. The heat capacity of the electron gas increases with the electron temperature in a linear fashion. In the limit of C_e approaching infinity, the value of B approaches one-half ($1/2$). The extreme position of the branch point at $(-1/B)$ is thus at -2. For the fast-transient process in metals, therefore, the branch point at $-1/B$ is always between the other two at 0 and -2.

Bromwich contour integration has been performed for obtaining the Laplace inversion according to equation (2.42) (Tzou and Zhang, 1995). The result is

$$\theta(\delta,\beta) = 1 - \frac{1}{\pi}\int_2^\infty \sin\left[\delta\sqrt{\frac{\rho(\rho-2)}{B\rho-1}}\right] e^{-\beta\rho}\frac{d\rho}{\rho} -$$

$$\frac{2}{\pi}\int_0^\infty \sin\left[\rho\delta\sqrt{\frac{2+(2B-1)\rho^2}{1+B\rho^2}}\right] \frac{e^{-\left(\frac{\rho^2}{1+B\rho^2}\right)\beta}}{\rho(1+B\rho^2)}d\rho. \tag{2.43}$$

The two integrands are bounded everywhere in their domains of integration. Both are finite at the lower bounds and exponentially decay to zero as the integral variable ρ approaches infinity. They are thus integrable in a strict sense.

Owing to the complicated integrands involved in equation (2.43), no analytical solution is possible. A numerical method for evaluating the integrals is still unavoidable. The upper bound being infinity in these integrals can be first approximated by a finite number in the numerical computation. Its value is then enlarged until the difference between two successive integrations is smaller than a prescribed threshold. Numerical integrations for equation (2.43) are postponed until the Riemann-sum approximation for the Laplace inversion is developed.

Riemann-sum approximation. Though having an analytical appearance, the improper integrals involved in equation (2.43) must be computed numerically. These two integrals do converge in all cases, but the rate of convergence is rather slow because of oscillations of the sinusoidal functions. This situation becomes more pronounced as the higher order effects of thermal lagging are further incorporated (see Sections 10.1 and 10.2). The numerical procedure for approximating the improper integrals, moreover, involves a series of summations to be performed. Recognizing this nature in numerical approximations, a special technique employing the Riemann-sum approximation is developed for the Laplace inversion (Tzou et al., 1994; Chiffelle, 1994). It performs the summation at an early stage, before the complicated Bromwich contour integration is evaluated.

The required Laplace inversion is given by equation (2.42). Introducing a variable transformation from p (complex) to ω (real),

$$p = \gamma + i\omega, \tag{2.44}$$

with γ being the real constant specifying the vertical segment in the Bromwich contour, equation (2.42) is reduced to a Fourier transform:

$$\theta(\delta,\beta) = \frac{e^{\gamma\beta}}{2\pi}\int_{-\infty}^\infty \overline{\theta}(\delta; p = \gamma + i\omega)e^{i\omega\beta}d\omega. \tag{2.45}$$

The Fourier integral thus obtained can be approximated by its Riemann sum. Denoting ω as the wave frequency and τ the half-period of its oscillations, i.e., $\omega = n\pi/\tau$ for the nth wave mode and $\Delta\omega_n = \pi/\tau$ for all modes,

$$\theta(\delta,\beta) = \frac{e^{\gamma\beta}}{2\tau} \sum_{n=-\infty}^{\infty} \overline{\theta}\left(\delta, p = \gamma + \frac{in\pi}{\tau}\right) e^{i(n\beta\pi/\tau)}. \tag{2.46}$$

Noticing further that the wave modes with positive and negative n values appear in pairs and

$$\overline{\theta}\left(\delta, \gamma + \frac{in\pi}{\tau}\right) e^{i(n\beta\pi/\tau)} + \overline{\theta}\left(\delta, \gamma - \frac{in\pi}{\tau}\right) e^{-i(n\beta\pi/\tau)}$$

$$= 2\operatorname{Re}\left[\overline{\theta}\left(\delta, \gamma + \frac{in\pi}{\tau}\right) e^{i(n\beta\pi/\tau)}\right], \tag{2.47}$$

equation (2.46) can be expressed as

$$\theta(\delta,\beta) = \frac{e^{\gamma\beta}}{\tau}\left[\frac{1}{2}\overline{\theta}(\delta,\gamma) + \operatorname{Re}\sum_{n=1}^{\infty}\overline{\theta}\left(\delta, \gamma + \frac{in\pi}{\tau}\right) e^{i(n\beta\pi/\tau)}\right] \tag{2.48}$$

where Re represents the real part of the summation. Since the function $e^{i(n\beta\pi/\tau)}$ has a fundamental period of 2π, the physical domain of β in equation (2.48) is $0 \le \beta \le 2\tau$. At $\beta = \tau$, more precisely, equation (2.48) yields

$$\theta(\delta,\beta) \cong \frac{e^{\gamma\beta}}{\beta}\left[\frac{1}{2}\overline{\theta}(\delta,\gamma) + \operatorname{Re}\sum_{n=1}^{N}\overline{\theta}\left(\delta, \gamma + \frac{in\pi}{\beta}\right)(-1)^n\right] \tag{2.49}$$

which is the inverse solution for $\theta(\delta, \beta)$. Unlike the improper integrals in equation (2.43), the Riemann-sum approximation for the Laplace inversion, equation (2.49), involves a single summation in the numerical process. Its accuracy depends on the value of γ and the truncation error dictated by N. The value of γ must be selected so that the Bromwich contour encloses all the branch points. With regard to the transformed solution shown by equation (2.33), therefore, any positive number can be selected for γ because the branch points are all negative. For faster convergence, however, numerous numerical experiments have shown that the value of γ satisfying the relation (Tzou et al., 1994; Chiffelle, 1994; Tzou, 1995a - c)

$$\gamma\beta \cong 4.7 \tag{2.50}$$

gives the most satisfactory results. The appropriate value of γ for faster convergence, in other words, depends on the instant of time (β) at which the lagging phenomenon is studied. The criterion shown by equation (2.50) is independent of the dimension of β. Should the energy equation with dimensions, namely, equation (2.10) or (2.20), be solved by the method of Laplace transform, β in equation (2.50) is replaced by the real time t while the constant 4.7 remains. The quantity γ in this case, of course, has a dimension of 1/s. Selection for the number of terms used in the Riemann sum, N, is straightforward. At given values of γ, δ, and β, the summation in equation (2.49) should continue until a prescribed threshold for the accumulated partial sums is satisfied.

The Riemann-sum approximation represented by equation (2.49) for the Laplace inversion has been rigorously examined, including the fundamental trigonometric, exponential, and hyperbolic functions (Chiffelle, 1994) whose Laplace inversions are known. With equation (2.50), which guides the "optimal" values of γ at various times β, satisfactory convergence is usually achieved within several tens of terms in the Riemann sum. The same numerical inversion technique has also been examined by the solution of Love's wave equation in elasticity and some complicated solutions of thermo-mechanical coupling in the fast-transient process.

The unit-step (heaviside) function displayed in Figure 2.3 provides the most critical challenge to the Riemann-sum approximation for the Laplace inversion. Should the Riemann-sum approximation, equation (2.49), be efficient, it should capture the sharp discontinuity located at $t = 1$, as well as the two constants, 0 and 1, jumping from one side of the discontinuity to another through a zero distance. Laplace transform of the unit-step function defined by

$$h(t - 1) = \begin{cases} 1 & \text{for } t \geq 1 \\ 0 & \text{for } t \leq 1 \end{cases} \tag{2.51}$$

is

$$L[h(t - 1)] = \frac{e^{-p}}{p}, \tag{2.52}$$

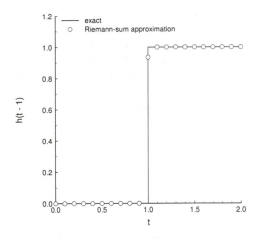

Figure 2.3 The heaviside function obtained by the Riemann-sum approximation for the Laplace inversion, equation (2.49). Here, $\gamma = 4.7/t$.

which is used as the function $\overline{\theta}$ in equation (2.49). The value of γ varies with time, $\gamma = 4.7/t$, with t being the physical time at which the Laplace inversion is being sought. The time domain from 0 to 2, with a sharp discontinuity existing at $t = 1$, is decomposed into 100 increments in the numerical computation. The "near perfect" agreement shown in Figure 2.3 results from an extremely small Cauchy's norm,

$$\frac{|\theta_N - \theta_{N-1}|}{\theta_{N-1}} \leq 10^{-10} . \tag{2.53}$$

This is to overcome the sensitive oscillations around the constants of 0 (for $t \in [0, 1)$) and 1 ($t \in (1, \infty)$). Such a small threshold, however, is not needed for most problems in thermal lagging because a sharp discontinuity seldom exists. Satisfactory results can still be obtained for other values of γt, say, $\gamma t = 0.1$ or 10. However, the number of terms required for convergence approaches the order of 10^6.

Another critical examination for the efficiency of the Riemann-sum approximation is the solution for the temperature wave (*CV* wave model) obtained by Baumeister and Hamill (1969):

$$\theta(\delta,\beta) = \left[e^{-\delta} + \delta \int_{\delta}^{\beta} e^{-z} \frac{I_1\left(\sqrt{z^2 - \delta^2}\right)}{\sqrt{z^2 - \delta^2}} \, dz \right] h(\beta - \delta) \tag{2.54}$$

where I_1 is the modified Bessel function of the first kind of order zero. The dimensionless variables δ, β, and θ are the same as those defined in equation (2.24), with τ_q replaced by α/C^2. At three representative instants of time, $\beta = 1, 2$, and 3, Figure 2.4 compares the results obtained by equation (2.49) and the analytical solution shown by equation (2.54). The transformed solution $\overline{\theta}$ needed in equation (2.49) is given by equation (2.38). Evaluation of equation (2.54) is made by the use of a 15-point Gaussian quadrature formula. As clearly demonstrated, the Riemann-sum approximation accurately captures the analytical results in all cases, which possess sharp discontinuities at the thermal wavefront, $\delta = \beta$ or $x = Ct$.

The temperature response with lagging can now be examined by the results from the Bromwich contour integration, equation (2.43), and the Riemann-sum approximation, equation (2.49). The transformed solution for the latter is given by equation (2.33). For two typical values of $B = 50$ and 100 in phonon scattering and phonon-electron interaction, the temperature distributions are displayed in Figure 2.5 for $\beta = 0.1, 0.2$, and 0.3. At a certain position (δ) and time (β), the temperature level increases with the value of B and hence the ratio of τ_T/τ_q. Since τ_T reflects the delayed response due to microstructural interaction effects, phonon scattering and phonon-electron interactions occurring in microscales seem to promote the temperature level. Also, note that *no* thermally undisturbed zone exists in the lagging response because the heat-affected zone extends to infinity. In all cases shown in Figure 2.5, the heat-penetration depth increases with the value of B.

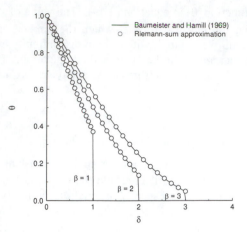

Figure 2.4 Comparison of the Riemann-sum approximation, equation (2.49), and the analytical solution obtained by Baumeister and Hamill (1969), equation (2.54).

The excellent agreement shown in Figure 2.5 not only validates the Riemann-sum approximation for the Laplace inversion, but also confirms the solution of equation (2.25) from two different approaches. This is desirable because the solution describing the lagging response in transient conditions has never been known before.

The Bromwich contour integration is generic but may be inefficient

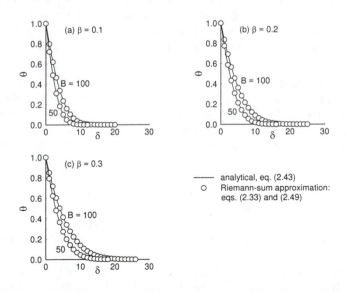

Figure 2.5 The temperature response with lagging for $B = 50$ and 100, typical values in phonon scattering and phonon-electron interaction. Here, $\beta = 0.1, 0.2$, and 0.3. The results of the Riemann-sum approximation are represented by circles.

because the branch points may shift intrinsically when different boundary or initial conditions are considered. It is simply impractical to repeat the procedure for all problems encountered in practice. With the high accuracy demonstrated in Figures 2.4 to 2.6, therefore, the Riemann-sum approximation for Laplace inversion shall be used throughout this book.

The dual-phase-lag model also describes the macroscopic behavior of both the diffusion and thermal waves in the same framework. For a consistent study, equation (2.25) is written in more general form:

$$\frac{\partial^2 \theta}{\partial \delta^2} + B \frac{\partial^3 \theta}{\partial \delta^2 \partial \beta} = A \frac{\partial \theta}{\partial \beta} + D \frac{\partial^2 \theta}{\partial \beta^2} \quad \text{with} \quad B = \frac{\tau_T}{2\tau_q} \qquad (2.55)$$

where A, D and B are three parameters describing the special behavior in limiting cases:
1. The lagging behavior caused by the interactions in microscale: $A = 2$, $D = 1$, and $B \neq 0$. Equation (2.55) reduces to equation (2.25) in this case.
2. The classical CV wave behavior: $A = 2$, $D = 1$, and $B = 0$. This is the case retrieving the solution of CV waves, equation (2.38).
3. The classical diffusion behavior: $A = 2$, $D = 1$, and $B = 1/2$ or $A = 2$, $D = 0$, and $B = 0$. Recalling that $\tau_T = \tau_q$ as $B = 1/2$, the two cases result in the same solution as that described from equations (2.29) to (2.31).

Adopting a *single* energy equation (2.55), Figure 2.6 describes diffusion, CV wave, and lagging behavior in the same framework of the dual-phase-lag model. The lagging response results in the highest temperature level among the three. The microstructural interaction effect, reflected by the delayed time for establishing the temperature gradient across a material volume (τ_T) significantly extends the physical domain of the thermal penetration depth. In fact, as indicated by Qiu and Tien (1992), the significantly larger heat-affected zone and the higher temperature level

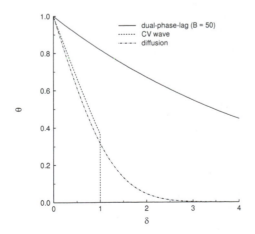

Figure 2.6 Comparison of the diffusion, wave, and lagging behaviors at the same instant of time ($\beta = 1.0$).

within the heat-affected zone are the main reasons that the parabolic two-step model agrees well with the experimental results of the short-pulse laser heating on metals (Brorson et al., 1987). The *CV* wave model does predict a higher temperature level in the heat-affected zone than the diffusion model, but the penetration depth is much shorter owing to the formation of the thermally undisturbed zone.

With all the macroscopic and microscopic behaviors described in the same framework, most important, the way(s) in which one behavior (such as diffusion or wave) *transits* into another (phonon scattering or phonon-electron interaction) can be described in full detail. Such transition may result from either shortening of the response time or shrinkage of the physical scale. Figure 2.7 gives an example of the way in which the microstructural interaction effect destroys the wave structure in a macroscopic response. To repeat, the *CV* wave model assumes $\tau = \tau_q$ and $\tau_T = 0$. When microscale interactions activate in heat transport, the value of τ_T starts to increase and the time at which a temperature gradient is established across the material volume, $t + \tau_T$, increases. A slight increase of τ_T from zero, exemplified by $B = 0.01$ in Figure 2.7, diminishes the sharp wavefront, and there exists no thermally undisturbed zone. When the value of B (and hence τ_T) further increases, the thermal penetration depth increases, and the temperature profile becomes smoother. This type of intrinsic transition may not be described by the existing macroscopic and microscopic models because they are established on different physical bases. The dual-phase-lag model is thus unique in this sense. The FORTRAN-77 code used for obtaining the solutions in Figures 2.7 and 2.8 is included in the Appendix. For applications with different transformed functions, namely, $\overline{\theta}(\delta; p)$, only the function subroutine FUNC(P) needs to be modified.

2.5.2 Method of Fourier Transform

With regard to the mathematical solution of equation (2.25), Fourier transform provides a parallel treatment to Laplace transform. The former transforms the space

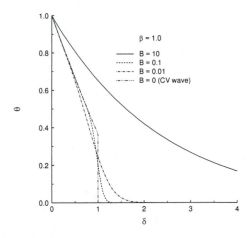

Figure 2.7 Diminution of the sharp thermal wavefront by the phase lag of the temperature gradient (τ_T).

variable (δ) to the mathematical space (ξ), and the latter transforms the time variable (β) to the mathematical space (p). For thermal lagging in a semi-infinite medium, equations (2.25) to (2.27), Fourier sine transform, defined as

$$\bar{\theta}_s(\beta;\xi) = \sqrt{\frac{2}{\pi}}\int_0^\infty \theta(\delta,\beta)\sin(\delta\xi)\,d\delta, \quad \theta(\delta,\beta) = \sqrt{\frac{2}{\pi}}\int_0^\infty \bar{\theta}_s(\beta;\xi)\sin(\delta\xi)\,d\xi$$

(2.56)

can be applied to reduce the partial differential equation (2.25) to an ordinary differential equation with respect to time. Multiplying equation (2.25) throughout by $\sin(\delta\xi)$ and integrating δ from 0 to ∞ results in

$$\frac{d^2\bar{\theta}_s}{d\beta^2} + (2 + B\xi^2)\frac{d\bar{\theta}_s}{d\beta} + \xi^2\bar{\theta}_s = \sqrt{\frac{2}{\pi}}\xi.$$

(2.57)

It has a general solution of the form

$$\bar{\theta}_s = C_1 e^{r_1\beta} + C_2 e^{r_2\beta} + \sqrt{\frac{2}{\pi}}\left(\frac{1}{\xi}\right), \quad r_{1,2} = \frac{-(2+B\xi^2)\pm\sqrt{4+B^2\xi^4+4\xi^2(B-1)}}{2}$$

(2.58)

with C_1 and C_2 determined from the transformed initial conditions (2.26):

$$\bar{\theta}_s = 0, \quad \frac{d\bar{\theta}_s}{d\beta} = 0 \quad \text{as} \quad \beta = 0.$$

(2.59)

The result is

$$C_1 = \sqrt{\frac{2}{\pi}}\frac{r_2}{\xi(r_1 - r_2)}, \quad C_2 = \sqrt{\frac{2}{\pi}}\frac{r_1}{\xi(r_2 - r_1)}$$

(2.60)

The transformed function $\bar{\theta}_s(\beta;\xi)$, obtained from equations (2.58) and (2.60), can then be substituted into equation (2.56) for the temperature in the physical domain:

$$\theta(\delta,\beta) = \sqrt{\frac{2}{\pi}}\int_0^\infty \left\{C_1 e^{r_1\beta} + C_2 e^{r_2\beta} + \sqrt{\frac{2}{\pi}}\left(\frac{1}{\xi}\right)\right\}\sin(\delta\xi)\,d\xi$$

(2.61)

No closed-form solution is possible for $\theta(\delta, \beta)$ owing to the complicated functions of ξ in C_1, C_2, r_1, and r_2. Numerical effort involved in the evaluation for the improper integral of equation (2.61), however, is the same as that for the Riemann-sum approximation.

Note that the method of separation of variables fails to apply because of the presence of the mixed-derivative term, $\partial^3\theta/\partial\delta^2\partial\beta$, in equation (2.25). Methods of integral transform, including both Laplace and Fourier transforms, are more suitable for this type of equation.

2.6 PRECEDENCE SWITCHING IN FAST-TRANSIENT PROCESSES

Characterization of the lagging response in terms of the parameter B, τ_T/τ_q, places the onset of precedence switching in the fast-transient process to diffusion. As discussed in equations (2.29) and (2.30), the case of $\tau_T = \tau_q$, or $B = 1/2$, in the dual-phase-lag model reduces to diffusion. For the case of $\tau_T > \tau_q$, the heat flux vector *precedes* the temperature gradient in the process of heat transfer, implying that the heat flux vector is the cause while the temperature gradient is the effect in heat flow.

For $\tau_T < \tau_q$, on the other hand, the temperature gradient precedes the heat flux vector, implying that the temperature gradient becomes the cause while the heat flux vector becomes the effect. Figure 2.8 illustrates this transition by varying the value of B from 0.1 to 0.9. Evidently, the temperature level is higher than that predicted by the diffusion model for $\tau_T > \tau_q$ (flux precedence) and is lower for $\tau_T < \tau_q$ (temperature gradient precedence). For fast-transient heating on metallic films the value of τ_T is roughly 1 to 2 orders of magnitude larger than that of τ_q (Tzou, 1995a, b), implying that the heat flux vector precedes the temperature gradient. As shown by Figure 2.8, the temperature level will be higher than that predicted by diffusion assuming $\tau_T = \tau_q$. Since the *CV* wave model assumes the precedence of the temperature gradient to the heat flux vector at all times, $\tau_q > \tau_T = 0$, the resulting temperature level deviates as much as diffusion from the experimental result. Assuming an inappropriate sequence of τ_T and τ_q, namely, $\tau_T = \tau_q$ in diffusion and τ_q

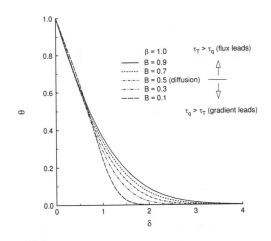

Figure 2.8 Precedence switching of the temperature gradient and the heat flux vector in the fast-transient process of heat transport. The curve of diffusion ($B = 0.5$) is at the onset of switching.

$> \tau_T \, (= 0)$ in *CV* waves, in a problem where τ_T is much greater than τ_q is the main reason that both diffusion and *CV*-wave models fail to describe the fast-transient response in the gold film experiment by Brorson et al. (1987).

2.7 RATE EFFECT

Although the wave structure in the dual-phase-lag model is destroyed by the microstructural interaction effect, the presence of the wave term in equation (2.20) (or (2.25)) allows *two* initial conditions to specify the lagging response of temperature. One example is the specification of the initial time-rate of change of temperature along with the initial temperature,

$$T(x,0) = T_0 \quad \text{and} \quad \frac{\partial T}{\partial t}(x,0) = \dot{T}_0 \quad \text{for} \quad x \in [0,\infty). \tag{2.62}$$

The boundary conditions remain the same, equations (2.21) and (2.22). The equations governing the lagging response are given by equations (2.25) and (2.27), but the initial conditions in equation (2.26) are replaced by

$$\theta(\delta,0) = 0 \quad \text{and} \quad \frac{\partial \theta}{\partial \beta}(\delta,0) = \dot{\theta}_0 \quad \text{with} \quad \dot{\theta}_0 = \frac{2\tau_q \dot{T}_0}{T_W - T_0}. \tag{2.63}$$

The emphasis, obviously, is placed on the effect of initial temperature rate on the lagging behavior. The transformed solution satisfying equations (2.25), (2.27), and (2.63) is straightforward:

$$\overline{\theta} = \left[\frac{p + 2 - \dot{\theta}_0}{p(p+2)}\right] e^{-\sqrt{\frac{p(p+2)}{1+Bp}}\,\delta} + \frac{\dot{\theta}_0}{p(p+2)}. \tag{2.64}$$

In the case of a zero initial temperature rate, $\dot{\theta}_0 = 0$, equation (2.64) reduces to equation (2.33). For a typical value of $B = 100$, Figure 2.9 shows the temperature distributions at various values of $\dot{\theta}_0$. The curve without a rate effect, $\dot{\theta}_0 = 0$, is the same as that shown in Figure 2.5(a) for $B = 100$ and $\beta = 0.1$. The initial time-rate of change of temperature reflects a uniform heating applied at $t = 0$, resulting in significant temperature rises throughout the entire body. The nonuniform distribution in the domain of $\delta < 10$ is the lagging response of major concern. When the initial temperature rate exceeds a certain value, evidenced by the distribution of $\dot{\theta}_0 = 10$ in Figure 2.9, the temperature in the vicinity of the boundary *exceeds* the boundary temperature of $\theta = 1$. This is the temperature *overshooting* phenomenon similar to that in the *CV* wave model that cannot be depicted by diffusion (Tzou et al., 1994).

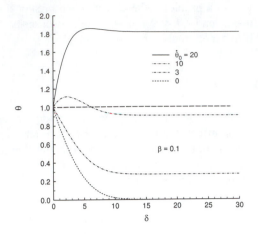

Figure 2.9 The effect of initial time-rate of change of temperature on the lagging response of temperature for $\dot{\theta}_0 = 0, 3, 10$, and 20.

In practice, however, temperature overshooting may be difficult to observe because of the extremely high rate of temperature change required to produce the exaggerated response. In metals, for example, the phase lag of the heat flux vector (τ_q) is of the order of picoseconds. A value of $\dot{\theta}_0 = 10$, according to equation (2.63), implies a temperature rate ($\dot{T}_0$) of the order of 10^{13} K/s to excite occurrence of the overshooting phenomenon. The physical domain in which temperature overshooting occurs, in addition, is down to the atomic level. For a value of $\delta \cong 5$ in Figure 2.9 where the local temperature exceeds the boundary temperature, equation (2.24) yields a value of x of the order of angstroms (10^{-10} m). An extremely high temperature rate occurring at an extremely small scale is the main difficulty in reproducing the temperature overshooting phenomenon in the laboratory.

2.8 PROBLEMS INVOLVING HEAT FLUXES AND FINITE BOUNDARIES

While the temperature formulation furnishes a simpler mathematical structure in revealing the fundamental characteristics of the lagging response, the flux formulation provides a more realistic simulation for applications. Owing to the complicated relationship between the temperature gradient and the heat flux vector as shown by equation (2.13), the problem involving a flux-specified boundary condition is nontrivial in the dual-phase-lag model. The mixed formulation in terms of both temperature and heat flux vector is a better approach in this situation.

The transient response in a semi-infinite medium subjected to thermal flux radiation at its boundary is considered for illustration. The one-dimensional forms of equations (2.5) and (2.7) are

$$-\frac{\partial q}{\partial x} = C_p \frac{\partial T}{\partial t} \tag{2.65a}$$

$$q + \tau_q \frac{\partial q}{\partial t} = -k \frac{\partial T}{\partial x} - k\tau_T \frac{\partial^2 T}{\partial x \partial t}. \tag{2.65b}$$

The initial conditions and the boundary condition at infinity remain the same, equation (2.22) and (2.23), while the boundary condition at $x = 0$ is replaced by

$$q = q_s\big[h(t) - h(t - t_s)\big] \quad \text{at } x = 0 \tag{2.66}$$

where q_s is the pulse intensity of the flux radiation and t_s denotes the pulse width in time. Since both temperature (equation (2.22)) and heat flux vector (equation (2.66)) are involved in the boundary condition, it is more appropriate to solve equation (2.65) in a coupled manner without further combination. Introducing the dimensionless variables,

$$\theta(x,\ t) = \frac{T(x,t) - T_0}{T_0}, \quad \eta = \frac{q}{T_0 C_p \sqrt{\alpha / \tau_q}}, \quad \beta = \frac{t}{2\tau_q}, \quad \delta = \frac{x}{2\sqrt{\alpha\tau_q}} \tag{2.67}$$

with β (dimensionless time) and δ (dimensionless space) identical to those in equation (2.24), the governing system becomes

$$-\frac{\partial \eta}{\partial \delta} = \frac{\partial \theta}{\partial \beta} \tag{2.68a}$$

$$2\eta + \frac{\partial \eta}{\partial \beta} = -\frac{\partial \theta}{\partial \delta} - B \frac{\partial^2 \theta}{\partial \delta \partial \beta} \quad \text{with} \quad B = \frac{\tau_T}{2\tau_q} \tag{2.68b}$$

$$\theta(\delta, 0) = 0 \quad \text{and} \quad \frac{\partial \theta}{\partial \beta}(\delta, 0) = 0 \quad \text{for} \quad \delta \in [0, \infty) \tag{2.69}$$

$$\eta(0, \beta) = \eta_s\big[h(\beta) - h(\beta - \beta_s)\big] \quad \text{and} \quad \theta(\delta, \beta) \to 0 \quad \text{as} \quad \delta \to \infty. \tag{2.70}$$

where

$$\beta_s = \frac{t_s}{2\tau_q}, \quad \eta_s = \frac{q_s}{T_0 C_p \sqrt{\alpha / \tau_q}} \tag{2.71}$$

are the dimensionless pulse width and intensity. Taking the Laplace transform of equation (2.68) and solving for $\overline{\theta}$ and $\overline{\eta}$ gives

$$\bar{\theta} = D_1 e^{A\delta} + D_2 e^{-A\delta}, \quad \bar{\eta} = -\sqrt{\frac{p(1+Bp)}{p+2}}\left(D_1 e^{A\delta} + D_2 e^{-A\delta}\right), \quad \text{with}$$

$$A = \sqrt{\frac{p(p+2)}{1+Bp}}. \tag{2.72}$$

Two unknowns, D_1 and D_2, are determined from the transformed boundary conditions of equation (2.70):

$$\bar{\eta} = \eta_s \left(\frac{1-e^{-\beta_s p}}{p}\right) \quad \text{at} \quad \delta = 0 \quad \text{and} \quad \bar{\theta} \to 0 \quad \text{as} \quad \delta \to \infty. \tag{2.73}$$

Equations (2.72) and (2.73) result in

$$D_1 = 0, \quad D_2 = \eta_s \sqrt{\frac{p+2}{p(1+Bp)}}\left(\frac{1-e^{-\beta_s p}}{p}\right) \tag{2.74}$$

which furnishes the solutions of $\bar{\theta}$ and $\bar{\eta}$ in the Laplace transform domain. The Laplace inversion can then be made by the Riemann-sum approximation, equation (2.49). The FORTRAN code in the Appendix is readily applicable, with the function subroutine, FUNC(P), replaced by the transformed solution of $\bar{\theta}$ (or $\bar{\eta}$) in equation (2.72) and the coefficients D defined in equation (2.74).

In the mixed formulation shown above, note that the special cases of diffusion and CV waves are retrieved as $B = 1/2$ and $B = 0$, respectively. The situation is identical to that described in equation (2.55) owing to the use of the same dimensionless variables for space and time. At a representative instant of time $\beta = 1$, Figure 2.10 shows the temperature distributions at various values of B. The pulse width is taken as one-fifth of the observation time, $\beta_s = 0.2$. Figure 2.10 shows that, after the deposition of thermal energy from $\beta = 0$ to $\beta = 0.2$, the lagging temperature predicted by the dual-phase-lag model ($B \neq 0$) decays at a *faster* rate than the thermal wave ($B = 0$). When the value of B gradually deviates from zero, the microstructural interaction effect (τ_T) diminishes the sharp ripple of temperature depicted by the CV wave model. The response curve reduces to that of diffusion as the value of B reaches 0.5. Figure 2.10 thus describes in detail the way in which a wave behavior transits into diffusion associated with the activation of heat transport in microscale. The corresponding microscopic mechanisms, to reiterate, are given by equations (2.11) (for a phonon-scattering-dominated mechanism) and (2.12) (for phonon-electron-interaction-dominated mechanism). For gradually increasing values of B (and hence τ_T), they imply a larger value of τ_N (in phonon scattering) and a smaller value of G (in phonon-electron interaction).

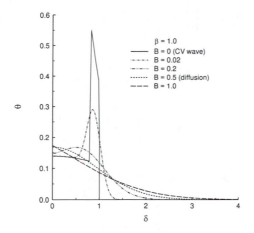

Figure 2.10 Distributions of lagging temperature in a semi-infinite solid for $B = 0$ (*CV* wave), 0.02, 0.2, 0.5 (diffusion), and 1.0. Here, $\beta = 1.0$, $\eta_s = 1.0$, and $\beta_s = 0.2$. The solid is heated from the front surface by heat flux.

Figure 2.11 shows the heat flux distributions in a semi-infinite medium. They are obtained by using the expression of $\bar{\eta}$ in equation (2.72) in the FUNC(P) subroutine in the FORTRAN code.

Equation (2.72) is in a convenient form to incorporate another boundary condition for a thin film with a finite thickness. Under the same flux condition at $x = 0$, equation (2.70) or (2.73), let us impose a zero slope of the temperature at L, the thickness of the film,

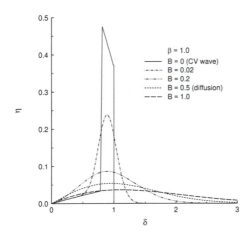

Figure 2.11 The flux distributions in the semi-infinite solid for $B = 0$ (*CV* wave), 0.02, 0.2, 0.5 (diffusion), and 1.0. Here, $\beta = 1.0$, $\eta_s = 1.0$, $\beta_s = 0.2$, and $\eta = q/T_0 C_p \times (\alpha/\tau_q)^{1/2}$.

$$\frac{\partial \theta}{\partial \delta} = 0 \quad \text{or} \quad \frac{d\overline{\theta}}{d\delta} = 0 \quad \text{at} \quad \delta = l \quad \text{with} \quad l = \frac{L}{2\sqrt{\alpha \tau_q}}. \tag{2.75}$$

It assumes negligible heat loss from the rear surface at short times (Qiu and Tien, 1992, 1993). In this case, the coefficients D_1 and D_2 become

$$D_1 = D_2 e^{-2Al}, \quad D_2 = \eta_s \sqrt{\frac{p+2}{p(1+Bp)}} \left[\frac{1-e^{-\beta_s p}}{p\left(1-e^{-2Al}\right)} \right]. \tag{2.76}$$

The temperature and flux solutions in the Laplace transform domain remain the same, equation (2.72). By the same procedure of Laplace inversion, the result for B = 50 at β = 0.5 is displayed in Figure 2.12 along with the results of diffusion and CV wave. The thermal wavefront depicted by the CV wave model ($B = 0$) is located at δ = β = 0.5, which yields a large temperature gradient owing to the presence of the thermally undisturbed zone. The temperature distribution becomes smoother when the value of B increases to 0.5 (the classical theory of diffusion). The dual-phase-lag model with B = 50, a typical value for metals (see Chapter 5), shows a unique pattern of *uniform* distribution in this case. The microstructural interaction effect (reflected by the large value of τ_T over τ_q) rapidly transports heat throughout the film, rendering a *constant* temperature across the film thickness. The temperature *gradient* becomes insignificant in the fast-transient process, which is an assumption made in the energy transport through the metal lattice in the microscopic phonon-electron interaction model (Qiu and Tien, 1992, 1993; see also Chapter 5).

Figure 2.13 shows the time history of lagging temperature at the rear surface of the film, δ = l = 1.0. A time domain in which there is no significant temperature rise exists in all cases. The thermal wave predicted by the CV wave model ($B = 0$) arrives last (at $\beta = \delta = 1$) owing to the finite speed of heat propagation

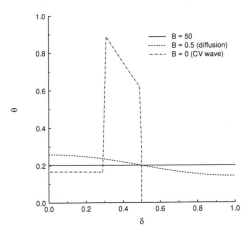

Figure 2.12 The uniform distributions of lagging temperature in a thin film for B = 50. Here, β = 0.5, l = 1.0, η_s = 1.0 and β_s = 0.2. The film is heated from the front surface and insulated at the rear surface.

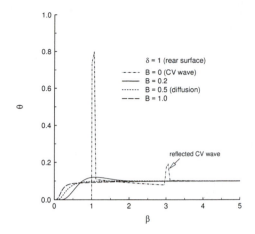

Figure 2.13 Time history of lagging temperature at the rear surface of the film. Here, $\delta = l = 1.0$, $\eta_s = 1.0$, and $\beta_s = 0.1$.

assumed in the model. The temperature rise time shortens as the value of B increases. Recalling that the parameter B $(\tau_T/2\tau_q)$ reflects the delayed time of the microstructural interaction effect (τ_T) relative to that of the fast-transient inertia (τ_q), this provides further evidence showing the enhancement of the overall heat transfer by the small-scale effect in the fast-transient process.

The classical theory of diffusion assumes an immediate equilibrium between the heat flux vector and the temperature gradient at any instant of time during the transient process, i.e.,

$$q(x,t) = -k\frac{\partial T}{\partial t}(x,t) \quad \text{or} \quad \eta(\delta,\beta) = -\frac{1}{2}\frac{\partial\theta}{\partial\beta}(\delta,\beta). \tag{2.77}$$

This is no longer the case in the dual-phase-lag model because of the presence of the phase lags.

Figure 2.14 shows the time histories of the heat flux vector and the temperature gradient at the midpoint of the film, $\delta = 0.5$. The value of B is taken as 0.2 $(\tau_T < \tau_q)$, implying that the temperature gradient precedes the heat flux vector in the transient heat transport. As a result, the temperature gradient increases at a faster rate than the heat flux vector.

For the reversed situation, where the heat flux vector precedes the temperature gradient in the transient response $(\tau_T > \tau_q)$, illustrated by Figure 2.15 at $B = 2.0$, the heat flux vector increases at a faster rate than the temperature gradient in the transient process. The precedence switch and, consequently, the interchange between cause and effect in the transient process of heat transport, is a salient feature in the dual-phase-lag model.

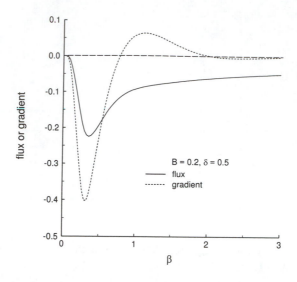

Figure 2.14 Time histories of the heat flux vector and the temperature gradient in the middle of the film. Here, $\delta = 0.5$, $l = 1.0$, $\eta_s = 1.0$, $\beta_s = 0.2$, and $B = 0.2$ (gradient precedence).

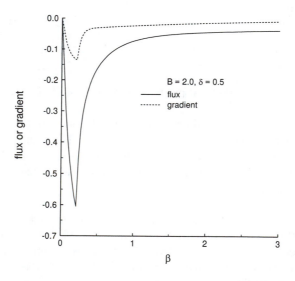

Figure 2.15 Time histories of the heat flux vector and the temperature gradient in the middle of the film. Here, $\delta = 0.5$, $l = 1.0$, $\eta_s = 1.0$, $\beta_s = 0.2$, and $B = 2.0$ (flux precedence).

2.9 CHARACTERISTIC TIMES DESCRIBING THE LAGGING RESPONSE

The constitutive equation (2.6) (or equation (2.7)) for the dual-phase-lag model is easy to confuse with the Cattaneo-Vernotte equation (2.2) (equation (2.3)) for the classical thermal waves. One argument is that equation (2.6) seems to be a direct result of a phase shift from t to $(t + \tau_T)$ in equation (2.2):

$$t^* = t + \tau_T. \tag{2.78}$$

Substituting equation (2.78) into (2.6) gives

$$\vec{q}(\vec{r}, t^* + \tau) = -k\nabla T(\vec{r}, t^*), \quad \text{with} \quad \tau = \tau_q - \tau_T, \tag{2.79}$$

which has an identical form to equation (2.2) in the time frame of t^*. Extra caution, however, must be taken when combining with the energy equation (2.5). A *consistent* shift of the timescale from t to t^* results in

$$-\nabla \bullet \vec{q}(\vec{r}, t^* - \tau_T) = C_p \frac{\partial T(\vec{r}, t^* - \tau_T)}{\partial t^*}, \tag{2.80}$$

which depicts heat transport occurring at $t^* - \tau_T$. In examining equations (2.79) and (2.80), *three* characteristic times remain in the framework: the characteristic time t^* at which the temperature gradient is established across a material volume, $(t^* + \tau)$ at which heat flow occurs and $(t^* - \tau_T)$ at which thermal energy is carried through. Expanding equations (2.79) and (2.80) with respect to t^* yields

$$\vec{q}(\vec{r}, t^*) + \tau \frac{\partial \vec{q}(\vec{r}, t^*)}{\partial t^*} = -k\nabla T(\vec{r}, t^*) \tag{2.81}$$

$$-\nabla \bullet \vec{q}(\vec{r}, t^*) + \tau_T \frac{\partial}{\partial t^*}[\nabla \bullet \vec{q}(\vec{r}, t^*)] = C_p\left[\frac{\partial T(\vec{r}, t^*)}{\partial t^*} - \tau_T \frac{\partial^2 T(\vec{r}, t^*)}{\partial t^{*2}}\right]. \tag{2.82}$$

Unlike the previous situation shown by equations (2.5) to (2.10), unfortunately, the heat flux vector *cannot* be eliminated from equations (2.81) and (2.82) in the time frame of t^*. Consequently, they have to be solved for $\vec{q}$ and T simultaneously. An informative expression, however, results from further substitution of equation (2.81) into (2.82):

$$\nabla^2 T(\vec{r}, t^*) = \frac{1}{\alpha}\frac{\partial T(\vec{r}, t^*)}{\partial t^*} - \frac{\tau_T}{\alpha}\frac{\partial^2 T(\vec{r}, t^*)}{\partial t^{*2}} - \frac{\tau_q}{\alpha}\frac{\partial}{\partial t^*}[\nabla \bullet q(\vec{r}, t^*)]. \tag{2.83}$$

The *negative* sign in front of the wave term and the additional term involving the heat flux vector intrinsically alter the fundamental characteristics in the *CV* wave equation. The negative sign, for example, completely *destroys* the wave structure in heat propagation, while the time derivative of the spatial gradient of the heat flux vector introduces some unexpected effects. Though bearing a great resemblance to the *CV* wave equation, therefore, equation (2.79) or (2.81) describing the lagging response in terms of two phase lags should *not* be confused with the approach of a *single* phase lag describing the wave behavior in heat conduction.

2.10 ALTERNATING DIFFUSION AND WAVE BEHAVIOR

In the absence of a general solution for equations (2.5) and (2.6), the unique feature of the lagging behavior can be illustrated by the Taylor series expansion. Including all the nonlinear terms of τ_T and τ_q, equation (2.7) becomes

$$\vec{q}(\vec{r},t) + \sum_{i=1}^{N} \frac{\tau_q^{(i)}}{(i)!} \frac{\partial^{(i)}\vec{q}}{\partial t^{(i)}}(\vec{r},t) = -k\left[\nabla T(\vec{r},t) + \sum_{j=1}^{M} \frac{\tau_T^{(j)}}{(j)!} \frac{\partial^{(j)}}{\partial t^{(j)}}\left(\nabla T(\vec{r},t)\right)\right], \quad M, N \geq 1.$$

(2.84)

For an exact representation, the number of terms in the Taylor series expansion, M and N, approach infinity. All the physical quantities in equation (2.84) occur at the same instant of time (t) as those in the energy equation (2.5), facilitating a further combination.   Because the volumetric heat source term does not alter the fundamental characteristic of the lagging response, we exclude this term in the following treatment. Eliminating the heat flux vector, by the same procedure, results in

$$\nabla^2 T + \sum_{i=1}^{M} \frac{\tau_T^{(i)}}{(i)!} \frac{\partial^{(i)}}{\partial t^{(i)}}\left(\nabla^2 T\right) = \frac{1}{\alpha}\left[\frac{\partial T}{\partial t} + \sum_{j=1}^{N} \frac{\tau_q^{(j)}}{(j)!} \frac{\partial^{(j+1)}T}{\partial t^{(j+1)}}\right],$$

(2.85)

which is in correspondence with equation (2.10) in the linearized version of the dual-phase-lag model. Equation (2.85) indicates a *progressive interchange* between diffusive and wave behaviors. When $N = M$, implying equal terms are taken in the Taylor series expansion on both sides of equation (2.84), equation (2.85) can be rearranged into the following form:

$$\frac{1}{N!} \frac{\partial^{(M)}}{\partial t^{(M)}}\left[\tau_T^M \nabla^2 T - \frac{\tau_q^M}{\alpha} \frac{\partial T}{\partial t}\right] + \text{lower-order terms} = 0.$$

(2.86)

The quantities enclosed by the brackets dictate the characteristics of the solution because they possess the *highest* order differentials. A particular solution is clearly seen,

$$\nabla^2 T - \frac{1}{\alpha^*}\frac{\partial T}{\partial t} = 0 \quad \text{with} \quad \alpha^* = \alpha\left(\frac{\tau_T}{\tau_q}\right)^M,\tag{2.87}$$

which is a *diffusion* equation with an equivalent thermal diffusivity α^*. The diffusive behavior depicted by equation (2.84) in the case of $M = N$ is thus clear. For $N = M + 1$, the next term taken after $N = M$ in the series approximation, the corresponding equation to equation (2.86) reads

$$\frac{\partial^{(M)}}{\partial t^{(M)}}\left[\frac{\tau_T^M}{M!}\nabla^2 T - \frac{1}{\alpha}\frac{\tau_q^{(M+1)}}{(M+1)!}\frac{\partial^2 T}{\partial t^2}\right] + \text{lower-order terms} = 0.\tag{2.88}$$

It results in a *wave* equation as the particular solution:

$$\nabla^2 T - \frac{1}{C^{*2}}\frac{\partial T}{\partial t} = 0 \quad \text{with} \quad C^* = \sqrt{\frac{(M+1)\alpha\tau_T^M}{\tau_q^{(M+1)}}} = C\sqrt{(M+1)\left(\frac{\tau_T}{\tau_q}\right)^M}\tag{2.89}$$

where $C = \sqrt{\alpha/\tau_q}$ is the speed of the *CV* wave, according to equation (2.4). Equation (2.89) demonstrates the subsequent *wave* behavior ($N = M + 1$) after diffusion ($N = M$). The equivalent thermal diffusivity, equation (2.87), and the equivalent thermal wave speed, equation (2.89), of the high-order modes may be smaller or greater than the preceding modes, depending on the ratio of (τ_T/τ_q).

2.11 DETERMINATION OF PHASE LAGS

The two phase lags characterizing the delayed response of the temperature gradient (τ_T) and the heat flux vector (τ_q) play decisive roles for the advancement of the dual-phase-lag model. Like the thermal conductivity and the thermal diffusivity, their values must be determined experimentally and well tabulated for engineering materials under various conditions, including the elevated temperature and the same medium with different microstructures.

In the absence of a thorough experiment determining the values of τ_T and τ_q at this stage of development, the analytical correlations to the existing modal parameters reflect such an attempt. In equation (2.12) correlating to the phonon-electron coupling factor, the thermal conductivity, and heat capacities of the electron gas and the metal lattice, for example, the values of τ_T and τ_q can be *calculated* from these experimentally determined parameters. The experimental facilities measuring the ultrafast transient response are very expensive, especially for the high-precision controlled short-pulse laser used for the determination of the phonon-electron coupling factor in picoseconds. The analytical correlation exemplified by equation

(2.12) not only provides an economical way to determine the values of τ_T and τ_q in this time frame, but also extends the previous effort for more productive results.

A successful experiment determining the values of τ_T and τ_q must achieve two goals. First, the experiment must be self-contained within the framework of the phase-lag concept. Correlation to other models is necessary but can hardly be sufficient. Second, based on the values of τ_T and τ_q thus determined and the use of the dual-phase-lag model, the resulting temperature curve must preserve all the salient features observed experimentally in the entire process of the short-time transient. For characterizing the linearized lagging behavior, one possibility is to use equation (2.7) for a one-dimensional specimen under separate conditions of a constant temperature gradient (for the determination of τ_q) and a constant heat flux (for τ_T).

For a one-dimensional solid subjected to a constant temperature gradient, as shown in Figure 2.16, the heat flux at the end of the coupon, $x = L$, can be recorded at various times in the transient process. The heat flux gauge used for this purpose must provide a direct measurement (such as the optical-fiber probe with low-impedance sapphire tips) and cannot be the one based on the measurement of temperature difference employing Fourier's law. The constant temperatures, T_0 and T_L maintained at the two ends at $x = 0$ and L, can be achieved by close contacts to different boiling fluids. The temperature gradient established over L increases with the difference of boiling temperatures of the two fluids. Under a constant temperature gradient, according to equation (2.7), the heat flux can be obtained as

$$q = k\nabla T \left(e^{-\frac{t}{\tau_q}} - 1 \right), \quad \text{with} \quad k\nabla T = \text{constant} . \tag{2.90a}$$

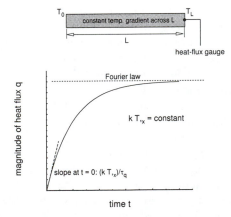

Figure 2.16 Determination of the phase lag of the heat flux vector, τ_q, from the initial slope of the heat flux versus time curve for a one-dimensional specimen subjected to a constant temperature gradient across L.

In a one-dimensional situation, $\nabla T = \partial T / \partial x \equiv T_{,x}$. The time-rate of change of the heat flux is thus

$$\frac{\partial q}{\partial t} = -\left(\frac{k\nabla T}{\tau_q}\right) e^{-\frac{t}{\tau_q}}. \tag{2.90b}$$

In the plot of heat flux (q) versus time (t), clearly, the slope of the q-t curve at $t = 0$ is

$$\left(\frac{\partial q}{\partial t}\right)_{t=0} = -\left(\frac{k\nabla T}{\tau_q}\right). \tag{2.90c}$$

Based on the known temperature gradient (∇T) and the thermal conductivity (k), therefore, the phase lag of the heat flux vector, τ_q, can be obtained from the initial slope in the q-t curve. This is illustrated in Figure 2.16, with arbitrary scales labeled on both axes for reference. The response curve assuming Fourier's law results in a constant heat flux (q = constant) in the time history. Under the assumption of an *instantaneous* response, a constant temperature gradient induces a constant heat flux vector in zero time, and the gradually increasing heat flux with time as shown in Figure 2.16 will not occur.

A similar procedure applies for the determination of τ_T. A constant heat flux is applied at one end of the coupon at $x = 0$, as illustrated in Figure 2.17, while a constant temperature T_L is maintained at $x = L$. A thermal gauge, including thermocouples for slower responses, infrared detectors for intermediate responses, or short-pulse lasers for fast responses, is attached to the heated end ($x = 0$) for measuring the temperature change in the time history. Integration of equation (2.7) in this case results in

$$T - T_L = \frac{q}{k}\left(e^{-\frac{t}{\tau_T}} - 1\right)(x - L), \quad \text{with} \quad q = \text{constant}, \tag{2.91a}$$

which gives

$$\frac{\partial T}{\partial t} = \left(\frac{q}{k\tau_T}\right) e^{-\frac{t}{\tau_T}}(x - L). \tag{2.91b}$$

At $x = 0$, clearly, the *initial* ($t = 0$) time-rate of change of temperature is

$$\left(\frac{\partial T}{\partial t}\right)_{t=0;\, x=0} = -\left(\frac{qL}{k\tau_T}\right). \tag{2.91c}$$

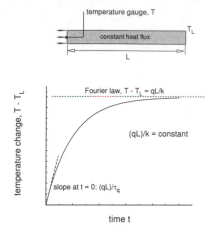

Figure 2.17 Determination of the phase lag of the temperature gradient, τ_T, from the initial slope of the temperature versus time curve at the heated end ($x = 0$) for a one-dimensional specimen subjected to a constant heat flux.

With the boundary heat flux (q) and the thermal conductivity (k) being known, the phase lag of the temperature gradient, τ_T, is obtained from the initial slope of the T-t curve recorded at the heated end at $x = 0$. This is illustrated in Figure 2.17. Again, the Fourier law raises the temperature change to a constant value, qL/k, in zero time.

The experimental setups shown in Figures 2.16 and 2.17 measuring the phase lags of τ_q and τ_T are feasible at least in principle. In practice, however, the thermal devices measuring the heat flux and the temperature change must respond sufficiently fast to capture the time-rising behavior of the heat flux and the temperature. For low-conducting porous media such as sand, referring to Chapter 6, the time frame for observing the time-rising behavior is of the order of seconds, which facilitates the use of traditional thermocouples. For rapid heating on metals, on the other hand, this time frame may shrink to the order of picoseconds to femtoseconds. The front-surface-pump, back-surface-probe technique developed by Brorson et al. (1987) and advanced recently by Qiu et al. (1994) is necessary to trace the rapid changes at short times.

The characteristic dimension of the specimen (L) also plays a decisive role in successful measurement. The coupon must be sufficiently small that the lagging behavior will not be averaged over by the internal structures. For sand as an example, this means involvement of several sand particles for a more pronounced lagging behavior. For metals, semiconductors, or dielectric crystals, on the other hand, a sufficiently small specimen implies the use of thin films containing only several grains. Evidently, the specimen size on microscale may be another difficulty in performing the experiment determining the phase lags from the short-time response.

Note that equation (2.7), and hence the initial slopes shown in Figures 2.16 and 2.17, results from the linear version of the dual-phase-lag model that neglects all the nonlinear effects of τ_T and τ_q. If the resulting values of τ_T and τ_q do not yield transient temperatures comparable to the experimental result, nonlinear effects of τ_T and τ_q may be needed in the analysis. The corresponding expressions for the heat flux vector (under a constant temperature gradient, equation (2.90a)) and the temperature (under a constant heat flux, equation (2.91a)) need to be reproduced to incorporate the high-order effect.

It is helpful to summarize the dual-phase-lag model presented in this chapter as follows:

1. The model captures the microstructural interaction effect in the fast-transient process. While the phase lag of the heat flux vector (τ_q) describes the thermal inertia in the short-time response, the phase lag of the temperature gradient (τ_T) describes the delayed time caused by the heat transport mechanisms occurring in microscale.

2. From a mathematical point of view, the lagging behavior introduces a new type of energy equation in conductive heat transfer. In addition to a wave term, it contains a mixed-derivative term, a second-order derivative in space and a first-order derivative in time. The mixed-derivative term reflects the delayed response due to microstructural interactions. It efficiently destroys the wave structure in heat propagation.

3. Compared with diffusion and CV waves, the lagging response of temperature results in a higher temperature level in the heat penetration zone. The larger the ratio of τ_T/τ_q (the parameter B), the larger the deviation from the diffusion and CV wave models. The lagging behavior is characterized by the ratio of τ_T/τ_q.

4. The case of diffusion with $\tau_T = \tau_q$ ($B = 0.5$) displays the onset for precedence switching in the process of heat transport. For $\tau_T > \tau_q$, the heat flux vector (cause) precedes the temperature gradient (effect), and the resulting temperature is higher than that predicted by diffusion. This is the case for microscopic phonon scattering (in dielectric films, insulators, and semi-conductors) and phonon-electron interaction (in metals). For $\tau_q > \tau_T$, the temperature gradient (cause) precedes the heat flux vector (effect), and the resulting temperature is lower than that predicted by diffusion. As shown by the experimental results in Chapters 6 and 7, this is the case for mesoscale heat transport in media with discrete internal structures. Also, at a fixed position inside the medium, the case of gradient precedence results in a faster rate of increase of the temperature gradient in the transient process. The case of flux precedence, on the other hand, results in a faster rate of increase of the heat flux vector in short times.

5. The lagging temperature responds to the initial time-rate of change of temperature, $\partial T/\partial t$ at $t = 0$. Such a rate effect may render a local temperature higher than the boundary temperature of the solid, a phenomenon called temperature overshooting, after Taitel (1972) in the wave theory of heat conduction. The initial temperature rate and the physical scale in which the temperature overshooting occurs, however, are difficult to achieve at this time. It is a mathematical result currently lacking experimental support.

6. The Riemann-sum approximation for the Laplace inversion, numerically, is stable and reliable. Perfect agreements with the analytical solutions are demonstrated in Figures 2.4 to 2.6. They support the use of this algorithm to explore other phenomena under various conditions of practical interest. Also, the use of the Laplace transform method minimizes the difficulty in problems

involving flux-specified boundary conditions. As shown in Section 2.8, the type of boundary condition really makes no difference when solving the boundary value problem in the Laplace transform domain.

7. The characteristic times describing the fast-transient process accounting for the small-scale effect, most important, dominate the lagging response. Only *two* characteristic times, t (for the temperature gradient and the energy equation for heat transport) and $t + \tau$ (for the heat flux vector), are involved in the *CV* wave model. While allowing the delayed response in the heat flux vector, it still assumes instantaneous heat transport right after the establishment of a temperature gradient. The temperature gradient, therefore, always precedes the heat flux vector in the *CV* wave model. The dual-phase-lag model, on the other hand, involves three characteristic times describing the process of heat transport. It relaxes the assumption of the precedence between the heat flux vector and the temperature gradient, allowing for a more flexible description when the microscale effect comes into the picture. As shown by the experimental support in Chapters 4 to 7, the additional phase lag is highly desirable to capture the delayed response due to the microscale interaction effect under various circumstances.

THREE

THERMODYNAMIC FOUNDATION

A constitutive equation depicts the way in which cause varies with effect in the transport process. Fourier's law in heat conduction is a typical example. In the case where the heat flux vector is viewed as the cause for heat flow and the temperature gradient is viewed as the effect, Fourier's law describes a linear relationship between the two, with the proportional constant defined as the thermal conductivity. Stoke's law of viscosity in momentum transfer is another example. Viewing the shear stress as the cause for fluid motion, the shear strain rate becomes the effect, and the proportional constant between the two is defined as the coefficient of viscosity. Cause and effect in these examples occur at the same instant of time, implying that their positions are interchangeable in physical phenomena. The difference between cause and effect has no physical significance in these cases. The CV wave model and the dual-phase-lag model display a more complicated situation. Assuming the temperature gradient is the cause and the heat flux vector is the effect, the CV wave model describes a wave phenomenon in heat propagation. The delay time between the two, called the relaxation time, renders a relaxation behavior that may be the basis for certain short-time responses. The dual-phase-lag model removes the strong assumption of precedence in heat flow. Either the temperature gradient or the heat flux vector may become the cause and the remaining one becomes the effect. The heat flow driven by the temperature gradient may have completely different characteristics from that driven by the heat flux vector. In many cases, most important, a different precedence sequence in heat flow implies a different substructural mechanism causing the delayed response.

No matter how fancy a constitutive behavior is, it must be admissible within the framework of the second law of thermodynamics. Physically, this means a positive-definite entropy production rate at all times in the thermodynamic process. In the estimate for the entropy production rate, however, it may be necessary to expand the classical framework assuming a quasi-stationary transition of thermodynamic states to account for the irreversibility occurring in the fast-transient process. This chapter is devoted to a detailed analysis of the lagging behavior with emphasis on the second law of thermodynamics. The expanded framework suitable for describing the nonequilibrium, irreversible transition of thermodynamic states in a fast-transient process is called *extended irreversible thermodynamics* (Jou et al., 1988; Tzou, 1993a).

3.1 CLASSICAL THERMODYNAMICS

Within the framework of classical thermodynamics, transition of the thermodynamic state in a physical process is assumed to occur at an extremely slow rate in the time history. It must be so slow that an equilibrium state, both thermodynamically and mechanically, is always achieved before the thermodynamic state moves on. From a microscopic point of view, as illustrated in Figure 3.1(a), this implies that the process time t is much longer than the relaxation time, the mean free time among successive collisions of molecules. At a representative instant of time t in the slowly varying process, the thermodynamic state is in equilibrium, and the thermodynamic properties such as pressure and temperature can be measured for defining the thermodynamic state. The pressure and temperature thus obtained, by eliminating the implicit variable of process time t, can be used to plot the p-T curve, the process diagram shown in Figure 3.1(b). Any two adjacent thermodynamic states on the process diagram, exemplified by A and B in Figure 3.1(b), require a sufficiently long time to travel through, although the physical time has been eliminated in preparing the process diagram. The thermodynamic transition possessing of such a long-time feature is defined as a *quasi-stationary* process.   Reversibility is a direct consequence of the quasi-stationary transition between two adjacent states, which implies an equal amount of effort needed to travel back and forth between A and B. Under this assumption, the incremental heat (dQ) added into a thermodynamic system at an instantaneous temperature T is equal to the product of the temperature and the resulting incremental change of entropy (ds). Mathematically, $dQ = Tds$. Absorption of heat may result in an incremental change of specific volume (dv)

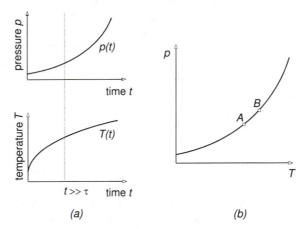

(a)   (b)

Figure 3.1 (*a*) Time histories of pressure and temperature measured in a slowly varying process of thermodynamic transition for $t \gg \tau$ and (*b*) process diagram of pressure versus temperature obtained by eliminating the process time.

under an instantaneous pressure (p). The resulting incremental work *done* by the system is denoted by pdv. The incremental change of internal energy of the thermodynamic system, $d\varepsilon$, according to the first law of thermodynamics, is thus

$$d\varepsilon = Tds - pdv \quad \text{or} \quad ds = \frac{d\varepsilon}{T} + \frac{pdv}{T}, \tag{3.1}$$

with temperature T measured in Kelvins, the absolute scale. Note that the specific quantities in equation (3.1), including $d\varepsilon$, ds, and dv, are defined per unit mass. In a quasi-stationary approach, the rate form corresponding to equation (3.1) is

$$\dot{s} = \frac{\dot{\varepsilon}}{T} + \frac{p\dot{v}}{T}, \tag{3.2}$$

with overdots denoting derivatives with respect to time. For converting the rate equation into a form on a basis per unit volume, we multiply the mass density (ρ) through the entire equation to give

$$\rho\dot{s} = \frac{\rho\dot{\varepsilon}}{T} + \frac{\rho p\dot{v}}{T}. \tag{3.3}$$

From a thermodynamic point of view, equation (3.3) expresses entropy (s) in terms of *two* thermodynamic properties, the internal energy (ε) and the specific volume (v). Mathematically, $s = s(\varepsilon, v)$. It is thus clear that the classical thermodynamics treats the conducting medium as a *simple* substance. From this point of view, the resulting time-rate of change of entropy per unit volume is

$$\rho\dot{s} = \rho\left(\frac{\partial s}{\partial \varepsilon}\right)_v \dot{\varepsilon} + \rho\left(\frac{\partial s}{\partial v}\right)_\varepsilon \dot{v}. \tag{3.4}$$

Comparing with equation (3.3), it gives

$$\frac{1}{T} = \left(\frac{\partial s}{\partial \varepsilon}\right)_v, \quad p = T\left(\frac{\partial s}{\partial v}\right)_\varepsilon. \tag{3.5}$$

The subscripts here represent the quantity held constant in partial differentiations. Equation (3.5) can be viewed as the definitions for temperature and pressure in terms of the derivatives of specific entropy with respect to internal energy and specific volume, respectively.

The time-rate of change of entropy per unit volume, $\rho\dot{s}$, is the sum of $\rho\dot{\varepsilon}/T$ (the internal energy per unit volume divided by temperature) and $\rho p\dot{v}/T$ (the work done by the thermodynamic system per unit volume divided by temperature). The internal energy per unit volume can be derived from the conservation of energy:

$$\rho\dot{\varepsilon} = -q_{i,i} - \sigma_{ij}\dot{e}_{ij} \quad \text{for} \quad i, j = 1, 2, 3 \tag{3.6}$$

where the repeated index represents summation. For example, $q_{i,i} = q_{1,1} + q_{2,2} + q_{3,3} \equiv \nabla \bullet \vec{q}$. The power produced by the stress tensor, $\sigma_{ij}\dot{e}_{ij}$ with $\dot{e}_{ij}$ representing the strain rate tensor in a *deformable* body, is included for a general treatment. To separate the work done by the hydrostatic pressure and the energy dissipation due to shear stresses, the stress tensor can be written as

$$\sigma_{ij} = p\delta_{ij} + \sigma'_{ij} \quad \text{for} \quad i, j = 1, 2, 3 \tag{3.7}$$

where δ_{ij} is the Kronecker delta. It has a value of zero for $i = j$, and unity otherwise. In terms of the matrix notation, alternatively, the Kronecker delta is a diagonal matrix with all the pivot elements being unity. For the system performing work under a hydrostatic tension, the normal stress components are identical, $\sigma_{11} = \sigma_{22} = \sigma_{33} = p$. The stress *deviator* σ'_{ij} defined in equation (3.7), therefore, contains only nonzero shear components (for $i \neq j$), while all the normal components are zero ($\sigma'_{11} = \sigma'_{22} = \sigma'_{33} = 0$). The viscous shear contributing only to the shape change (which does not perform work) is thus separated from the normal tension contributing to the volume change (which performs work) of the thermodynamic system. Substituting equation (3.7) into (3.6) yields

$$\frac{\rho\dot{\varepsilon}}{T} = -\frac{q_{i,i}}{T} - \frac{p\dot{e}}{T} - \frac{\sigma'_{ij}\dot{e}_{ij}}{T}, \quad \text{with} \quad e \equiv \delta_{ij}\dot{e}_{ij} = \dot{e}_{ii}, \tag{3.8}$$

which gives the first term in equation (3.3).

The second term in equation (3.3), the work performed by the thermodynamic system per unit volume, can be derived from the conservation of mass. Assuming a homogeneous density distribution throughout the system, the continuity equation requires that

$$\frac{\partial \rho}{\partial t} + \rho\nabla \bullet \vec{u} = 0 \quad \text{or} \quad \dot{\rho} + \rho u_{i,i} = 0 \tag{3.9}$$

where $\vec{u}$ denotes the velocity vector at a material point. Since the mass density ρ is reciprocal to the specific volume v, $\rho = 1/v$, the time-rate of change of density in equation (3.9) can be replaced by

$$\dot{\rho} = -\frac{\dot{v}}{v^2}. \tag{3.10}$$

Divergence of the velocity vector, in addition, is the time-rate of change of the divergence of the displacement vector ($\vec{w}$). Mathematically,

$$\nabla \bullet \vec{u} = \frac{d}{dt}(\nabla \bullet \vec{w}) = \frac{d}{dt}(w_{i,i}) = \dot{e} \tag{3.11}$$

where

$$e_{ij} = \frac{1}{2}\left(\frac{\partial w_i}{\partial x_j} + \frac{\partial w_j}{\partial x_i}\right), \quad \text{implying} \quad e_{11} = \frac{\partial w_1}{\partial x_1}, \quad e_{22} = \frac{\partial w_2}{\partial x_2}, \quad e_{33} = \frac{\partial w_3}{\partial x_3},$$

(3.12)

the normal components of Cauchy strain tensor describing small deformation. Substituting equations (3.10) and (3.11) into (3.9), the continuity equation becomes

$$\rho \dot{v} = \dot{e}, \quad \text{resulting in} \quad \frac{\rho p \dot{v}}{T} = \frac{p \dot{e}}{T}.$$

(3.13)

Equations (3.8) (for the specific internal energy) and (3.13) (for the specific work done) can now be substituted into equation (3.3) to obtain the time-rate of change of specific entropy:

$$\rho \dot{s} = -\frac{q_{i,i}}{T} - \frac{\sigma'_{ij} \dot{e}_{ij}}{T},$$

(3.14)

with the terms containing $\dot{e}$ in equations (3.8) and (3.13) canceling each other. Noting that

$$\frac{\partial}{\partial x_i}\left(\frac{q_i}{T}\right) = \frac{1}{T}\frac{\partial q_i}{\partial x_i} - \frac{q_i}{T^2}\frac{\partial T}{\partial x_i},$$

(3.15)

equation (3.14) can be arranged into a special form:

$$\rho \dot{s} + \frac{\partial}{\partial x_i}\left(\frac{q_i}{T}\right) = -\frac{q_i}{T^2}\frac{\partial T}{\partial x_i} - \frac{\sigma'_{ij}\dot{e}_{ij}}{T}.$$

(3.16)

Equation (3.16) is in a standard form comparable to the Reynolds transport equation for the entropy flux vector. Let us consider the thermodynamic system shown in Figure 3.2. The total volume of the thermodynamic system is denoted by V. The entropy flux vector, $\bar{J}$, flows out of the system through the system boundary represented by S. The unit normal of the representative surface area dS is denoted by $\bar{n}$. The entropy production rate per unit volume within the thermodynamic system is denoted by Σ. The time-rate of change of entropy in the entire volume is thus

$$\int_V \rho \dot{s}\, dV = \int_V \Sigma\, dV - \int_S \bar{J} \cdot \bar{n}\, dS.$$

(3.17)

For combining the surface to the volume integral, the divergence theorem is applied,

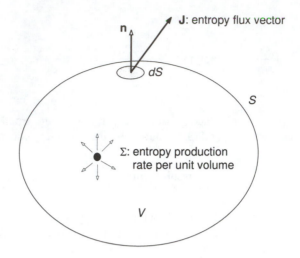

Figure 3.2 The entropy flux vector flowing out of the system boundary ($\bar{J}$) and the entropy production rate within the thermodynamic system.

$$\int_S \bar{J} \bullet \bar{n} \, dS = \int_V \nabla \bullet \bar{J} \, dV \, , \tag{3.18}$$

resulting in

$$\rho \dot{s} + \nabla \bullet \bar{J} = \Sigma \quad \text{or} \quad \rho \dot{s} + J_{i,i} = \Sigma \, . \tag{3.19}$$

This is the differential form of the Reynolds transport equation for the entropy flux vector. Comparing equations (3.16) and (3.19) gives

$$J_i = \frac{q_i}{T} \, , \quad \Sigma = -\frac{q_i}{T^2} \frac{\partial T}{\partial x_i} - \frac{\sigma'_{ij} \dot{e}_{ij}}{T} \, , \tag{3.20}$$

or, in terms of vector notations,

$$\bar{J} = \frac{\bar{q}}{T} \, , \quad \Sigma = -\left(\frac{1}{T^2}\right) \bar{q} \bullet \nabla T - \frac{\sigma'_{ij} \dot{e}_{ij}}{T} \, . \tag{3.21}$$

For a thermodynamically admissible process, the entropy production rate, Σ, must be *positive definite*. From the second expression in equation (3.21), this implies

$$\Sigma = -q_i \left(\frac{1}{T^2} \frac{\partial T}{\partial x_i}\right) - \sigma'_{ij} \left(\frac{\dot{e}_{ij}}{T}\right) \geq 0 \, . \tag{3.22}$$

One of the *many* possibilities for equation (3.22) to be valid is

$$\left(\frac{1}{T^2}\frac{\partial T}{\partial x_i}\right) = -Aq_i \quad \text{and} \quad \left(\frac{\dot{e}_{ij}}{T}\right) = -B\sigma'_{ij} \quad \text{with} \quad A, B \geq 0, \quad (3.23)$$

which warrants a quadratic form for the entropy production rate. Equation (3.23) results in two constitutive equations for the thermal and mechanical fields. For a rigid conductor where no deformation is possible, the strain rate tensor is zero and the second expression in equation (3.23) can be ignored. To the thermal field, equation (3.23) indicates

$$q_i = -\left(\frac{1}{AT^2}\right)\frac{\partial T}{\partial x_i} \quad \text{or} \quad \vec{q} = -k\nabla T \quad \text{with} \quad A = \frac{1}{kT^2}, \quad k \geq 0. \quad (3.24)$$

This is the familiar Fourier's law in heat conduction, which has been derived as a natural consequence in ensuring a positive-definite entropy production rate in the classical framework of thermodynamics. It is important to note, however, that the classical thermodynamic assumes a quasi-stationary transition of thermodynamic states, revealing that Fourier's law takes on this assumption. Fourier's law describes heat flow in transition of thermodynamic states; in other words, it describes a situation in thermo*statics*. For the mechanical field, on the other hand, equation (3.23) indicates

$$\sigma'_{ij} = -\left(\frac{1}{BT}\right)\dot{e}_{ij} = -2\mu\dot{e}_{ij} \quad \text{with} \quad B = \frac{1}{2\mu T}, \quad \mu \geq 0. \quad (3.25)$$

This is the familiar Stoke's law of viscosity in Newtonian fluids. Again, it results from the natural consequence ensuring a positive-definite entropy production rate in transition of the thermodynamic states, taken from the same quasi-stationary assumption. Stoke's law of viscosity in momentum transfer appears as a counterpart of Fourier's law in energy transfer in the classical framework of thermodynamics.

3.2 EXTENDED IRREVERSIBLE THERMODYNAMICS

When the process time becomes comparable to the relaxation time in molecular collisions or the thermalization time between electrons and phonons in microscale, transition of the thermodynamic states is defined to take place at a "fast" pace. The thermodynamic "state" in these processes may not have sufficient time to achieve equilibrium before it moves on in the time history. As a result, the thermodynamic process becomes *irreversible*. Losing support for making these important assumptions, the classical framework of thermodynamics must be extended to account for the fast-transient effect and the irreversibility in the short-time response. Existing effort includes extension of the classical framework (Coleman, 1964;

Coleman et al., 1982, 1986; Bai and Lavine, 1995) and development of the extended irreversible thermodynamics (Jou et al., 1988; Tzou, 1993a).

The quasi-stationary assumption, first, has to be removed in the fast-transient process. In addition to the specific internal energy (ε) and the specific volume (v), *additional* fundamental state variables need to be introduced in defining the specific entropy (s) to reflect the irreversible nature in thermodynamic transitions. The thermodynamic state is highly *nonequilibrium* in a fast-transient process. The heat flux vector (q_i, for the thermal field) and the deviatoric stress tensor (σ'_{ij}, for the mechanical field) are natural consequences in nonequilibrium, irreversible transitions of thermodynamic states. Physically, therefore, it is simply intuitive to use them as the additional *fundamental* state variables in defining the specific entropy. Mathematically, this implies $s \equiv s(\varepsilon, v, q_i, \sigma'_{ij})$. Here, for a consistent representation with the stress tensor, a tensor notation is used for the heat flux vector, $\vec{q} \equiv q_i$ for i = 1, 2, and 3. The specific entropy depends on four state variables, indicating that the conducting medium is no longer a simple substance. The rate equation corresponding to equation (3.4) is

$$\rho \dot{s} = \rho \left(\frac{\partial s}{\partial \varepsilon} \right)_{v, q_i, \sigma'_{ij}} \dot{\varepsilon} + \rho \left(\frac{\partial s}{\partial v} \right)_{\varepsilon, q_i, \sigma'_{ij}} \dot{v} + \rho \left(\frac{\partial s}{\partial q_i} \right)_{\varepsilon, v, \sigma'_{ij}} \dot{q}_i + \rho \left(\frac{\partial s}{\partial \sigma'_{ij}} \right)_{\varepsilon, v, q_i} \dot{\sigma}'_{ij} .$$

(3.26)

Extending the results of equation (3.5) from classical thermodynamics, equation (3.26) becomes

$$\rho \dot{s} = \frac{\rho \dot{\varepsilon}}{T} + \frac{\rho p \dot{v}}{T} + \rho \left(\frac{\partial s}{\partial q_i} \right)_{\varepsilon, v, \sigma'_{ij}} \dot{q}_i + \rho \left(\frac{\partial s}{\partial \sigma'_{ij}} \right)_{\varepsilon, v, q_i} \dot{\sigma}'_{ij} .$$

(3.27)

In a general situation, the specific entropy (s) can be expanded into a Taylor series of both q_i and σ_{ij}' around an equilibrium state,

$$s = s_0 + \left(\frac{\partial s}{\partial q_i} \right)_0 q_i + \left(\frac{\partial s}{\partial \sigma'_{ij}} \right)_0 \sigma'_{ij} + \frac{1}{2} \left\{ \left(\frac{\partial^2 s}{\partial q_k \partial q_k} \right)_0 q_i q_i + 2 \left(\frac{\partial^2 s}{\partial q_k \partial \sigma'_{kj}} \right)_0 q_i \sigma'_{ij} + \left(\frac{\partial^2 s}{\partial \sigma'_{kl} \partial \sigma'_{kl}} \right)_0 \sigma'_{ij} \sigma'_{ij} \right\} + \cdots$$

(3.28)

with the subscript "0" referring to the quantities at the equilibrium state. Only the second-order terms in equation (3.28) will be considered in the following derivations, implying a moderate disturbance of specific entropy from that at the equilibrium state. The first-order derivatives in equation (3.27) are thus obtained by direct differentiations,

$$\frac{\partial s}{\partial q_i} = \left(\frac{\partial s}{\partial q_i}\right)_0 + \left(\frac{\partial^2 s}{\partial q_k \partial q_k}\right)_0 q_i + \cdot \cdot$$

$$\frac{\partial s}{\partial \sigma'_{ij}} = \left(\frac{\partial s}{\partial \sigma'_{ij}}\right)_0 + \left(\frac{\partial^2 s}{\partial \sigma'_{kl} \partial \sigma'_{kl}}\right)_0 \sigma'_{ij} + \cdot \cdot$$

(3.29)

where the coupling terms between the thermal and mechanical fields in equation (3.28) are further neglected for the time being. The thermomechanical coupling effect is discussed separately in Section 3.4 for a more concentrated study. At an equilibrium state, the specific entropy of the thermodynamic system reaches a maximum, implying that

$$\left(\frac{\partial s}{\partial q_i}\right)_0 = \left(\frac{\partial s}{\partial \sigma'_{ij}}\right)_0 = 0.$$

(3.30)

The coefficients of q_i and σ'_{ij} in equation (3.29) depend on the second-order derivatives of specific entropy with respect to the heat flux vector and the stress deviator, respectively. For the estimate of their signs, consider a convex, monotonically increasing curve of entropy shown in Figure 3.3(a). As the heat flux or stress deviator increases in the thermodynamic process, the slope (the first-order derivative of specific entropy) on the entropy curve decreases, as shown in Figure 3.3(b). The slope on the *first*-order derivative of the entropy curve shown in Figure 3.3(b), namely the *second*-order derivatives of specific entropy with respect to the heat flux vector and stress deviator, consequently, *decreases* as the heat flux vector or stress deviator increases. As a result, both the second-order derivative of specific entropy with respect to the heat flux vector, $\partial^2 s / \partial q_i \partial q_i$, and that with respect to the stress deviator, $\partial^2 s / \partial \sigma'_{ij} \partial \sigma'_{ij}$, are *negative*. Along with equation (3.30), therefore,

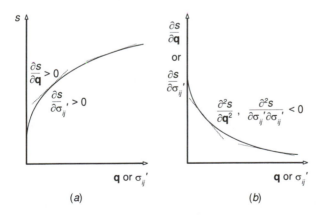

(a)   (b)

Figure 3.3 Negative second-order derivatives of specific entropy with respect to the heat flux vector and the stress deviator. (*a*) The positive first-order derivatives, and (*b*) the negative second-order derivatives.

equation (3.29) becomes

$$\frac{\partial s}{\partial q_i} \cong -C_1 q_i, \quad \frac{\partial s}{\partial \sigma'_{ij}} \cong -C_2 \sigma'_{ij}, \quad \text{with} \quad C_1 = -\left(\frac{\partial^2 s}{\partial q_i \partial q_i}\right)_0 \geq 0; \quad C_2 = -\left(\frac{\partial^2 s}{\partial \sigma'_{ij} \partial \sigma'_{ij}}\right)_0 \geq 0.$$

(3.31)

Substituting equation (3.31) into (3.27), the time-rate of change of entropy per unit mass is obtained:

$$\rho \dot{s} = \frac{\rho \dot{\varepsilon}}{T} + \frac{\rho p \dot{v}}{T} - \rho\left(C_1 q_i\right)\dot{q}_i - \rho\left(C_2 \sigma'_{ij}\right)\dot{\sigma}'_{ij}.$$

(3.32)

Owing to negligence of the high-order terms in the Taylor series expansions, to repeat, equation (3.32) bears the assumption of slight disturbances of q_i and σ_{ij} from their equilibrium values. The first and second terms on the right side of equation (3.32), defined in equations (3.8) and (3.13), can be treated in the same manner. The corresponding equation to equation (3.16) is

$$\rho \dot{s} + \frac{\partial}{\partial x_i}\left(\frac{q_i}{T}\right) = -q_i\left[\frac{1}{T^2}\frac{\partial T}{\partial x_i} + \rho C_1 \dot{q}_i\right] - \sigma'_{ij}\left[\frac{\dot{e}_{ij}}{T} + \rho C_2 \dot{\sigma}'_{ij}\right].$$

(3.33)

Equation (3.33) is now ready for comparison with the Reynolds transport equation for the entropy flux vector, equation (3.19). While the entropy flux vector (J_i), the first expression in equation (3.20), remains the same, the entropy production rate Σ is generalized into the following form:

$$\Sigma = -q_i\left[\frac{1}{T^2}\frac{\partial T}{\partial x_i} + \rho C_1 \dot{q}_i\right] - \sigma'_{ij}\left[\frac{\dot{e}_{ij}}{T} + \rho C_2 \dot{\sigma}'_{ij}\right].$$

(3.34)

To ensure a positive-definite value for Σ, a sufficient (but necessary) condition is

$$\frac{1}{T^2}\frac{\partial T}{\partial x_i} + \rho C_1 \dot{q}_i = -A q_i, \quad A \geq 0$$

$$\frac{\dot{e}_{ij}}{T} + \rho C_2 \dot{\sigma}'_{ij} = -B \sigma'_{ij}, \quad B \geq 0,$$

(3.35)

in an alternative form,

$$\vec{q} + \tau \frac{\partial \vec{q}}{\partial t} = -k \nabla T, \quad \text{with} \quad k = \frac{1}{AT^2} \quad \text{and} \quad \tau = \frac{\rho C_1}{A} \geq 0$$

$$\sigma'_{ij} + \tau_M \frac{\partial \sigma'_{ij}}{\partial t} = -2\mu \dot{e}_{ij} \quad \text{with} \quad \mu = \frac{1}{2BT} \quad \text{and} \quad \tau_M = \frac{\rho C_2}{B} \geq 0.$$

(3.36)

The vector notations are retrieved in the first expression for a more conventional appearance. The first expression describes the process of heat transport with a nonequilibrium and irreversible transition of thermodynamic states. Comparing to equation (2.3) for the CV wave model, it has exactly the same form. The second equation coincides with the constitutive equation describing Maxwell's fluid (Flügge, 1967), with τ_M being the relaxation time characterizing the relaxation behavior of stresses under a constant strain. Again, they result from the natural consequence of ensuring a positive-definite entropy production rate.

In the framework of extended irreversible thermodynamics, to reiterate, the CV wave model results from two major assumptions. First, the nonequilibrium and irreversible states describing the CV wave behavior do not deviate much from the equilibrium state. The second-order terms incorporated in the Taylor series expansion for the specific entropy in equation (3.28), however, do allow for a moderate disturbance from the equilibrium state in comparison with the first-order expansions. The second-order expansions do not include the thermomechanical coupling effect, which is perfectly legitimate for heat transport in *rigid* conductors. Second, and most important, the CV wave behavior is a direct result of a monotonically increasing, but *convex* entropy curve shown in Figure 3.3(a). A convex entropy curve seems to result from the assumption of precedence of the temperature gradient (cause) to the heat flux vector (effect) in the process of heat transport ($\tau_q = \tau$, $\tau_T = 0$; see the discussion of the CV wave model in Chapter 2). Such a gradient precedence appears to be a strong assumption, but this special mode of heat transport is justified within the framework of the extended irreversible thermodynamics.

Rather than a general formulation including all the high-order and thermomechanical coupling effects, note that our purpose here is to find a special case that supports the CV wave equation in the framework of the second law of thermodynamics. The two major assumptions summarized above, namely, the second-order disturbances from an equilibrium state and negligence of the thermomechanical coupling effect, outline the possible physical conditions for this special behavior to exist. Should the high-order terms and the thermomechanical coupling effect be further incorporated in the formulation, equations (3.28) and (3.29), the resulting equations governing heat and momentum transport in the thermodynamic system (equation (3.36)) become complicated. The CV wave model only appears as a special case in the generalized equation describing coupled heat transport. The coupling effect, in addition, renders a complicated situation, in that the energy equation involves mechanical stress and strain rates while the momentum equation involves heat flux vector and temperature gradient. Mathematically, this implies the necessity of solving the energy and momentum equations simultaneously for obtaining the temperature and stress distributions in the deformable body. The constitutive equations thus obtained are more general, but the emphasis would be on the effect of thermomechanical coupling, a complicated subject in thermomechanics postponed to Chapter 9.

3.3 LAGGING BEHAVIOR

Inspired by the success shown in Section 3.2, implementation of the generalized fluxes, including both the heat flux vector in heat transport and the mechanical stress in momentum transport, into the fundamental state variables seems to be a plausible approach for studying the nonequilibrium and irreversible transition of thermodynamic states. In extending this procedure to study the lagging behavior that covers the small-scale response in both space and time in the same framework, note that the resulting *CV* wave model (small-scale response in time only) describing the process of heat transport is a direct consequence assuming precedence of the temperature gradient to the heat flux vector in the transient process. Absorbing the *effect* of heat flow, namely, the heat flux vector, in the list of fundamental variables defining a nonequilibrium, the irreversible thermodynamic state is a mathematical way of reflecting such special behavior. Since the dual-phase-lag model does allow interchange between cause and effect to reflect different delayed responses in various situations, it is simply intuitive to incorporate the generalized *gradients*, including both the *temperature gradient* for heat transport and Cauchy *strain tensor* (displacement gradient in general, referring to equation (3.12)) for momentum transport, into the fundamental state variables for a rigorous treatment.

      In the case of gradient precedence, the constitutive equations for heat and momentum transfer, equation (3.35), are repeated here for continuity:

$$\frac{1}{T^2}\frac{\partial T}{\partial x_i} + \rho C_1 \dot{q}_i = -Aq_i, \quad A \geq 0$$

for gradient-precedence, $\tau_T < \tau_q$.   (3.35′)

$$\frac{\dot{e}_{ij}}{T} + \rho C_2 \dot{\sigma}'_{ij} = -B\sigma'_{ij}, \quad B \geq 0$$

They result from implementation of the *effect* in heat and momentum transport, namely, the generalized fluxes q and σ'_{ij}, in the list of fundamental variables defining the nonequilibrium, irreversible entropy, $s \equiv s(\varepsilon, v, q_i, \sigma'_{ij})$. In the conjugate situation where the generalized fluxes precede the generalized gradients, the generalized *gradients* become the effect, and the functional form of specific entropy becomes $s \equiv s \equiv s(\varepsilon, v, \nabla T, e_{ij})$, with the Cauchy strain tensor e_{ij} incorporated for the same reason that the case of heat transport is included in deformable bodies. The time-rate of change of specific entropy evolves into the following form:

$$\rho \dot{s} = \rho \left(\frac{\partial s}{\partial \varepsilon}\right)_{v,q_i,\nabla T, \atop \sigma'_{ij},e_{ij}} \dot{\varepsilon} + \rho \left(\frac{\partial s}{\partial v}\right)_{\varepsilon,q_i,\nabla T, \atop \sigma'_{ij},e_{ij}} \dot{v} + \rho \left[\frac{\partial s}{\partial(\nabla T)}\right]_{\varepsilon,v,e_{ij}} \nabla \dot{T} + \rho \left[\frac{\partial s}{\partial e_{ij}}\right]_{\varepsilon,v,\nabla T} \dot{e}_{ij}$$

(3.37)

With the assistance of equation (3.5), similarly, the rate equation becomes

$$\rho \dot{s} = \frac{\rho \dot{\varepsilon}}{T} + \frac{\rho p \dot{v}}{T} + \rho \left[\frac{\partial s}{\partial (\nabla T)} \right]_{\varepsilon, v, e_{ij}} \nabla \dot{T} + \rho \left[\frac{\partial s}{\partial e_{ij}} \right]_{\varepsilon, v, \nabla T} \dot{e}_{ij}. \tag{3.38}$$

Retaining the *second*-order terms in ∇T and e_{ij} in the Taylor series expansions for s,

$$s = s_0 + \left(\frac{\partial s}{\partial T_{,i}} \right)_0 T_{,i} + \left(\frac{\partial s}{\partial e_{ij}} \right)_0 e_{ij} + \frac{1}{2} \left\{ \left(\frac{\partial^2 s}{\partial T_{,k} \, \partial T_{,k}} \right)_0 T_{,i} \, T_{,i} \right.$$

$$\left. + 2 \left(\frac{\partial^2 s}{\partial T_{,k} \, \partial e_{kj}} \right)_0 T_{,i} \, e_{ij} + \left(\frac{\partial^2 s}{\partial e_{kl} \partial e_{kl}} \right)_0 e_{ij} e_{ij} \right\} + \cdots. \tag{3.39}$$

With the tensor notation introduced in the temperature gradient for a consistent representation, $T_{,i} \equiv \nabla T$, the parallel expression to equation (3.27) is

$$\rho \dot{s} = \frac{\rho \dot{\varepsilon}}{T} + \frac{\rho p \dot{v}}{T} - \rho (C_3 \nabla T) \nabla \dot{T} - \rho \left(C_4 e_{ij} \right) \dot{e}_{ij} \tag{3.40}$$

where

$$C_3 = - \left(\frac{\partial^2 s}{\partial T_{,k} \, \partial T_{,k}} \right) > 0 \quad C_4 = - \left(\frac{\partial^2 s}{\partial e_{kl} \partial e_{kl}} \right) > 0. \tag{3.41}$$

The entropy equation in this case, similarly, reads as

$$\rho \dot{s} + \frac{\partial}{\partial x_i} \left(\frac{q_i}{T} \right) = -T_{,i} \left(\frac{q_i}{T^2} + \rho C_3 \dot{T}_{,i} \right) - \dot{e}_{ij} \left(\frac{\sigma'_{ij}}{T} + \rho C_4 e_{ij} \right) \tag{3.42}$$

where the results of equations (3.8) for the specific internal energy and (3.13) for the specific work done under hydrostatic tension have been used. While the entropy flux vector remains the same, $J_i = q_i/T$, the entropy production rate is

$$\Sigma = -T_{,i} \left(\frac{q_i}{T^2} + \rho C_3 \dot{T}_{,i} \right) - \dot{e}_{ij} \left(\frac{\sigma'_{ij}}{T} + \rho C_4 e_{ij} \right). \tag{3.43}$$

To ensure a positive-definite value for Σ, again, a sufficient (but necessary) condition is

$$\frac{q_i}{T^2} + \rho C_3 \dot{T}_{,i} = (A_2 - A_1)T_{,i}, \quad A_1, A_2 \geq 0, \quad A_1 > A_2$$

$$\frac{\sigma'_{ij}}{T} + \rho C_4 e_{ij} = (B_2 - B_1)\dot{e}_{ij}, \quad B_1, B_2 \geq 0, \quad B_1 > B_2$$

$$\text{for flux-precedence, } \tau_q < \tau_T. \qquad (3.44)$$

They describe the constitutive behavior governing heat and momentum transport in the case of precedence of the fluxes to the gradients.

Equation (3.36) describes the transition of thermodynamic states with a gradient precedence ($\tau_T < \tau_q$), while equation (3.44) describes the transition of thermodynamic states with a flux precedence ($\tau_q < \tau_T$). The combination of the two thus describes a full picture of the lagging behavior where cases of both flux and gradient precedence are included in the same framework. The first expression in equation (3.44) results in

$$T_{,i} = \left(\frac{A_1}{A_2}\right)T_{,i} + \left(\frac{1}{A_2 T^2}\right)q_i + \left(\frac{\rho C_3}{A_2}\right)\dot{T}_{,i}. \qquad (3.45)$$

Replacing the term of $T_{,i}$ in the first expression of equation (3.35) by equation (3.45) then gives

$$q_i + \frac{\rho C_1}{\left[A + 1/\left(A_2 T^4\right)\right]}\dot{q}_i = -\frac{\left[A_1/\left(A_2 T^2\right)\right]}{\left[A + 1/\left(A_2 T^4\right)\right]}T_{,i} - \frac{\left[\rho C_3/\left(A_2 T^2\right)\right]}{\left[A + 1/\left(A_2 T^4\right)\right]}\dot{T}_{,i}, \qquad (3.46)$$

in terms of the more familiar vector notations,

$$\vec{q} + \tau_q \frac{\partial \vec{q}}{\partial t} = -K\left[\nabla T + \frac{\partial}{\partial t}(\nabla T)\right] \quad \text{with}$$

$$K = \frac{\left[A_1/\left(A_2 T^2\right)\right]}{\left[A + 1/\left(A_2 T^4\right)\right]} > 0, \quad \tau_q = \frac{\rho C_1}{\left[A + 1/\left(A_2 T^4\right)\right]} > 0, \quad \tau_T = \frac{\rho C_3}{A_1} > 0. (3.47)$$

Equation (3.47) has exactly the same form as equation (2.7) describing the linearized lagging behavior. It is now derived as a natural consequence ensuring a positive-definite entropy production rate on the basis of the second law of extended irreversible thermodynamics.

The counterpart in mechanical deformation is obtained in the same manner. The second expression in equation (3.44) gives

$$\dot{e}_{ij} = \left(\frac{B_2}{B_1}\right)\dot{e}_{ij} - \left(\frac{1}{TB_1}\right)\sigma'_{ij} - \left(\frac{\rho C_4}{B_1}\right)e_{ij} \quad \text{for } \tau_q < \tau_T. \qquad (3.48)$$

Substituting equation (3.48) into the second expression of (3.36) (for $\tau_T < \tau_q$) results in

$$\sigma'_{ij} + \tau_M \frac{\partial \sigma'_{ij}}{\partial t} = \lambda e_{ij} - 2\mu \dot{e}_{ij} \quad \text{with}$$

$$\tau_M = \frac{\rho C_2}{\left[B - \left(1 / B_1 T^2 \right) \right]} > 0, \quad \lambda = \frac{\rho C_4}{B_1 T \left[B - \left(1 / B_1 T^2 \right) \right]} > 0,$$

$$2\mu = \frac{B_2}{B_1 T \left[B - \left(1 / B_1 T^2 \right) \right]} > 0, \quad B_1 B > \frac{1}{T^2} \tag{3.49}$$

which resembles the combined behavior of the Maxwell fluid (equation (3.36)) and the Hookean solid.

3.4 THERMOMECHANICAL COUPLING

Equations (3.32) and (3.40), and hence equations (3.36) and (3.47), are particular forms neglecting the effect of thermomechanical coupling in production of entropy in nonequilibrium, irreversible thermodynamic transitions. In a more rigorous treatment including the thermomechanical coupling effect, *both* the generalized fluxes and the generalized gradients can be implemented into the list of fundamental variables defining the entropy. The mathematics is more tedious, but the resulting constitutive equations describing coupled heat and momentum transport allow both cases of gradient- and flux-precedence in the same framework without separate consideration.

Mathematically, the functional form of specific entropy in this more rigorous treatment can be written as $s \equiv s(\varepsilon, v, q_i, T_{,i}, \sigma'_{ij}, e_{ij})$, giving

$$\rho \dot{s} + \frac{\partial}{\partial x_i}\left(\frac{q_i}{T} \right) = -\frac{q_i}{T^2}\frac{\partial T}{\partial x_i} - \frac{\sigma'_{ij}\dot{e}_{ij}}{T} + \rho\left(\frac{\partial s}{\partial q_i} \right)_{\substack{\varepsilon, v, \nabla T, \\ \sigma'_{ij}, e_{ij}}}\dot{q}_i + \rho\left[\frac{\partial s}{\partial(\nabla T)} \right]_{\substack{\varepsilon, v, q_i, \\ \sigma'_{ij}, e_{ij}}}\nabla \dot{T}$$

$$+ \rho\left(\frac{\partial s}{\partial \sigma'_{ij}} \right)_{\substack{\varepsilon, v, q_i, \\ \nabla T, e_{ij}}}\dot{\sigma}'_{ij} + \rho\left[\frac{\partial s}{\partial e_{ij}} \right]_{\substack{\varepsilon, v, q_i, \\ \nabla T, \sigma'_{ij}}}\dot{e}_{ij}, \tag{3.50}$$

in correspondence with equations (3.33) and (3.42). The entropy flux vector remains the same, $J_i = q_i/T$, but the entropy production rate per unit volume evolves into

$$\Sigma = -\frac{q_i}{T^2}\frac{\partial T}{\partial x_i} - \frac{\sigma'_{ij}\dot{e}_{ij}}{T} + \rho\left(\frac{\partial s}{\partial q_i}\right)_{\varepsilon,v,\nabla T,\ \sigma'_{ij},e_{ij}}\dot{q}_i + \rho\left[\frac{\partial s}{\partial(\nabla T)}\right]_{\varepsilon,v,q_i,\ \sigma'_{ij},e_{ij}}\nabla\dot{T}$$

$$+\rho\left(\frac{\partial s}{\partial\sigma'_{ij}}\right)_{\substack{\varepsilon,v,q_i,\\ \nabla T,e_{ij}}}\dot{\sigma}'_{ij} + \rho\left[\frac{\partial s}{\partial e_{ij}}\right]_{\substack{\varepsilon,v,q_i,\\ \nabla T,\sigma'_{ij}}}\dot{e}_{ij}, \tag{3.51}$$

in comparison with equation (3.19). The value of Σ depends on the rate changes of specific entropy with respect to the heat flux vector q_i, the temperature gradient $T_{,i} \equiv \nabla T$, the stress deviator σ'_{ij}, and the deformation rate tensor $\dot{e}_{ij}$. The moderately disturbed specific entropy from the equilibrium state, the second-order Taylor series expansion in correspondence with equations (3.28) and (3.39), is

$$s(q_i,T_{,i},\sigma'_{ij},e_{ij}) = s_0 + \left(\frac{\partial s}{\partial q_i}\right)_0 q_i + \left(\frac{\partial s}{\partial T_{,i}}\right)_0 T_{,i} + \left(\frac{\partial s}{\partial\sigma'_{ij}}\right)_0 \sigma'_{ij} + \left(\frac{\partial s}{\partial e_{ij}}\right)_0 e_{ij}$$

$$+\frac{1}{2}\left\{\left(\frac{\partial^2 s}{\partial q_k\partial q_k}\right)_0 q_i q_i + \left(\frac{\partial^2 s}{\partial T_{,k}\partial T_{,k}}\right)_0 T_{,i}\,T_{,i} + \left(\frac{\partial^2 s}{\partial\sigma'_{kl}\partial\sigma'_{kl}}\right)_0 \sigma'_{ij}\sigma'_{ij} + \left(\frac{\partial^2 s}{\partial e_{kl}\partial e_{kl}}\right)_0 e_{ij}e_{ij}\right.$$

$$+2\left[\left(\frac{\partial^2 s}{\partial\sigma'_{kl}\partial e_{kl}}\right)_0 \sigma'_{ij}e_{ij} + \left(\frac{\partial^2 s}{\partial T_{,k}\partial e_{kj}}\right)_0 T_{,i}\,e_{ij} + \left(\frac{\partial^2 s}{\partial T_{,k}\partial\sigma'_{kj}}\right)_0 T_{,i}\,\sigma'_{ij}\right.$$

$$\left.\left.+\left(\frac{\partial^2 s}{\partial q_k\partial e_{kj}}\right)_0 q_i e_{ij} + \left(\frac{\partial^2 s}{\partial q_k\partial\sigma'_{kj}}\right)_0 q_i\sigma'_{ij} + \left(\frac{\partial^2 s}{\partial T_{,k}\partial q_k}\right)_0 T_{,i}\,q_i\right]\right\}$$

$$+ \text{third-order terms.} \tag{3.52}$$

The lengthy terms in the second-order derivatives result from the involvement of four independent variables in the present case. The first-order derivatives involving the thermomechanical coupling effect are thus

$$\left(\frac{\partial s}{\partial q_i}\right)_{\substack{\varepsilon,v,T_{,i}\\ \sigma'_{ij},e_{ij}}} = \left(\frac{\partial s}{\partial q_i}\right)_0 + \left(\frac{\partial^2 s}{\partial q_k\partial q_k}\right)_0 q_i + \left(\frac{\partial^2 s}{\partial q_k\partial e_{kj}}\right)_0 e_{ij}$$

$$+\left(\frac{\partial^2 s}{\partial q_k\partial\sigma'_{kj}}\right)_0 \sigma'_{ij} + \left(\frac{\partial^2 s}{\partial q_k\partial T_{,k}}\right)_0 T_{,i} \tag{3.53a}$$

$$\left(\frac{\partial s}{\partial T_{,i}}\right)_{\substack{\varepsilon,v,q_i \\ \sigma'_{ij},e_{ij}}} = \left(\frac{\partial s}{\partial T_{,i}}\right)_0 + \left(\frac{\partial^2 s}{\partial T_{,k}\,\partial T_{,k}}\right)_0 T_{,i} + \left(\frac{\partial^2 s}{\partial T_{,k}\,\partial e_{kj}}\right)_0 e_{ij}$$

$$+ \left(\frac{\partial^2 s}{\partial T_{,k}\,\partial \sigma'_{kj}}\right)_0 \sigma'_{ij} + \left(\frac{\partial^2 s}{\partial T_{,k}\,\partial q_k}\right)_0 q_i \qquad (3.53b)$$

$$\left(\frac{\partial s}{\partial \sigma'_{ij}}\right)_{\substack{\varepsilon,v,q_i \\ T_{,i},e_{ij}}} = \left(\frac{\partial s}{\partial \sigma'_{ij}}\right)_0 + \left(\frac{\partial^2 s}{\partial \sigma'_{kl}\,\partial \sigma'_{kl}}\right)_0 \sigma'_{ij} + \left(\frac{\partial^2 s}{\partial \sigma'_{kl}\,\partial e_{kl}}\right)_0 e_{ij}$$

$$+ \left(\frac{\partial^2 s}{\partial T_{,k}\,\partial \sigma'_{kj}}\right)_0 T_{,i} + \left(\frac{\partial^2 s}{\partial q_k\,\partial \sigma'_{kj}}\right)_0 q_i \qquad (3.53c)$$

$$\left(\frac{\partial s}{\partial e_{ij}}\right)_{\substack{\varepsilon,v,q_i \\ T_{,i},\sigma'_{ij}}} = \left(\frac{\partial s}{\partial e_{ij}}\right)_0 + \left(\frac{\partial^2 s}{\partial e_{kl}\,\partial e_{kl}}\right)_0 e_{ij} + \left(\frac{\partial^2 s}{\partial \sigma'_{kl}\,\partial e_{kl}}\right)_0 \sigma'_{ij}$$

$$+ \left(\frac{\partial^2 s}{\partial T_{,k}\,\partial e_{kj}}\right)_0 T_{,i} + \left(\frac{\partial^2 s}{\partial q_k\,\partial e_{kj}}\right)_0 q_i. \qquad (3.53d)$$

A direct substitution of equation (3.53) into (3.51) gives the entropy production rate. Collecting terms led by q_i, $T_{,i}$, σ'_{ij}, and e_{ij} into four groups gives

$$\Sigma = q_i Q_i + T_{,i}\,G_i + \sigma'_{ij} S_{ij} + e_{ij} E_{ij}, \text{ for } i, j = 1, 2, 3 \qquad (3.54)$$

where

$$Q_i = -\left(\frac{1}{T^2}\right) T_{,i} + \rho\left(\frac{\partial^2 s}{\partial q_k\,\partial q_k}\right)_0 \dot{q}_i + \rho\left(\frac{\partial^2 s}{\partial q_k\,\partial T_{,k}}\right)_0 \dot{T}_{,i}$$

$$+ \rho\left(\frac{\partial^2 s}{\partial q_k\,\partial \sigma'_{kj}}\right)_0 \dot{\sigma}'_{ij} + \rho\left(\frac{\partial^2 s}{\partial q_k\,\partial e_{kj}}\right)_0 \dot{e}_{ij} \qquad (3.55)$$

$$G_i = \rho\left(\frac{\partial^2 s}{\partial q_k\,\partial T_{,k}}\right)_0 \dot{q}_i + \rho\left(\frac{\partial^2 s}{\partial T_{,k}\,\partial T_{,k}}\right)_0 \dot{T}_{,i}$$

$$+\rho\left(\frac{\partial^2 s}{\partial T_{,k}\,\partial\sigma'_{kj}}\right)_0 \dot{\sigma}'_{ij} + \rho\left(\frac{\partial^2 s}{\partial T_{,k}\,\partial e_{kj}}\right)_0 \dot{e}_{ij} \qquad (3.56\ Cont.)$$

$$S_{ij} = -\left(\frac{1}{T}\right)\dot{e}_{ij} + \rho\left(\frac{\partial^2 s}{\partial q_k\,\partial\sigma'_{kj}}\right)_0 \dot{q}_i + \rho\left(\frac{\partial^2 s}{\partial T_{,k}\,\partial\sigma'_{kj}}\right)_0 \dot{T}_{,i}$$

$$+\rho\left(\frac{\partial^2 s}{\partial\sigma'_{kl}\,\partial\sigma'_{kl}}\right)_0 \dot{\sigma}'_{ij} + \rho\left(\frac{\partial^2 s}{\partial\sigma'_{kl}\,\partial e_{kl}}\right)_0 \dot{e}_{ij} \qquad (3.57)$$

$$E_{ij} = \rho\left(\frac{\partial^2 s}{\partial q_k\,\partial e_{kj}}\right)_0 \dot{q}_i + \rho\left(\frac{\partial^2 s}{\partial T_{,k}\,\partial e_{kj}}\right)_0 \dot{T}_{,i}$$

$$+\rho\left(\frac{\partial^2 s}{\partial\sigma'_{kl}\,\partial e_{kl}}\right)_0 \dot{\sigma}'_{ij} + \rho\left(\frac{\partial^2 s}{\partial e_{kl}\,\partial e_{kl}}\right)_0 \dot{e}_{ij} . \qquad (3.58)$$

A *quadratic* form of Σ, again, is one of the many possibilities ensuring a positive-definite entropy production rate. From equation (3.54), this implies

$$Q_i = A_3 q_i - B_3 T_{,i}; \quad G_i = A_4 T_{,i} + B_3 q_i; \quad S_{ij} = A_5\sigma'_{ij} - B_4 e_{ij}; \quad E_{ij} = A_6 e_{ij} + B_4\sigma'_{ij},$$

with

$$A_3, A_4, A_5, A_6 \geq 0 \quad \text{and} \quad B_3, B_4 \geq 0. \qquad (3.59)$$

Combining with equation (3.58), more explicitly, the admissible forms of heat and momentum transport equations, namely, the *constitutive* equations, ensuring a positive-definite entropy production rate are

$$-\left(\frac{1}{T^2}\right)\nabla T + \rho\left(\frac{\partial^2 s}{\partial q_k\,\partial q_k}\right)_0 \dot{\vec{q}} + \rho\left(\frac{\partial^2 s}{\partial q_k\,\partial T_{,k}}\right)_0 \nabla\dot{T}$$

$$+\rho\left(\frac{\partial^2 s}{\partial q_k\,\partial\sigma'_{kj}}\right)_0 \dot{\sigma}'_{ij} + \rho\left(\frac{\partial^2 s}{\partial q_k\,\partial e_{kj}}\right)_0 \dot{e}_{ij} = A_3\vec{q} - B_3\nabla T , \qquad (3.60a)$$

$$\rho\left(\frac{\partial^2 s}{\partial q_k\,\partial T_{,k}}\right)_0 \dot{\vec{q}} + \rho\left(\frac{\partial^2 s}{\partial T_{,k}\,\partial T_{,k}}\right)_0 \nabla\dot{T}$$

$$+\rho\left(\frac{\partial^2 s}{\partial T_{,k}\,\partial\sigma'_{kj}}\right)_0 \dot{\sigma}'_{ij} + \rho\left(\frac{\partial^2 s}{\partial T_{,k}\,\partial e_{kj}}\right)_0 \dot{e}_{ij} = A_4\nabla T + B_3\bar{q}, \qquad (3.60b\ Cont.)$$

$$-\left(\frac{1}{T}\right)\dot{e}_{ij} + \rho\left(\frac{\partial^2 s}{\partial q_k\,\partial\sigma'_{kj}}\right)_0 \dot{\bar{q}} + \rho\left(\frac{\partial^2 s}{\partial T_{,k}\,\partial\sigma'_{kj}}\right)_0 \nabla\dot{T}$$

$$+\rho\left(\frac{\partial^2 s}{\partial\sigma'_{kl}\,\partial\sigma'_{kl}}\right)_0 \dot{\sigma}'_{ij} + \rho\left(\frac{\partial^2 s}{\partial\sigma'_{kl}\,\partial e_{kl}}\right)_0 \dot{e}_{ij} = A_5\sigma'_{ij} - B_4 e_{ij}, \qquad (3.60c)$$

$$\rho\left(\frac{\partial^2 s}{\partial q_k\,\partial e_{kj}}\right)_0 \dot{\bar{q}} + \rho\left(\frac{\partial^2 s}{\partial T_{,k}\,\partial e_{kj}}\right)_0 \nabla\dot{T}$$

$$+\rho\left(\frac{\partial^2 s}{\partial\sigma'_{kl}\,\partial e_{kl}}\right)_0 \dot{\sigma}'_{ij} + \rho\left(\frac{\partial^2 s}{\partial e_{kl}\,\partial e_{kl}}\right)_0 \dot{e}_{ij} = A_6 e_{ij} + B_4\sigma'_{ij} \qquad (3.60d)$$

where vector notations are recovered whenever possible for a more conventional appearance, $\bar{q} \equiv q_i$ and $\nabla T \equiv T_{,i}$ for $i = 1, 2, 3$. Within the framework of extended irreversible thermodynamics, equation (3.60) describes the thermomechanical coupling in nonequilibrium irreversible transition of thermodynamic states. Coupled with the conservation equations for energy and momentum in thermodynamic transitions, they depict a complicated system to be solved for the four major unknowns, $\bar{q}$, T, σ_{ij}, and e_{ij}. Note that the second-order derivatives of specific entropy calculated at the equilibrium state in equation (3.60) are equivalent to the coefficients C used in equations (3.31) and (3.41). They are assumed to be negative *constants* for establishing the correlation to the linearized dual-phase-lag model in the *simplest* situation.

3.4.1 Rigid Conductors

Equations (3.60a) and (3.60b) describing the coupling behavior in heat transport are repetitive. They are actually identical. To illustrate this property, consider heat transport in a *rigid conductor* where no deformation occurs ($e_{ij} = 0$) and that the stress rates ($\dot{\sigma}'_{ij}$) are all zero. Equations (3.60a) and (3.60b) in this case reduce to

$$-\left(\frac{1}{T^2}\right)\nabla T + \rho\left(\frac{\partial^2 s}{\partial q_k\,\partial q_k}\right)_0 \dot{\bar{q}} + \rho\left(\frac{\partial^2 s}{\partial q_k\,\partial T_{,k}}\right)_0 \nabla\dot{T} = A_3\bar{q} - B_3\nabla T \qquad (3.61a)$$

$$\rho\left(\frac{\partial^2 s}{\partial q_k\,\partial T_{,k}}\right)_0 \dot{\bar{q}} + \rho\left(\frac{\partial^2 s}{\partial T_{,k}\,\partial T_{,k}}\right)_0 \nabla\dot{T} = A_4\nabla T + B_3\bar{q}. \qquad (3.61b)$$

Both equations can be cast into the same form:

$$\bar{q} + \tau_q \frac{\partial \bar{q}}{\partial t} = -k\left[\nabla T + \tau_T \frac{\partial}{\partial t}(\nabla T)\right],$$ (3.62)

with

$$k = \frac{1}{A_3}\left(\frac{1}{T^2} - B_3\right), \quad \tau_T = \frac{-\rho}{\frac{1}{T^2} - B_3}\left(\frac{\partial^2 s}{\partial q_k \partial T_{,k}}\right)_0, \quad \tau_q = -\frac{\rho}{A_3}\left(\frac{\partial^2 s}{\partial q_k \partial q_k}\right)_0$$ (3.63a)

in equation (3.61a), and

$$k = \frac{A_4}{B_3}, \quad \tau_T = -\frac{\rho}{A_4}\left(\frac{\partial^2 s}{\partial T_{,k}\,\partial T_{,k}}\right)_0, \quad \tau_q = -\frac{\rho}{B_3}\left(\frac{\partial^2 s}{\partial q_k \partial T_{,k}}\right)_0$$ (3.63b)

in equation (3.61b).   Equation (3.62), ensuring a positive-definite entropy production rate, has exactly the same form as equation (2.7), the constitutive equation depicting the linearized lagging response. For positive values of thermal conductivity k, phase lag of the temperature gradient τ_T, and phase lag of the heat flux vector τ_q, additional constraints are imposed:

$$0 < B_3 < \frac{1}{T^2}, \quad \left(\frac{\partial^2 s}{\partial q_k \partial q_k}\right)_0 < 0, \quad \left(\frac{\partial^2 s}{\partial T_{,k}\,\partial T_{,k}}\right)_0 < 0, \quad \left(\frac{\partial^2 s}{\partial q_k \partial T_{,k}}\right)_0 < 0.$$ (3.64)

Note that the first two conditions in the second-order derivatives of specific entropy at the equilibrium state repeat the results in equation (3.41) for C_3 and C_4. Along with the additional mixed-derivative term, they describe the similar convex behavior of the entropy curve shown in Figure 3.3.   In the framework of the linearized lagging behavior, it seems that a *negative* second-order derivative of specific entropy with respect to the state variables is a general trend.

3.4.2 Isothermal Deformation

A similar situation exists in isothermal deformation where the effect from the thermal field is absent, $\nabla T = 0$ and $\dot{\bar{q}} = \bar{0}$. Equations (3.60c) and (3.60d) in this case reduce to

$$-\left(\frac{1}{T}\right)\dot{e}_{ij} + \rho\left(\frac{\partial^2 s}{\partial \sigma'_{kl}\partial \sigma'_{kl}}\right)_0 \dot{\sigma}'_{ij} + \rho\left(\frac{\partial^2 s}{\partial \sigma'_{kl}\partial e_{kl}}\right)_0 \dot{e}_{ij} = A_5\sigma'_{ij} - B_4 e_{ij}$$ (3.65a)

$$\rho\left(\frac{\partial^2 s}{\partial\sigma'_{kl}\partial e_{kl}}\right)_0 \dot{\sigma}'_{ij} + \rho\left(\frac{\partial^2 s}{\partial e_{kl}\partial e_{kl}}\right)_0 \dot{e}_{ij} = A_6 e_{ij} + B_4 \sigma'_{ij}. \tag{3.65b}$$

Again, equations (3.65a) and (3.65b) can be cast into an identical form,

$$\sigma'_{ij} + \tau_\sigma \dot{\sigma}'_{ij} = \lambda_1 e_{ij} + \tau_e \dot{e}_{ij}, \quad \text{with} \tag{3.66}$$

$$\tau_\sigma = -\frac{\rho}{A_5}\left(\frac{\partial^2 s}{\partial\sigma'_{ij}\partial\sigma'_{ij}}\right)_0, \quad \tau_e = -\frac{1}{A_5}\left[\rho\left(\frac{\partial^2 s}{\partial\sigma'_{ij}\partial e_{ij}}\right)_0 - \frac{1}{T}\right], \quad \lambda_M = \frac{B_4}{A_5} \tag{3.67a}$$

in equation (3.65a), and

$$\tau_\sigma = -\frac{\rho}{B_4}\left(\frac{\partial^2 s}{\partial\sigma'_{ij}\partial e_{ij}}\right)_0, \quad \tau_e = -\frac{\rho}{B_4}\left(\frac{\partial^2 s}{\partial e_{ij}\partial e_{ij}}\right)_0, \quad \lambda_M = \frac{A_6}{B_4} \tag{3.67b}$$

in equation (3.65b). For positive values of the τ and λ_M, the following conditions are further imposed:

$$\left(\frac{\partial^2 s}{\partial\sigma'_{ij}\partial\sigma'_{ij}}\right)_0 < 0, \quad \left(\frac{\partial^2 s}{\partial\sigma'_{ij}\partial e_{ij}}\right)_0 < 0, \quad \left(\frac{\partial^2 s}{\partial e_{ij}\partial e_{ij}}\right)_0 < 0, \tag{3.68}$$

which requires the same type of convex entropy curve shown in Figure 3.3. As a counterpart of equation (3.62), describing the lagging behavior in heat transport, equation (3.66) resembles the combined behavior of the Maxwell fluid (Tzou, 1993a) and the Hookean solid in nonequilibrium, irreversible thermodynamic transition. Both equations (3.62) and (3.66) result from the natural consequence ensuring a positive-definite entropy production rate, supporting their admissibility within the framework of extended irreversible thermodynamics.

The framework of extended irreversible thermodynamics established in this work, evidently, is not restricted to the derivation of the constitutive equation for the thermal field. It may include the complicated thermomechanical coupling shown by equation (3.60), the thermal field alone, equation (3.62), or only the mechanical deformation under an isothermal condition, equation (3.66). The procedure presented in this chapter that justifies a positive-definite entropy production rate in thermodynamic transition provides a general tool for *deriving* the thermodynamically admissible constitutive equations for both thermal and mechanical responses.

When nonlinear effects of τ_T and τ_q gradually come into the picture, such as the T wave structure discussed in Section 10.1, the same procedure may be extended to ensure a positive entropy production rate in the fast-transient process, with a careful selection of the proportional constants such as C_1 to C_4 in equations (3.31) and (3.41). The quadratic form of the entropy production rate may not exist for an easier recognition of a positive-definite value, implying the need for other special forms of the entropy production rate to satisfy the second law of extended irreversible thermodynamics. Prior to the lengthy exercise involving all the high order effects of phase lags in the thermodynamic framework, however, it is my personal belief that finding the experimental support and the physical conditions in which the *linearized* dual-phase-lag model applies are far more important than a full expansion into the nonlinear regimes of τ_T and τ_q. Reviewing the history in the development of the thermal wave theory, the extensive effort devoted to the linearized *CV* wave model, equation (2.3), prior to the full accommodation of nonlinear effects based on the *single*-phase-lag model, equation (2.2), bears the same merit.

TEMPERATURE PULSES IN SUPERFLUID LIQUID HELIUM

Superfluidity refers to flow without viscosity, a quantum effect for fluids at a temperature close to 0 K. The isotope of helium ^{4}He may be the only known substance that remains fluid at such a low temperature. At a temperature below the λ point, 2.19 K, where a second-order phase transition occurs, superfluidity in liquid helium (helium II) becomes pronounced. It results in several remarkable features that do not exist in regular fluids with viscosity effects. The theory of superfluid was developed by Landau in 1941, and is summarized in a concise chapter in the book by Landau and Lifshitz (1959).

This chapter starts with a brief review of the superfluid concept in helium II. With emphasis on the resulting second sound wave in heat propagation, the classical CV wave and the dual-phase-lag models follow to address the thermal wave effect from a phenomenological approach. The experimental results obtained by Bertman and Sandiford (1970) for heat propagation in liquid helium are carefully examined to reflect the refined structure of thermal waves. Though not conclusive, the dual-phase-lag concept is shown to provide more detailed resolution than the classical CV wave. It is thus hypothesized that the additional delay (in *time*) between the heat flux vector and the temperature gradient is due to the finite time required for activating the low-temperature molecules to transport heat.

4.1 SECOND SOUND IN LIQUID HELIUM

At a temperature close to absolute zero, thermomechanical behavior of liquid helium can be described in terms of the combined motion of the superfluid component and the normal viscous-fluid component (Landau and Lifshitz, 1959). Owing to the absence of viscosity in the superfluid component, no friction, and hence no momentum transfer, could result between these components in their relative motion. Denoting the mass density of liquid helium by ρ,

$$\rho = \rho_s + \rho_n \qquad (4.1)$$

where the subscripts s and n refer to the superfluid and normal components, respectively. Likewise, the mass flux density ($\bar{j}$) can be decomposed into

$$\bar{j} = \rho_s \bar{v}_s + \rho_n \bar{v}_n \qquad (4.2)$$

with $\bar{v}$ denoting the flow velocity. The linearized equations of fluid dynamics include the mass conservation for the combined flow,

$$\frac{\partial \rho}{\partial t} + \nabla \bullet \bar{j} = 0, \qquad (4.3)$$

conservation of momentum,

$$\frac{\partial \bar{j}}{\partial t} + \nabla p = 0, \qquad (4.4)$$

conservation of entropy flux induced by the normal viscous-fluid component (since the superfluid component has no viscosity) along

$$\frac{\partial (\rho s)}{\partial t} + \rho s \nabla \bullet \bar{v}_n = 0, \qquad (4.5)$$

and the flow acceleration of superfluid induced by the gradient of chemical potential (μ),

$$\frac{\partial \bar{v}_s}{\partial t} + \nabla \mu = 0. \qquad (4.6)$$

Convective terms in these equations have been neglected owing to linearization.

Fundamental characteristics in heat propagation can be studied by neglecting the effect of thermal expansion, $\bar{j} = \bar{0}$, resulting in

$$\rho_s \bar{v}_s + \rho_n \bar{v}_n = 0 \quad \text{and} \quad \bar{v}_s = -\left(\frac{\rho_n}{\rho_s}\right) \bar{v}_n. \qquad (4.7)$$

from equation (4.2). In the absence of overall mass flux in liquid helium, clearly, the superfluid and normal viscous-fluid components flow in opposite directions, resulting in a stationary mass center in any specified volume element.

The absence of thermal expansion in liquid helium, $\bar{j} = \bar{0}$, implies a constant overall mass density from equation (4.3), ρ = constant. Equation (4.5) reduces to

$$\frac{\partial s}{\partial t} + s\nabla \bullet \bar{v}_n = 0 \tag{4.8}$$

in this case. From equation (4.4), moreover, the pressure gradient vanishes, which implies a constant pressure (p = constant) throughout the flow field. The gradient of chemical potential in equation (4.6), $\nabla\mu$, relates to the gradients of temperature and pressure by the thermodynamic identities

$$\nabla\mu = -s\nabla T + \frac{\nabla p}{\rho}. \tag{4.9}$$

In the absence of a pressure gradient, $\nabla p = 0$, implying that $\nabla\mu = -s\nabla T$, equation (4.6) becomes

$$\frac{\partial \bar{v}_s}{\partial t} = s\nabla T. \tag{4.10}$$

For a better focus on the thermodynamic behavior, flow velocities are eliminated from equations (4.8) and (4.10). Taking the time derivative on the divergence of equation (4.7) results in

$$\rho_s \frac{\partial}{\partial t} \nabla \bullet \bar{v}_s + \rho_n \frac{\partial}{\partial t} \nabla \bullet \bar{v}_n = 0. \tag{4.11}$$

The first terms can be obtained from the divergence of equation (4.10),

$$\frac{\partial}{\partial t} \nabla \bullet \bar{v}_s = \nabla s \bullet \nabla T + s\nabla^2 T. \tag{4.12}$$

The second term, on the other hand, can be obtained from the time derivative of equation (4.8),

$$\frac{\partial}{\partial t} \nabla \bullet \bar{v}_n = \frac{\left(\frac{\partial s}{\partial t}\right)^2 - s\frac{\partial^2 s}{\partial t^2}}{s^2}. \tag{4.13}$$

Substituting equations (4.12) and (4.13) into equation (4.11), the equation containing entropy and temperature is obtained:

$$\frac{\left[\left(\frac{\partial s}{\partial t}\right)^2 - s\frac{\partial^2 s}{\partial t^2}\right]}{s^2} + \left(\frac{\rho_s}{\rho_n}\right)\left(\nabla s \bullet \nabla T + s\nabla^2 T\right) = 0 \tag{4.14}$$

Even though the equations of fluid dynamics have been linearized, the resulting equation governing heat propagation in liquid helium, equation (4.14), is highly nonlinear.

Consider now disturbances of entropy (s') and temperature (T') imposed on the existing equilibrium fields, $s_0(\bar{r}, t)$ and $T_0(\bar{r}, t)$, respectively:

$$s(\bar{r},t) = s_0(\bar{r},t) + s'(\bar{r},t), \quad T(\bar{r},t) = T_0(\bar{r},t) + T'(\bar{r},t). \tag{4.15}$$

The disturbances are assumed small so that the second-order terms can be neglected. Substituting equation (4.15) into equation (4.14) and linearizing results in

$$\frac{1}{s_0^2}\left(\frac{\partial s_0}{\partial t}\right)^2 - \frac{1}{s_0}\left(\frac{\partial^2 s_0}{\partial t^2}\right) + \left(\frac{\rho_s}{\rho_n}\right)\left(\nabla s_0 \bullet \nabla T_0 + s_0\nabla^2 T_0\right) = 0, \tag{4.16}$$

which governs the primary equilibrium configurations, and

$$\left(\frac{2}{s_0^2}\right)\left(\frac{\partial s_0}{\partial t}\right)\frac{\partial s'}{\partial t} - \frac{1}{s_0}\left(\frac{\partial^2 s'}{\partial t^2}\right) + \left[\frac{1}{s_0^2}\left(\frac{\partial^2 s_0}{\partial t^2}\right) - \frac{2}{s_0^3}\left(\frac{\partial s_0}{\partial t}\right)^2\right]s'$$

$$+\left(\frac{\rho_s}{\rho_n}\right)\left[\nabla s_0 \bullet \nabla T' + \nabla s_0 \bullet \nabla s' + \left(\nabla^2 T_0\right)s' + s_0\nabla^2 T'\right] = 0, \tag{4.17}$$

which governs the disturbance fields. Since the disturbance is assumed to be small, implying a small deviation from the equilibrium state,

$$s' = \frac{cT'}{T_0} \tag{4.18}$$

with c being a proportional constant, in J/kg K. Substituting equation (4.18) into equation (4.17), a single equation governing the propagation of temperature disturbance in liquid helium is obtained:

$$\left(\frac{2c}{s_0^2 T_0}\right)\left(\frac{\partial s_0}{\partial t}\right)\frac{\partial T'}{\partial t} - \left(\frac{c}{s_0 T_0}\right)\frac{\partial^2 T'}{\partial t^2} + \left[\frac{c}{s_0^2 T_0}\left(\frac{\partial^2 s_0}{\partial t^2}\right) - \frac{2c}{s_0^3 T_0}\left(\frac{\partial s_0}{\partial t}\right)^2\right]T'$$

$$+\left(\frac{\rho_s}{\rho_n}\right)\left[\nabla s_0 \bullet \nabla T' + \nabla T_0 \bullet \frac{c}{T_0^2}\left(T_0\nabla T' - T'\nabla T_0\right) + \left(\frac{c\nabla^2 T_0}{T_0}\right)T' + s_0\nabla^2 T'\right] = 0 \tag{4.19}$$

Complexity in this equation results from the unsteady, nonuniform equilibrium fields of s_0 and T_0. Should the equilibrium configurations be steady and uniform, s_0 ≡ constant and T_0 ≡ constant, equation (4.19) is greatly simplified:

$$\nabla^2 T' = \frac{1}{u_2^2} \frac{\partial^2 T'}{\partial t^2}, \quad \text{with} \quad u_2 = \sqrt{\frac{s_0^2 T_0 \rho_s}{c \rho_n}} . \tag{4.20}$$

Distinguishing from the speed of sound in normal fluid, $u_1 = (\partial p / \partial \rho)^{1/2}$, the quantity u_2 is the speed of *second* sound in superfluid liquid helium. Propagation of a temperature disturbance in helium II, as clearly depicted by equation (4.20), is a *wave* phenomenon rather than the conventional diffusion. The pure wave equation shown by equation (4.20) should not be viewed as *exact*, owing to negligence of nonlinear effects and presumed steady and homogeneous states of temperature and entropy. The wave behavior describing the propagation of temperature disturbance, however, is a salient feature that revolutionizes the classical concept of diffusion.

The *CV* wave model aims to describe the same type of wave behavior by taking into account the relaxation time in heat transport. Should the *CV* wave speed be correlated to the second-sound wave speed in liquid helium, the relaxation time results from

$$\tau_q \sim \left(\frac{\rho_n}{\rho_s} \right) \frac{c \alpha}{s_0^2 T_0} . \tag{4.21}$$

This identity becomes exact at extremely short times when the effect of diffusion is negligible in comparison with the effect of the thermal wave.

Retaining the same relaxation behavior in heat transport, the dual-phase-lag model continues to explore the effect of microstructural interactions on thermal wave propagation. For a temperature disturbance propagating in liquid helium below the λ point, the additional time delay, τ_T, is attributed to the finite time required for activating the low-temperature molecules to conduct heat in an efficient manner. Incorporating such generalized lagging behavior in heat propagation, the experimental results obtained by Bertman and Sandiford (1970) are revisited to reveal salient features of thermal lagging at extremely low temperatures.

It will be shown that thermal lagging results in several phenomena that are compatible with the experimental results. Phase lag of the heat flux vector, τ_q, describes the delayed response of temperature at the probed position. Phase lag of the temperature gradient, τ_T, on the other hand, smoothes the temperature pulse into a bell-shaped curve.

4.2 EXPERIMENTAL OBSERVATIONS

Bertman and Sandiford (1970) performed transient experiments to illustrate the "second-sound" (wave) effect for heat propagation through superfluid liquid helium at 1 K. Their experimental setup is briefly sketched in Figure 4.1. The time history

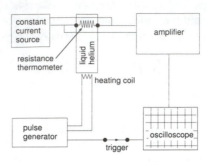

Figure 4.1 The oscilloscope for measuring the time history of temperature at a fixed position in the liquid helium sample (Bertman and Sandiford, 1970).

of temperature at a *fixed* (probed) position in liquid helium is recorded by an oscilloscope. An electric pulse from a generator applies a thermal pulse to the liquid helium sample, which simultaneously triggers the oscilloscope. Temperature increases at the probed position are recorded by a resistance thermometer. A typical oscilloscope trace is shown in Figure 4.2. There exists a significant time delay before a significant temperature rise is detected at the probed position. The local temperature then rises and falls rapidly, causing a bell-shaped temperature-time response curve. The temperature ripple travels at a constant speed in liquid helium. When changing the probed position, in other words, only the arrival time of the thermal signal changes, which is a characteristic in wave propagation. The delayed response and the rapid rise-and-fall behavior in temperature are salient features that cannot be observed in diffusion-dominated phenomena.

Note that the pure wave behavior depicted by equation (4.20) or that by the *CV* wave equation possesses a sharp wavefront. The infinite temperature-rate at the wavefront would result in a more exaggerated response than that shown in Figure

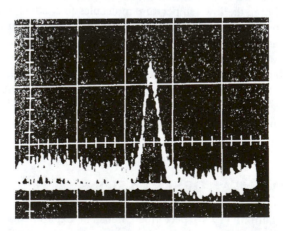

Figure 4.2 A typical oscilloscope trace of temperature measured at a fixed position inside the liquid helium sample (reproduced from Bertman and Sandiford, 1970).

4.2. In addition to the wave behavior, there probably exist some micromechanisms that spread out the temperature response.

The experimental result shown in Figure 4.2 has often been quoted as experimental evidence supporting the wave behavior in heat conduction. Two special features can be observed. First, before arrival of the thermal pulse at the probed position, there exists a time domain in which no significant temperature rise is detected. Second, the temperature response displays a rapid rise-and-fall behavior that is not observed in the traditional behavior of diffusion. Owing to the lack of scales provided in Bertman and Sandifords' experiment, unfortunately, these special behaviors can only be studied in a qualitative sense.

4.3 LAGGING BEHAVIOR

The energy equation describing the generalized lagging response is equation (2.20):

$$\frac{\partial^2 T}{\partial x^2} + \tau_T \frac{\partial^3 T}{\partial x^2 \partial t} = \frac{1}{\alpha} \frac{\partial T}{\partial t} + \frac{\tau_q}{\alpha} \frac{\partial^2 T}{\partial t^2}. \qquad (4.22)$$

Reduction of the dual-phase-lag model, equation (4.22), to the classical diffusion and wave models is discussed in Section 2.3, following equation (2.10). To reiterate, the case of $\tau_T = \tau_q$ reduces to the diffusion model, while the case of $\tau_T = 0$ and $\tau_q \neq 0$ reduces to the CV wave model. Before the problem involving a heat pulse is studied, let us first analyze a simpler problem involving a *temperature pulse* applied at the boundary of a finite medium. The way in which the pulse structure is affected by the lagging behavior is better focused in this manner. The initial temperature of the sample is assumed constant, T_0. Observing the large temperature difference between the heater (at room temperature) and the liquid helium sample (at 1 K) shown in Figure 4.2, the Dirac-delta function, $\Delta(t)$, is assumed for describing the boundary pulse,

$$T = T_w \Delta(t) \quad \text{at} \quad x = 0. \qquad (4.23)$$

Equation (4.23) depicts a temperature pulse that rises and falls in zero time. Though nonphysical, the resulting response provides an asymptotic behavior when the heating duration becomes extremely short. The physical observation is assumed to be made in such a short time that the temperature at the other end of the sample (L) is undisturbed:

$$T = T_0 \quad \text{at} \quad x = L. \qquad (4.24)$$

Equation (4.24) also excludes the effect of wave reflection. The system is assumed to be disturbed from a "stationary" state, implying that

$$T = T_0 \quad \text{and} \quad \frac{\partial T}{\partial t} = 0 \quad \text{as} \quad t = 0. \qquad (4.25)$$

Equations (4.22) to (4.25) furnish the formulation in terms of temperature.

The analysis in terms of dimensionless variables is desirable, especially in the absence of a definite scale in Figure 4.2. In equation (2.24) in Chapter 2, note that the constant factor "2" was introduced to retrieve the variables defined by Baumeister and Hamill (1969). This is no longer necessary for the present study. Introducing

$$\theta = \frac{T}{T_w}, \quad \beta = \frac{t}{\tau_q}, \quad \delta = \frac{x}{\sqrt{\alpha \tau_q}}, \tag{4.26}$$

equations (4.22) to (4.25) become

$$\frac{\partial^2 \theta}{\partial \delta^2} + B \frac{\partial^3 \theta}{\partial \delta^2 \partial \beta} = \frac{\partial \theta}{\partial \beta} + \frac{\partial^2 \theta}{\partial \beta^2} \quad \text{with} \quad B = \frac{\tau_T}{\tau_q} \tag{4.27}$$

$$\theta(\delta, \beta) = \Delta(\beta) \quad \text{at} \quad \delta = 0 \quad \text{and} \quad \theta \to 0 \quad \text{as} \quad \delta \to \infty \tag{4.28}$$

$$\theta = 0 \quad \text{and} \quad \frac{\partial \theta}{\partial \beta} = 0 \quad \text{as} \quad \beta = 0. \tag{4.29}$$

Note that the constant coefficient "1/2" in the parameter B disappears compared to that in equation (2.25). The Laplace transform solution satisfying equations (4.27) to (4.29) is easily obtained:

$$\overline{\theta}(\delta; p) = \frac{\sinh[(l - \delta)A]}{\sinh(Al)} \quad \text{with} \quad A = \sqrt{\frac{p(1+p)}{1 + Bp}} \quad \text{and} \quad l = \frac{L}{\sqrt{\alpha \tau_q}}. \tag{4.30}$$

It can be used directly in the Riemann-sum approximation for the Laplace inversion, with the function subroutine FUNC(P) in the Appendix replaced by equation (4.30).

The results of diffusion and CV wave are retrieved under the conditions $B = 0$ ($\tau_T = 0$, the CV wave model) and $B = 1$ ($\tau_T = \tau_q$, the diffusion model) and are shown in Figure 4.3. For the CV waves the thermal wavefront arrives at $\beta = \delta$, resulting from $x = \left(\sqrt{\alpha / \tau_q}\right) t$. The value of l, the dimensionless length of the medium, is normalized to unity while the temperature response at the midpoint ($\delta = 0.5$) is taken to assure no effect from wave reflection. The diffusion model with $B = 1$ predicts a sharp rise of temperature at early times, approximately in $0 \le \beta \le 0.04$. The temperature level then tapers off in the rest of the transient stage and vanishes at $\beta \cong 0.5$. Referring to the oscilloscope trace of transient temperature shown in Figure 4.2, the diffusion model does not capture the salient features. The CV wave model, an approach capturing the fast-transient effect (in time) but neglecting the small-scale consequence (in space), results in an *infinite* temperature upon impingement of the sharp wavefront. The infinite temperature results from the temperature pulse simulated by the Dirac-delta function, equation (4.23) or (4.28). The dual-phase-lag

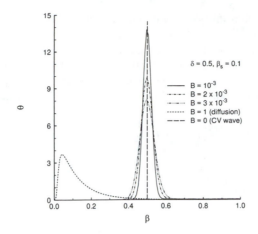

Figure 4.3 Time-histories of local temperature at $\delta = 0.5$ predicted by the *CV* wave model ($B = 0$), diffusion model ($B = 1$) and the dual-phase-lag model with $B = 0.001$, 0.002, and 0.003.

model, even in the absence of a realistic model for the boundary pulse, predicts profiles of transient temperature similar to the bell-shaped profile observed in Figure 4.2. The peak temperature decreases and the pulse duration spreads wider as the value of B (τ_T/τ_q) increases. Recalling that the phase lag of the temperature gradient, τ_T, measures the delayed time caused by the inert behavior of molecules at extremely low temperature, the lower peak temperature and the wider pulse duration are attributed to the microscale effect of heat transport in space.

The reason for considering a delta function in the boundary pulse now becomes clear. The boundary pulse results in a delta-function response in the *CV* wave model accounting *only* for the small-scale effect in time. When the small-scale effect in space activates, it is reflected by the gradually increasing value of τ_T, the delayed time caused by the inert behavior of molecules at low temperature. Figure 4.3 clearly shows the way in which such an effect in microscale destroys the sharp wavefront resulting from the thermal inertia in the transient process. None of the other types of boundary conditions could provide such a distinct resolution for the fast-transient and the small-scale effects in heat transport.

4.4 HEATING PULSE IN TERMS OF FLUXES

On the basis of the fundamental understanding thus developed, we are now ready to consider a more realistic, but more complicated boundary pulse. As depicted in Figure 4.1, the heating provided by the heater coil at the boundary of the liquid helium sample is more appropriately described by a heat flux with a time duration of t_s:

$$q(x,t) = q_s[h(t) - h(t - t_s)] \quad \text{at} \quad x = 0 \qquad (4.31)$$

where $h(\bullet)$ denotes the unit-step function.   Equation (4.31) describes a constant intensity of heat flux (q_s) for $t \in [0, t_s]$ and remains zero otherwise. Owing to the presence of a flux-dependent boundary condition, the temperature representation, equation (4.22), is no longer suitable to use because of the complicated relationship between the heat flux vector and the temperature gradient shown in equation (2.13). The mixed formulation, equations (2.5) and (2.7), shall be used instead:

$$-\frac{\partial q}{\partial x}(x,t) = C_p \frac{\partial T}{\partial t}(x,t) \tag{4.32}$$

$$q(x,t) + \tau_q \frac{\partial q}{\partial t}(x,t) = -k \frac{\partial T}{\partial x}(x,t) - k\tau_T \frac{\partial^2 T}{\partial x \partial t}. \tag{4.33}$$

Since the effect of wave reflection is excluded, let us consider a *semi-infinite* medium in this case.   The local temperature in the neighborhood of the flux-irradiated boundary is the same as that in a finite body without the effect of reflection. The regularity condition at a distance far from the boundary is thus

$$T(x,t) \to 0 \quad \text{as} \quad x \to \infty. \tag{4.34}$$

The initial conditions remain the same, equation (4.25).
    Introducing the following dimensionless variables,

$$\eta = \frac{q}{q_s}, \quad \theta = \frac{T - T_0}{q_s\sqrt{\alpha\tau_q}/k}, \quad \beta = \frac{t}{\tau_q}, \quad \delta = \frac{x}{\sqrt{\alpha\tau_q}}, \tag{4.35}$$

equations (4.31) to (4.34) and (4.25) become

$$\eta + \frac{\partial \eta}{\partial \beta} = -\frac{\partial \theta}{\partial \delta} - B \frac{\partial^2 \theta}{\partial \delta \partial \beta} \quad \text{with} \quad B = \frac{\tau_T}{\tau_q}, \tag{4.36}$$

$$\frac{\partial \eta}{\partial \delta} = -\frac{\partial \theta}{\partial \beta}$$

$$\eta(\delta,\beta) = h(\beta) - h(\beta - \beta_s) \quad \text{at} \quad \delta = 0, \quad \theta \to 0 \quad \text{as} \quad \delta \to \infty \tag{4.37}$$

$$\theta = 0 \quad \text{and} \quad \frac{\partial \theta}{\partial \beta} = 0 \quad \text{as} \quad \beta = 0 \tag{4.38}$$

where $\beta_s = t_s/\tau_q$. The Laplace transform solution satisfying equations (4.36) to (4.38) can be obtained in a straightforward manner,

$$\overline{\eta}(\delta; p) = \overline{\eta}_b(p)e^{-A\delta}, \quad \overline{\theta}(\delta; p) = \overline{\eta}_b(p)\sqrt{\frac{1+p}{p(1+Bp)}}\, e^{-A\delta} \qquad (4.39)$$

with

$$\overline{\eta}_b(p) = \frac{1-e^{-\beta_s p}}{p} \qquad (4.40)$$

being the Laplace transform of the boundary heat flux, equation (4.37). Equation (4.39) can now be used in the subroutine FUNC(P) in the Appendix to obtain the inverse solution.

Figure 4.4 shows the temperature pulses predicted by the diffusion and *CV* wave models at different locations inside the medium. The pulse width of the boundary flux is taken as $\beta_s = 0.1$. The classical diffusion model with $B = 1$, like the previous case, predicts early responses, a result that is not observed in the oscilloscope trace shown in Figure 4.2. The *CV* wave model with $B = 0$ results in block pulses that resemble the experimental results. Owing to the sharp wavefront implemented in the model, however, the pulse width of the temperature is exactly the same as the pulse width of the boundary heat flux ($\beta_s = 0.1$). The sharp edges on arrival and departure of the thermal wavefront and the small oscillations on the tops of the block pulses, in addition, are somewhat awkward when compared to the oscilloscope trace shown in Figure 4.2.

These situations hold when changing the pulse widths of the heat flux applied at the boundary, as shown in Figure 4.5. As the pulse width of the boundary heat flux shortens from 0.1 to 0.08 to 0.05, exemplified by the response curves of *CV* waves at $\delta = 0.5$, the pulse width of the temperature shrinks accordingly. The oscillatory profiles on the tops of the block pulses remain. From Figures 4.3 to 4.5,

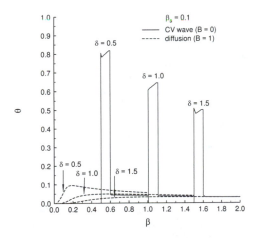

Figure 4.4 The temperature pulses at different locations in the medium, $\delta = 0.5, 1.0$, and 1.5, for $\beta_s = 0.1$ predicted by the classical diffusion ($B = 1$) and *CV* wave ($B = 0$) models.

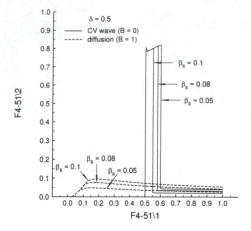

Figure 4.5 The temperature pulses at δ = 0.5 produced by the boundary heat flux with various pulse widths, β_s = 0.1, 0.08, and 0.05.

it is clear that the *CV* wave model inherits every detail from the boundary conditions, including the pulse shape and the pulse width. This result is attributed to the sharp wavefront depicted by the model.

The temperature pulses predicted by the dual-phase-lag model are displayed in Figure 4.6. As a typical example, the pulse width is chosen as β_s = 0.1, and the value of δ is taken as 0.5 (the closest distance in Figure 4.4). When the value of *B* gradually deviates from zero, it efficiently destroys the sharp boundaries of the thermal wavefront. The microscale effect (the inert behavior of molecules, to

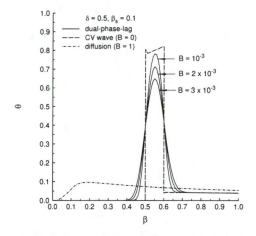

Figure 4.6 The temperature pulses predicted by the dual-phase-lag model, B = 0.001, 0.002, and 0.003. Here, δ = 0.5 and β_s = 0.1.

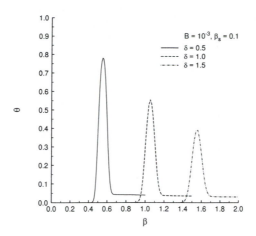

Figure 4.7 The temperature pulses at δ = 0.5, 1.0, and 1.5 predicted by the dual-phase-lag model with B = 0.001 and β_s = 0.1.

reiterate) in heat transport nicely spreads the pulse into a bell shape, just like the oscilloscope trace shown in Figure 4.2. If the CV wave model, called the second-sound phenomenon by Bertman and Sandiford (1970), is acceptable for describing the temperature ripples in liquid helium, evidenced by the close pulse shapes shown in Figures 4.2 (experiment) and 4.6 (analysis), the dual-phase-lag model seems to provide an even closer result to the experimental observation.

Figure 4.7 shows the evolution of temperature pulses as a function of distance from the boundary. The peak value of temperature decreases owing to the energy dispersion, but the bell shape of the pulse remains. At a fixed position of δ = 0.5, Figure 4.8 shows the effect of the pulse width of the boundary heat flux. As the

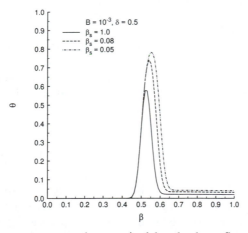

Figure 4.8 The temperature pulses excited by the heat flux with different pulse widths, β_s = 0.1, 0.08, and 0.05, predicted by the dual-phase-lag model with B = 0.001.

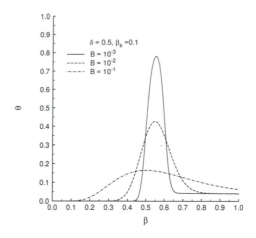

Figure 4.9 Diminution of the bell-shaped pulse at larger values of B predicted by the dual-phase-lag model with $B = 0.001, 0.01$, and 0.1.

pulse width (β_s) reduces from 0.1 to 0.08 to 0.05, the pulse width of the temperature decreases accordingly. The bell shape of the pulse, however, remains.

In passing, note that not all the values of B in the dual-phase-lag model may yield close agreement to the experimental observation. As the value of B further increases, implying a more pronounced effect in microscale heat transport, the temperature pulse will spread further and the peak temperature will continuously decrease. This situation is illustrated in Figure 4.9. When the value of B increases to 0.1, the delayed response due to the inert behavior of molecules becomes pronounced. It yields an earlier response time and destroys the localized structure of the temperature pulse. The peak temperature is significantly reduced, rendering a more uniform temperature rise in the time history.

In Figures 4.4 to 4.9, it is desirable to know the domain of real time in which the rapid rise-and-fall behavior of temperature occurs. According to the threshold values of $\alpha = 1.083 \times 10^{-6}$ m^2/s for the thermal diffusivity (Eckert and Drake, 1972) and $C = 19$ m/s for the thermal wave speed (Peshkov, 1944), a unity value of β $(\beta = 1)$ corresponds to approximately 3 ns. Although the timescale was not provided in Bertman and Sandiford's experiment, the rise-and-fall behavior should occur on the order of nanoseconds.

4.5 OVERSHOOTING PHENOMENON OF TEMPERATURE

In addition to the resemblance of the pulse shape, *overshooting* of temperature upon impingement of lagging waves may be clear evidence for obtaining experimental support. Temperature overshooting, in brief, is concerned with the *excessive* temperature established in a conducting medium when two thermal wavefronts confront each other. In the absence of a time delay due to the microscale effect, $\tau_T = 0$, this phenomenon was first discussed by Taitel (1972) in the framework of the classical CV wave model. An objection was raised because the case of the field temperature exceeding the boundary temperature of excitation is somewhat unacceptable from the viewpoint of diffusion. Recently, the phenomenon of temperature overshooting was revisited by Tzou et al. (1994) and Özisik and Tzou (1994), confirming that the temperature overshooting indeed results from the *time-rate effect* of temperature in the wave theory of heat conduction. The physical domain in which temperature overshooting occurs, however, was found to be on *microscale*, which is excluded in the framework of the CV wave model. A detailed understanding of the overshooting phenomenon, therefore, must incorporate the microstructural interaction effect in consideration of the fast-transient process.

The dual-phase-lag model seems to be suitable for further exploration of the temperature overshooting phenomenon because it accounts for the microscale interaction effect in the phase lag of the temperature gradient (τ_T). Let us consider a one-dimensional specimen of liquid helium with a finite length L, as shown in Figure 4.10. The initial temperature is assumed to be uniform, $T = T_0$ at $t = 0$. The temperature at $x = 0$ and $x = L$ suddenly raises to a constant T_w, which produces two thermal signals propagating toward the center of the specimen. The present problem involves all temperature-specified boundary conditions, supporting the use of equation (2.25) for the sake of convenience. The dimensionless variables are defined in equation (2.24). The initial condition remains the same, equation (2.26), while the boundary conditions are modified:

$$\theta(0,\beta) = \theta(l,\beta) = 1, \quad \text{with} \quad l = \frac{L}{2\sqrt{\alpha\tau_q}}. \tag{4.41}$$

The Laplace transform solution satisfying equations (2.25), (2.26), and (4.41) is straightforward:

$$\overline{\theta}(\delta; p) = C_1 e^{Al} + C_2 e^{-Al} \tag{4.42a}$$

where

$$A = \sqrt{\frac{p(p+2)}{1+Bp}}, \quad B = \frac{\tau_T}{2\tau_q}, \quad C_1 = \frac{e^{-Al} - e^{-2Al}}{p\left(1 - e^{-2Al}\right)}, \quad C_2 = \frac{1 - e^{-Al}}{p\left(1 - e^{-2Al}\right)}. \tag{4.42b}$$

Equation (4.42), likewise, is implemented into the FUNC(P) subroutine in the

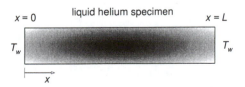

Figure 4.10 The one-dimensional specimen of liquid helium prepared for producing the phenomenon of temperature overshooting.

Appendix to obtain the inverse solution. At representative instants of time, $\beta = 0.4$, 0.7, 1.0, and 1.7, the temperature distributions in the one-dimensional specimen are shown in Figure 4.11. Without loss of generality, the dimensionless length l is assumed to be unity ($l = 1$), and the position $\delta = 0.5$ is at the midpoint of the specimen. Only distributions in the half-span ($0 \leq \delta \leq 0.5$) are displayed, owing to the symmetry of the problem. According to the CV wave model ($\tau_T = 0$, $B = 0$), the thermal wavefront is located at $x = Ct$ or $\delta = \beta$ according to equation (2.24). At $\beta = 0.4$ shown in Figure 4.11(a), therefore, the thermal wavefronts are located at $\delta = 0.4$ (for the thermal wave emanating from the left boundary at $x = 0$) and $\delta = 0.6$ (the thermal wave emanating from the right boundary at $x = l = 1$). The physical domain of $0.4 \leq \delta \leq 0.6$ is thus the thermally undisturbed zone according to the CV wave model. The time delay due to the inert behavior of molecules, namely, the phase lag of the temperature gradient τ_T, smooths the sharp wavefront and extends the physical domain of the heat-affected zone. To the left side of the thermal wavefront, the temperature level predicted by the dual-phase-lag model increases as the value of τ_T (B) decreases. Penetrating to the right side of the thermal wavefront, on the contrary, the temperature level increases with the value of τ_T (B). The physical domain of the heat affected zone, conceivably, increases as the time delay due to the inert behavior of molecules increases. The classical theory of diffusion ($\tau_T = \tau_q$ or $B = 0.5$), assuming an instantaneous response, predicts the highest temperature among the three. At $\beta = 0.4$, the field temperature already reached about 90% of the boundary temperature ($\theta = 1$).

At $\beta = 0.7$, the two wavefronts first confront each other at $\delta = 0.5$ (at the midpoint, $\beta = 0.5$) and arrive later at $\delta = 0.7$ (for the thermal wavefront emanating

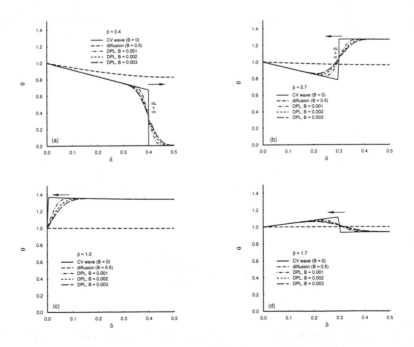

Figure 4.11 Overshooting of temperature due to the time-rate effect. Temperature distributions in the one-dimensional specimen with $l = 1$ are shown at (a) $\beta = 0.4$, (b) $\beta = 0.7$, (c) $\beta = 1.0$, and (d) $\beta = 1.7$.

from the left boundary) and $\delta = 0.3$ (for the thermal wavefront emanating from the right boundary). This is shown in Figure 4.11(b), with the arrow head indicating the direction of wave propagation. The diffusion model already predicts a *steady-state* temperature ($\theta = 1$) at this instant of time, while both the *CV* wave model and dual-phase-lag model predict *higher* temperatures ($\theta > 1$) than the boundary temperature ($\theta = 1$). This is the temperature overshooting phenomenon discussed by Taitel (1972). Figure 4.12, discussed below, demonstrates that the time-rate of change of temperature, $\partial T/\partial t$ or $\partial \theta/\partial \beta$, is the cause for this unusual behavior. At $\beta = 1$, Figure 4.11(c), the thermal wavefront emanating from the right boundary confronts the left boundary, inducing a slightly higher overshooting temperature. The dual-phase-lag model basically predicts the same behavior, but the amount of overshooting decreases as the value of τ_T (B) increases. The temperature overshooting gradually diminishes as the transient time lengthens. At $\beta = 1.7$, shown in Figure 4.11(d), the amount of overshooting reduces to about 40% of the amount shown in Figure 4.11(c).

In searching for the physical interpretation for the temperature overshooting, recall the rate effect shown by Figure 2.9. The field temperature within the conducting medium may exceed the boundary temperature if the time-rate of change of temperature exceeds a certain value. The initial time-rate of change of temperature shown in Figure 2.9 appears as an external source exciting the conducting medium in addition to the suddenly raised temperature at the boundary. In the present problem, shown in Figure 4.11, the interior of the conducting medium experiences a heating rate (a nonzero time-rate of increase of temperature) as the thermal disturbances propagate toward the center. In the dual-phase-lag model, such a time-rate of change of temperature depends on the phase lags of the temperature gradient and the heat flux vector. When the heating rate reaches a certain level, as shown in Figure 2.9, the field temperature in the local area may become exaggerated and exceed the temperature at the boundary. Figure 4.12 demonstrates this behavior explicitly. The time-rate of change of temperature is obtained by the Laplace inversion of

$$L\left[\frac{\partial \theta}{\partial \beta}(\delta,\beta)\right] = p\overline{\theta}(\delta;p),\qquad(4.43)$$

which can be easily obtained in conjunction with equation (4.42) for $\overline{\theta}(\delta;p)$. The *CV* wave model results in an *infinite* time-rate of change of temperature at the thermal wavefront ($\delta = \beta$), evidenced by the time derivative of equation (2.54):

$$\theta(\delta,\beta) = \left[e^{-\delta} + \delta\int_{\delta}^{\beta} e^{-z}\frac{I_1\left(\sqrt{z^2 - \delta^2}\right)}{\sqrt{z^2 - \delta^2}}dz\right]\Delta(\beta - \delta)$$

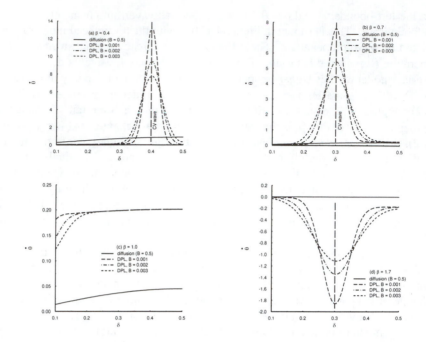

Figure 4.12 Distributions of the time-rate of change of temperature in the one-dimensional specimen with $l = 1$ at (a) $\beta = 0.4$, (b) $\beta = 0.7$, (c) $\beta = 1.0$, and (d) $\beta = 1.7$.

$$+ \left[\delta\, e^{-\beta}\, \frac{I_1\!\left(\sqrt{\beta^2 - \delta^2}\right)}{\sqrt{\beta^2 - \delta^2}} \right] h(\beta - \delta) \qquad (4.44 \; Cont.)$$

where the Dirac-delta function $\Delta(\beta - \delta)$ results from the time derivative of the unit-step function $h(\beta - \delta)$. On removal of the unrealistic assumption of an infinite speed of heat propagation from the diffusion model, unfortunately, the classical *CV* wave model introduces another paradox of an *infinite* time-rate of change of temperature at the thermal wavefront. The dual-phase-lag model, as shown in Figure 4.12(a), predicts a high rate of change in the neighborhood of the thermal wavefront (at $\delta = \beta = 0.4$), but the temperature rate remains finite. Upon impingement of the two lagging "waves",[1] the exaggerated time-rate of increase of temperature ($\dot{\theta} > 0$) heats the local area, resulting in a temperature level higher than the temperature at the boundary ($\theta = 1$). The time-rate of change of temperature increases as the ratio of τ_T to τ_q (the value of B) decreases, referring to Figure 4.12(a) to 4.12(c). As the

---

[1] The lagging behavior depicted by the linear dual-phase-lag model, equation (2.25), is *parabolic* in nature. The term "wave" used here is only for convenience.

transient time lengthens, the temperature rate decreases (Figure 4.12(a) to 4.12(c)), and reflection of lagging "waves" may induce a negative temperature rate ($\dot{\theta} < 0$) upon impingement on the incident waves (Figure 4.12(d)). Such a cooling rate reduces the amount of temperature overshooting (Figures 4.11(b) and 4.11(c)) to a nominal level (Figure 4.11(d)). The classical theory of diffusion results in a minor temperature rate at all times. The time-rate of change of temperature rapidly decreases to zero as the transient time lengthens.

The temperature overshooting phenomenon strongly depends on the specimen size. For $l = 1$, as shown in Figure 4.11(c) at $\beta = 1$, the amount of overshooting may reach as much as 38% over the boundary temperature. As the specimen size increases, however, the local time-rate of change of temperature dramatically decreases, resulting in a rapid diminution of the overshooting phenomenon. This is explicitly shown in Figure 4.13, where the specimen size (l) increases from 1 to 7. The temperature overshooting phenomenon diminishes as the specimen size exceeds 3 ($l \geq 3$), and the field temperature becomes lower than the boundary temperature at all times. According to the same values of $\alpha = 1.083 \times 10^{-6}$ m^2/s (thermal diffusivity for liquid helium) and $C = 19$ m/s (thermal wave speed), a value of $l = 1$ implies a real dimension of L on the order of submicrons (10^{-7} m). Again, a specimen size on microscale and a typical response time of the order of nanoseconds constitute a serious challenge for a successful production of the overshooting phenomenon in the laboratory.

The maximum amount of overshooting being 38% results from impingement of *two* lagging "waves" emanating from two opposite boundaries. For producing a more pronounced amplitude of temperature overshooting, it is desirable to produce a higher time-rate of change of temperature by sending more lagging waves into the conducting medium at the same time. One possibility is to use a cylindrical geometry instead of a one-dimensional specimen, as shown in Figure 4.14. The cylindrical specimen has a uniform initial temperature T_0. The circumference of the cylinder is suddenly raised to a constant temperature T_w,

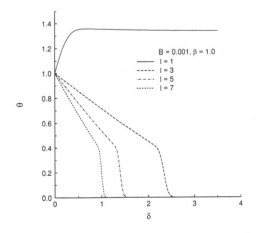

Figure 4.13 Effect of size on the amount of temperature overshooting for $\beta = 1.0$, $B = 0.001$, and $l = 1.0, 3.0, 5.0$, and 7.0.

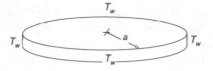

Figure 4.14   A cylindrical specimen for producing pronounced temperature overshooting on impingement of an infinite number of lagging "waves."

resulting in an *infinite* number of lagging waves propagating toward the center of the specimen.   A much higher time-rate of increase of temperature results from impingement of these lagging waves at the same time, resulting in temperature overshooting about 2 orders of magnitude higher than that in the one-dimensional specimen. Satisfying the regular condition at the central axis of the specimen, i.e., the temperature remains bounded at $r = 0$, the Laplace transform solution for $\bar{\theta}$ in correspondence with equation (4.42) is

$$\bar{\theta}(\delta; p) = \frac{J_0\left(-\delta\sqrt{\dfrac{p(p+2)}{1+Bp}}\right)}{pJ_0\left(-D\sqrt{\dfrac{p(p+2)}{1+Bp}}\right)}, \quad D = \frac{a}{2\sqrt{\alpha\tau_q}} \tag{4.45}$$

with J_0 being the Bessel function of the first kind of order zero.   All the dimensionless variables remain the same as those defined in equation (2.24), with the space variable x replaced by the radial distance r.

The temperature overshooting represented by equation (4.45) is more pronounced than that occurring in the one-dimensional specimen,   as shown in Figure 4.15 at $\beta = 0.4, 0.9, 1.0, 1.02$ (the most exaggerated stage), 1.4, and 1.8. The dimensionless radius D is taken as unity. Owing to a more pronounced effect of the temperature rate, temperature overshooting activates when all the lagging waves approach the center ($\delta = 0$). This is shown in Figure 4.15(b) at $\beta = 0.9$. Note that according to the *CV* wave model, the thermal wavefront at this instant of time is located at $\delta = 0.1$ and all the front ends of the thermal waves have not yet met. The amount of temperature overshooting reaches a maximum at $\beta = 1.02$, as shown in Figure 4.15(d). The pronounced rate effect exaggerates the temperature level to approximately 3.1 times the temperature imposed at the circumference ($\theta = 1$). When the transient time lengthens, Figures 4.15(d) to 4.15(f), the temperature level gradually tapers off, but the overshooting phenomenon persists.   A higher overshooting temperature is a desirable feature for producing the overshooting phenomenon in the laboratory.

The overshooting phenomenon is mainly attributed to the fast-transient effect in time. As the time delay due to the microstructural effect (the inert behavior of molecules at low temperatures in this case) increases, the value of τ_T (and hence the ratio of B) increases, which diminishes the overshooting behavior. This is shown in Figure 4.16, where the result at the critical instant, $\beta = 1.02$, for the most

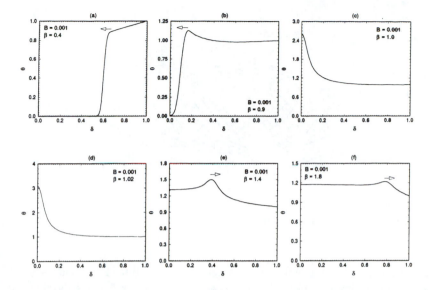

Figure 4.15 Temperature overshooting in a cylindrical specimen at (a) $\beta = 0.4$, (b) $\beta = 0.9$, (c) $\beta = 1.0$, (d) $\beta = 1.02$ (the most exaggerated stage), (e) $\beta = 1.4$, and (f) $\beta = 1.8$ at $D = 1.0$ and $B = 0.001$.

exaggerated response is displayed. The temperature overshooting disappears ($\theta < 1$) as the value of B reaches approximately 0.3. Note that the result for diffusion is retrieved for $B = 0.5$. The temperature overshooting phenomenon, therefore, is pertinent to the CV wave model and the dual-phase-lag model with a small value of τ_T (B).

For developing an overall picture in space (δ) and time (β), Figure 4.17 displays a three-dimensional surface describing the overshooting phenomenon. It shows the physical domains of space and time in which temperature overshooting is

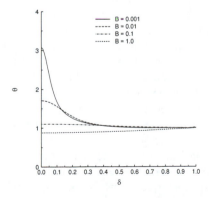

Figure 4.16 The effect of τ_T / τ_q (B) on temperature overshooting. Here, $\beta = 1.02$ (the most exaggerated stage).

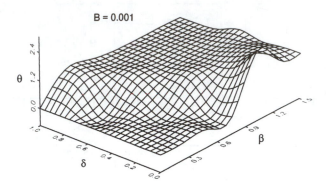

Figure 4.17 Temperature overshooting in the space and time domain.

more pronounced. They are in the neighborhood of $\delta = 1 - \beta$ for $\beta \leq 1$ and $\delta = \beta - 1$ for $\beta \geq 1$. The maximum temperature occurs at $\beta = 1.02$ and in the neighborhood of $\delta = 0$ (the central axis of the cylindrical specimen).

The temperature overshooting phenomenon, at least in terms of the relaxation-time behavior absorbed in τ_q, is not restricted to superfluid liquid helium. This section illustrates this unusual behavior by using the threshold value of B (τ_T $/2\tau_q$) for liquid helium. The dimensionless analysis provided herewith is generally applicable to other types of media, but the physical time domain in which temperature overshooting occurs may become even shorter. For metals, for example, a value of β (the dimensionless time) being unity corresponds to several picoseconds. For producing the overshooting phenomenon in the laboratory, this implies that the thermal device used for recording the temperature must respond accurately in such a short time. The pump-and-probe technique (Brorson et al., 1987; Qiu et al., 1994), described in Chapter 5, seems to be applicable for this purpose, but the feasibility needs to be proven experimentally. Note that temperature overshooting, like the thermal resonance phenomenon (Tzou, 1991b, c, 1992d, e) in solids under high-frequency excitations, may provide an efficient means in thermal processing of materials. Should the excessively high temperature, including the location at which it occurs, be well controlled, the exaggerated response in temperature provides additional heating at the desired location.

It is also important to note that the rapid rise-and-fall behavior described in Figures 4.6 to 4.9 and the temperature overshooting phenomenon described in Figures 4.11 and 4.15 are the direct result of $\tau_q \gg \tau_T$ (very small values of B). This condition supports the fact that the temperature gradient is the cause, driving the heat flow (the heat flux vector, being the effect) in the transient process. Evidenced by the failure of the classical theory of diffusion (which assumes an *instantaneous* response between the temperature gradient and the heat flux vector) shown in Figures 4.4 and 4.5, physical interpretations for these special behaviors must rely on the lagging behavior in the transient response. The classical *CV* wave model assuming $\tau_T = 0$ and $\tau_q > 0$ appears as a special case in the general condition of $\tau_q \gg$

τ_T. The importance of the dual-phase-lag model is not limited to the satisfactory comparison with the experimental result. It also demonstrates that the prolonged response in time at a fixed position and the localized temperature response (the rapid rise-and-fall behavior) observed in Bertman and Sandiford's experiment may *not* be a monopoly of the thermal wave behavior. In fact, the energy equation in the dual-phase-lag model, evidenced by equation (4.22), is *parabolic* in nature. There exists no thermal wavefront in the history of heat propagation, and the heat-affected zone extends to *infinity* in the physical domain. The temperature rise may be *insignificant* in a certain time domain (see Figures 4.3 and 4.6 to 4.9), but the thermal disturbance arrives instantaneously. As revealed by the great similarity of the pulse shape *regardless* of the type of the boundary condition, the dual-phase-lag model, which accounts for the delayed response due to the microscale effect in the fast-transient process, seems to more precisely describe the transient phenomena for heat propagation through superfluid liquid helium.

The great similarity of the pulse shape shown in Figure 4.9 to the oscilloscope trace shown in Figure 4.2 reveals the threshold value of τ_T/τ_q for liquid helium. For retaining the bell shape of the pulse, the ratio of τ_T to τ_q should be smaller than 10^{-2}. This is the closest estimate we could get in the absence of a scale in the experimental result. Should the refined scales be provided in Figure 4.2, the values of τ_T and τ_q could be determined precisely. This is illustrated in Chapter 6, where a detailed transient experiment in casting sand is examined for this purpose.

In the dual-phase-lag model, to reiterate, the prolonged response time for a significant temperature rise at a fixed position is mainly caused by the fast-transient effect absorbed in the phase lag of the heat flux vector, τ_q. The relatively smooth rise-and-fall behavior, on the other hand, is attributed to the phase lag of the temperature gradient, τ_T. While retaining these salient features in the transient response at extremely low temperatures, the dual-phase-lag model uses much less computation time. Under the same prescribed Cauchy norm for convergence, 10^{-10}, for example, the curves employing the dual-phase-lag model shown in Figure 4.6 are obtained in 0.84 to 0.86 second. The curve employing the *CV* wave model, on the other hand, requires 1032 seconds, owing to the intensive iterations in capturing the sharp wavefronts. These data are obtained based on a 486-66 MHz personal computer. When applied to the prediction of the transient response in multi-dimensional bodies, economy of computation time may be another attractive feature of the dual-phase-lag model.

FIVE
ULTRAFAST PULSE-LASER HEATING ON METAL FILMS

Ultrafast lasers with a pulse duration of the order of picoseconds provide an efficient means to study the electron-phonon dynamics in extremely short times. To date, the shortest pulse duration reaches 96 picoseconds (Brorson et al., 1987) and 100 picoseconds (Qiu et al., 1994). For high-conducting metals such as gold, the time frame of primary concern is only several picoseconds, and the penetration depth is of the order of submicrons. The extremely small scales in both space and time challenge the concepts of thermal diffusion (in the classical theory of diffusion) and thermal relaxation (in the *CV* wave model) when used alone for describing the heat transport process in this type of problem. In addition to the thermal diffusivity and the relaxation time, the phonon-electron coupling factor is a dominating parameter in the short-time energy exchange between the hot electron gas and the metal lattice; see Section 1.2. Along with diffusion, the resulting *thermalization* behavior interferes with the relaxation behavior in the fast-transient process, resulting in complicated phenomena on a small scale that cannot be depicted by the existing macroscopic models. Chapter 5 is devoted to the lagging behavior in the ultrafast process of laser heating on metals. Though necessary but not sufficient, the purpose is to determine the possible values of the phase lags that result in close transient temperatures in the picosecond domain. The physical mechanisms for the lagging behavior, to reiterate, are the finite time required for the phonon-electron interaction to take place (the phase lag of the temperature gradient) and the fast-transient effect of thermal inertia (the phase lag of the heat flux vector).

5.1 EXPERIMENTAL OBSERVATIONS

Continuing the novel experiment by Brorson et al. (1987), Qiu et al. (1994) produced femtosecond, ultrafast laser heating on gold films. Figure 5.1 summarizes their experimental system for measuring the femtosecond transient response. The

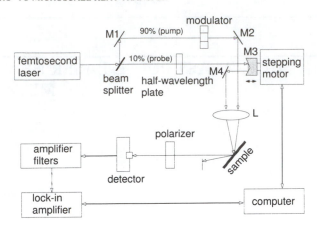

Figure 5.1 The experimental system for measuring the femtosecond transient response in gold film. Reproduced from Qiu et al. (1994).

laser pulse from the synchronously pumped linear-cavity dye laser driven by a continuously pumped, actively mode locked Nd-YAG laser was separated by a beam splitter into two beams with an intensity ratio of 9:1. The high-intensity beam, called the pump beam, is used to heat the specimen, while the low-intensity beam, called the probe beam, is used to measure the temperature change of the specimen. The 9-to-1 ratio ensures that measurement of the temperature change by the probe beam does not interfere with the heating process provided by the pump beam. The mirrors in Figure 5.1, labeled M1 to M4 in Figure 5.1, are used to change the light path of the laser beam. The pump laser passes through a light modulator, modulating the beam intensity at 1.2 MHz. Reflected by another mirror, M2, the modulated light beam was then focused onto the heated specimen by a microscope lens (denoted by L). The induced reflectivity change (proportional to the normalized temperature change of the specimen after the first few picoseconds, to be discussed below) on the specimen was detected by the probe beam, which passes mirror M3 mounted on a stepping motor for a precise control of delay times in probing. Every motor step moves the mirror M3 by 5 μm, resulting in a change of the light path of the probe beam and causing a time delay of 33 femtoseconds. By placing mirror M3 at different locations, the reflectivity changes can be detected at different delay times. Every mirror position yields a single data point for the reflectivity change at the resulting single delay time. A series of tests with different mirror positions (and hence different delay times) is thus needed to depict the entire transient process. The polarization of the probe beam is rotated to be orthogonal to the pump beam by a half-wavelength plate, so that the scattered pump beam can be blocked by a linear polarizer in front of a PIN-type detector.

The femtosecond laser heating experiment involves many other details for the precision control and analysis of the transient response in extremely short times. Figure 5.1 only summarizes the most fundamental principles in this type of experiment. The paper by Qiu et al. (1994) provides a very detailed illustration for the entire system, including detailed experimental conditions for both pump and probe laser beams, stepping motor ranges, functionality and response times of high-speed photodiodes detecting the reflected probe pulses, and conversion diagrams from heating pulses to the measurement result of reflectivity changes in the entire laser heating process.

Aging of the laser dyes varies the laser pulse duration on a day-to-day basis. Determination of the pulse duration is thus as important a task as measurement of the transient reflectivity change itself. In the experiment by Qiu et al. (1994), the pulse duration used in the femtosecond heating is measured by the laser beam itself. As shown in Figure 5.2, a laser pulse is separated by a 50:50 beam splitter. Each pulse undergoes a different optical path, which causes a time delay denoted by τ, before they were focused on a nonlinear frequency-doubling crystal. In consistency with the notation used by Qiu et al. (1994), note that τ used here refers to the delay time of the probe beam (for measuring the reflectivity change of the sample) relative to the pump beam. It is a *real time* in the short-time transient and should not be confused with the relaxation time used elsewhere in this book. After the frequency is doubled, the intensity of the laser beam is

$$I_s(t,\tau) = C_0\, I(t)\, I(t+\tau) \tag{5.1}$$

where $I(t)$ is the light intensity of the preceding beam, $I(t+\tau)$ is the light intensity of the delayed beam, and C_0 is a crystal constant. The measured light intensity $I_s(\tau)$ is obtained by summing $I_s(t, \tau)$ over the entire time domain, called the autocorrelation of the laser pulse:

$$I_s(\tau) = C_0 \int_{-\infty}^{\infty} I(t)\, I(t+\tau)\, dt \tag{5.2}$$

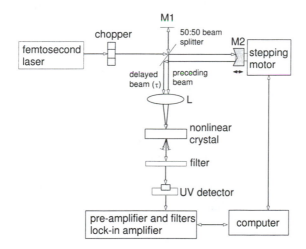

Figure 5.2 Experimental system determining the pulse autocorrelation and pulse duration. Reproduced from Qiu et al. (1994).

Again, every position of mirror M2 (controlled by the stepping motor) results in a single path of light for the delayed beam. A single path of light, in turn, gives a fixed value of the delay time (τ) between the two beams in the transient process. For obtaining the autocorrelation function of the light intensity shown by equation (5.2), therefore, a series of tests with various mirror positions needs to be performed for developing the entire picture at various delay times.

5.2 LASER LIGHT INTENSITY

Light intensity of the laser beam, $I(t)$, appears as the volumetric heat source term (the energy absorption rate in the sample) in the analytical modeling. Its analytical form in the time history can be arbitrarily chosen, but the resulting autocorrelation of the laser pulse, defined in equation (5.2), must be close to that measured experimentally. Based on the light intensity thus determined, the volumetric heating in the sample is

$$S(x,t) = S_0 e^{-\frac{x}{\delta}} I(t),\tag{5.3a}$$

with δ denoting the penetration depth of laser radiation, and S_0 the intensity of laser absorption (Qiu and Tien, 1992, 1993, 1994):

$$S_0 = 0.94 J \left(\frac{1-R}{t_p \delta} \right).\tag{5.3b}$$

In equation (5.3b), J is laser fluence, R the radiative reflectivity of the sample to the laser beam, and t_p the full-width-at-half-maximum pulse duration. Equation (5.3a) assumes an exponentially decayed heating intensity in the thickness direction of the sample.

5.2.1 Gaussian Distribution

Traditionally, the Gaussian profile is used to simulate the light intensity of laser pulses:

$$I(t) = I_0 e^{-\psi \left(\frac{t}{t_p} \right)^2}\tag{5.4a}$$

with ψ being a constant, $\psi = 4 \ln(2) \cong 2.77$. Substituting equation (5.3) into (5.2), the autocorrelation of the laser pulse is

$$I_s(\tau) = C_0 \int_{-\infty}^{+\infty} e^{-\psi\left(\frac{t}{t_p}\right)^2} e^{-\psi\left(\frac{t+\tau}{t_p}\right)^2} dt = \left[C_0 I_0^2 \sqrt{\frac{\pi t_p^2}{2\psi}} \right] e^{-\frac{\psi}{2}\left(\frac{\tau}{t_p}\right)^2}.$$ (5.4b)

The bracketed quantity is the autocorrelation at $\tau = 0$, $I_s(0)$. The *normalized* autocorrelation of the laser pulse is thus

$$\frac{I_s(\tau)}{I_s(0)} = e^{-\frac{\psi}{2}\left(\frac{\tau}{t_p}\right)^2} = e^{-2\ln 2\left(\frac{\tau}{t_p}\right)^2} \qquad \text{(Gaussian)}.$$ (5.5)

Qiu et al. (1994) showed that the Gaussian distribution resulting from equation (5.5) agrees well with the experimental result for $t_p = 100$ femtoseconds (fs).

Selection of a particular form for the light intensity function, $I(t)$ in equation (5.4a), is relatively arbitrary as long as it results in an autocorrelation function comparable to the experimental result. In the finite difference solution obtained by Qiu et al. (1994), for example, the Gaussian profile shown by equation (5.4a) was directly implemented into the semi-implicit Crank-Nicholson scheme with no special difficulty in discretizing the volumetric heat source. For the Laplace transform technique to be used later for studying the transient reflectivity change on the film surfaces, however, the Gaussian distribution results in a Laplace transformed solution involving an error function. When performing the Riemann-sum approximation for the Laplace inversion, equation (2.55), the error function with a complex argument is needed, which unfortunately, is very inconvenient from a numerical point of view. To overcome this difficulty, an alternate form for the light intensity, $I(t)$, is considered.

5.2.2 Alternate Form of Light Intensity

A particular form of light intensity will be explored that not only yields an autocorrelation of laser pulse comparable to the experimental result but also facilitates the direct use of the Riemann-sum approximation for the Laplace inversion. The ideal light intensity must possess a maximum at $t = 0$, a characteristic in the Gaussian profile shown in equation (5.4a). A possibility possessing this property is

$$I(t) = I_0 e^{-a\left|\frac{t}{t_p}\right|} \qquad \text{with} \quad a \geq 0, \quad t_p \geq 0$$ (5.6a)

The autocorrelation of the laser pulse corresponding to equation (5.4b) is, from equation (5.2),

$$I_s(\tau) = C_0 I_0^2 \int_{-\infty}^{+\infty} e^{-\psi|t|} e^{-\psi|t+\tau|} dt \qquad \text{with} \quad \psi = \frac{a}{t_p} > 0.$$ (5.6b)

Further evaluation of $I_s(\tau)$ needs to be performed in different domains of τ, owing to the absolute values. For $\tau \geq 0$, equation (5.6b) can be decomposed into

$$I_s(\tau) = C_0 I_0^2 \left[\int_{-\infty}^{-\tau} e^{\psi t}\, e^{\psi(t+\tau)} dt + \int_{-\tau}^{0} e^{\psi t}\, e^{-\psi(t+\tau)} dt + \int_{0}^{\infty} e^{-\psi t}\, e^{-\psi(t+\tau)} dt \right],$$

$$\text{for } \tau \geq 0. \qquad (5.6c)$$

For $\tau < 0$, on the other hand, equation (5.6b) gives

$$I_s(\tau) = C_0 I_0^2 \left[\int_{-\infty}^{0} e^{\psi t}\, e^{\psi(t+\tau)} dt + \int_{0}^{-\tau} e^{-\psi t}\, e^{\psi(t+\tau)} dt + \int_{-\tau}^{\infty} e^{-\psi t}\, e^{-\psi(t+\tau)} dt \right],$$

$$\text{for } \tau < 0. \qquad (5.6d)$$

Integrations of the right sides of equations (5.6c) and (5.6d) are straightforward, resulting in

$$I_s(\tau) = \begin{cases} C_0 I_0^2 \left(\dfrac{e^{-\psi\tau}}{\psi} + \tau\, e^{-\psi\tau} \right), & \text{for } \tau \geq 0 \\[3mm] C_0 I_0^2 \left(\dfrac{e^{\psi\tau}}{\psi} - \tau\, e^{\psi\tau} \right), & \text{for } \tau < 0 \end{cases} \qquad (5.7a)$$

or, in a normalized form,

$$\frac{I_s(\tau)}{I_s(0)} = \begin{cases} e^{-\psi\tau} + \psi\tau e^{-\psi\tau}, & \text{for } \tau \geq 0 \\ e^{\psi\tau} - \psi\tau\, e^{\psi\tau}, & \text{for } \tau < 0 \end{cases} \qquad (5.7b)$$

Figure 5.3 compares the autocorrelation of the laser pulse obtained in equation (5.7b) with the experimental result obtained by Qiu et al. (1994). The excellent agreement supports the use of equation (5.6a) for describing the light intensity of the laser beam. The use of equation (5.6a), to repeat, is for the convenience in obtaining the Riemann-sum approximation for the Laplace inversion. The Gaussian profile shown in equation (5.5) yields satisfactory agreement with the experimental result, as shown in Figure 5.3. Note that the maximum amplitude of autocorrelation occurs at zero time delay. In analytical modeling for the short-time response, therefore, the "initial" condition is advanced to the instant at which the autocorrelation function is zero. In the use of the Gaussian profile, referring to Figure 5.3, this implies an "initial" instant of time at $t = -2t_p \cong -200$ ps, the initial time used by Qiu and Tien (1992, 1993, 1994) in their analytical model.

With the pulse structure in laser heating thus understood, Figure 5.4 shows the experimental results of reflectivity change measured in the front surface of a gold film of thickness 0.1 μm. Results obtained by both Qiu et al. (1994) and

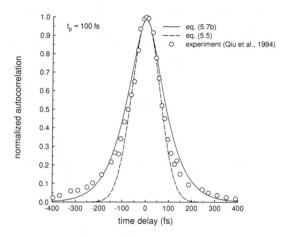

Figure 5.3 Comparison of the normalized autocorrelation of the laser pulse, equation (5.7b), equation (5.5) (Gaussian distribution), and the experimental result (reproduced from Qiu et al., 1994) at $a = 1.88$, $t_p = 100$ fs.

Brorson et al. (1987) are displayed for comparison. Most remarkable are the close results between the two independent experiments obtained in the picosecond response. Receiving the photon energy from the laser beam, the electron gas develops an extremely high temperature in an extremely short period of time, of the order of several tenths of a picosecond. The lattice temperature and the electron-gas temperature are in a highly nonequilibrium state in this stage, with the lattice temperature being several times lower. Through the energy exchange between electrons and phonons (quanta of the metal lattice), thermalization between phonons

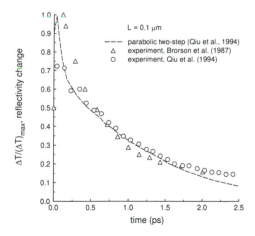

Figure 5.4 Experimental results of reflectivity change by Brorson et al. (1987) and Qiu et al. (1994) at the front surface of a gold film of thickness 0.1 μm, and predictions from the parabolic two-step model (Qiu et al., 1994) in Section 5.3.

and electrons occurs, rendering a slow process of temperature decay in the electron gas. It is important to note that in the first few hundred femtoseconds in the transient response, the reflectivity change does *not* reflect the real temperature change in the electron gas (Fann et al., 1992a, b; Qiu et al., 1994). When comparing the results of temperature change of the electron gas (calculated from the parabolic two-step model, see Section 5.3) with the experimental results of reflectivity change, coincidence should *not* be expected in the first few hundred femtoseconds.

5.3 MICROSCOPIC PHONON-ELECTRON INTERACTION MODEL

Qiu and Tien (1994) accounted for the temperature-dependent heat capacity of electron gas and thermal conductivity in their effort to model the two-step process (parabolic) of energy transport between phonons and electrons in the short-time transient:

$$C_e(T_e)\frac{\partial T_e}{\partial t} = \frac{\partial}{\partial x}\left(\frac{T_e}{T_l}k\frac{\partial T_e}{\partial x}\right) - G(T_e - T_l) + S \qquad (5.8)$$

$$C_l\frac{\partial T_l}{\partial t} = G(T_e - T_l). \qquad (5.9)$$

The short-time laser heating on thin films is modeled by a one-dimensional problem, with x being the spatial distance measured from the front surface of the film subjected to the short-pulse laser irradiation. The quantity G is the phonon-electron coupling factor that dictates the refined mechanism of phonon-electron interaction in transporting heat. Its value depends on a group of microscopic properties, which will be discussed in more detail in Section 5.4. While assuming diffusive behavior for heat transport in the electron gas, equation (5.8), equation (5.9) neglects the effect of heat diffusion in the metal lattice because heat transport by free electrons, at least for pure metals, is much greater than that by lattice vibrations. The thermal conductivity k is the equilibrium value when thermalization between electrons and phonons is achieved ($T_e = T_l$). As a result, shown by equation (5.8), thermal conductivity of the electron gas is assumed to be proportional to the temperature of the electron gas and inversely proportional to the temperature of the metal lattice. Heat capacity of the electron gas, $C_e(T_e)$, is assumed to be proportional to the temperature of the electron gas. Employing the Gaussian profile, the laser heating source S in equation (5.8) results from the substitution of equations (5.3b) and (5.4a) into equation (5.3a):

$$S(x,t) = 0.94J\left(\frac{1-R}{t_p\delta}\right)e^{-\frac{x}{\delta}-2.77\left(\frac{t}{t_p}\right)^2}, \qquad (5.10)$$

with δ being the radiation penetration depth on the order of nanometers. Also, a typical value of $R = 0.93$ for visible light is selected in the analysis.

During the short period of laser heating, heat losses from the front and back surfaces of the film are assumed negligible, implying

$$\frac{\partial T_e}{\partial x}(0,t) = \frac{\partial T_e}{\partial x}(L,t) = \frac{\partial T_l}{\partial x}(0,t) = \frac{\partial T_l}{\partial x}(L,t) = 0 \qquad (5.11)$$

with L denoting the film thickness. The initial conditions for both the electron and the metal lattice at $t = -2t_p$ are

$$T_e(x,-2t_p) = T_l(x,-2t_p) = T_0. \qquad (5.12)$$

From a physical point of view, equation (5.12) describes the laser heating of the electron-lattice system from a thermalization state.

In an attempt to solve the two-step system, equations (5.8) and (5.9) subject to the boundary condition (5.11) and initial condition (5.12), Qiu and Tien employed the semi-implicit Crank-Nicholson finite difference scheme owing to the strong nonlinearity in the energy transport equations. To capture the temperature-varying (and hence the space- and time-varying) properties, they used 400 grid points in the numerical computation for the electron temperature (T_e) and the lattice temperature (T_l).

Reflectivity changes of metals result from variations of electron distributions. A description of the reflectivity change in terms of the temperature change of the electron gas is only appropriate when electrons follow thermal equilibrium distributions, a condition that occurs *after* the first few hundred femtoseconds in the transient response. Imperfection in the electron-phonon system, such as voids and lattice distortion, might have significant effects on the transient response in this initial period. After the first few hundred femtoseconds, the electron distribution becomes thermally equilibrated, and the normalized reflectivity change is proportional to the normalized temperature change of the electron gas (Rosei and Lynch, 1972; Rosei, 1974; Guerrisi and Rosei, 1975; Juhasz et al., 1992):

$$\frac{\Delta T_e}{\left(\Delta T_e\right)_{max}} \cong \frac{\Delta R_e}{\left(\Delta R_e\right)_{max}} \qquad (5.13)$$

with "max" denoting the maximum value of ΔT_e or ΔR occurring in the transient process. Such a proportional relation holds between room temperature and approximately 700 K in the electron gas. The normalized temperature change in the electron gas during the transient stage, the left side of equation (5.13), can be calculated from the parabolic two-step model, equations (5.8) and (5.9). The right side of equation (5.13), on the other hand, can be measured experimentally as described in Section 5.2. For the same case of a thin film of thickness 0.1 μm (L), Figure 5.4 shows excellent agreement between the parabolic two-step model and the experimental result for the normalized reflectivity change (normalized temperature change in the electron gas) at the front surface of the film. The parabolic two-step model accurately captures the temperature drop at the front surface of the film in the thermalization process, a special feature in the fast-transient process that cannot be depicted by the conventional diffusion and thermal wave models (Qiu and Tien, 1992, 1993; Qiu et al., 1994).

5.4 CHARACTERISTIC TIMES — THE LAGGING BEHAVIOR

Although diffusive behavior is assumed in the parabolic two-step model, the phonon-electron interaction in microscale requires a *finite period of time* to take place; see Figure 2.1. The resulting time delay in heat transport through the electron-phonon system may not be significant if only a few phonons and electrons are involved. However, when hundreds and thousands of these energy carriers interact in the process of heat transport, the *cumulated* time delay for a *significant* rise of temperature in the electron-phonon system may become comparable to the phonon-electron relaxation time. The characteristic times governing the phonon-electron heat transport described by equations (5.8) and (5.9), in other words, need to be further explored.

Because the temperature-varying properties do not alter the fundamental characteristics in heat transport, evidenced by the persistent Fourier behavior in conventional heat conduction regardless of the temperature-dependent or temperature-independent thermal conductivity, we first reveal the time-concept by considering *constant* thermal properties in equations (5.8) and (5.9). The two-step heat transport equations in this case become

$$C_e \frac{\partial T_e}{\partial t} = k \nabla^2 T_e - G(T_e - T_l) \tag{5.14a}$$

$$C_l \frac{\partial T_l}{\partial t} = G(T_e - T_l). \tag{5.14b}$$

For metals, externally supplied photons (laser) first increase the temperature of the electron gas, as represented by equation (5.14a). Through the phonon-electron interactions, as the second step of heating, the hot electron gas then increases the temperature of the metal lattice as represented by equation (5.14b). The energy exchange between phonons and electrons is characterized by the coupling factor G:

$$G = \frac{\pi^4 (n_e v_s \kappa)^2}{k}. \tag{5.15}$$

The coupling factor (G) depends on the number density of free electrons per unit volume (n_e), the Boltzmann constant (κ), and the speed of sound v_s:

$$v_s = \frac{\kappa}{2\pi h}\left(6\pi^2 n_a\right)^{-\frac{1}{3}} T_D.$$ (5.16)

Through the speed of sound, therefore, the phonon-electron coupling factor further depends on the Planck constant (h), the atomic number density per unit volume (n_a), and the Debye temperature (T_D). The s-band approximation employed by Qiu and Tien (1992) provides an accurate estimate for the number density of free electrons in pure metals.

Equations (5.14a) and (5.14b) can be combined by eliminating the electron-gas temperature (T_e). First, from equation (5.14b), the electron-gas temperature can be expressed in terms of the lattice temperature and its time derivative:

$$T_e = T_l + \frac{C_l}{G}\frac{\partial T_l}{\partial t}.$$ (5.17a)

Second, substituting equation (5.17a) into (5.14a) and using the result of $G(T_e - T_l)$ from equation (5.14b) results in

$$\nabla^2 T_l + \frac{\alpha_e}{C_E^2}\frac{\partial}{\partial t}\left(\nabla^2 T_l\right) = \frac{1}{\alpha_E}\frac{\partial T_l}{\partial t} + \frac{1}{C_E^2}\frac{\partial^2 T_l}{\partial t^2},$$ (5.17b)

where

$$\alpha_E = \frac{k}{C_e + C_l} \quad \text{and} \quad C_E = \sqrt{\frac{kG}{C_e C_l}}$$ (5.18)

represent the equivalent thermal diffusivity (α_E) and thermal wave speed (C_E). Equation (5.18) is a *single* energy equation governing heat transport through the metal lattice. The parallel equation governing heat transport through the electron gas can be obtained in the same manner. From equation (5.14a), the lattice temperature can be expressed in terms of the electron-gas temperature:

$$T_l = T_e - \frac{k}{G}\nabla^2 T_e + \frac{C_e}{G}\frac{\partial T_e}{\partial t}.$$ (5.19a)

Substituting equation (5.19a) into (5.14b) and using the result of $G(T_e - T_l)$ from equation (5.14a) results in

$$\nabla^2 T_e + \frac{\alpha_e}{C_E^2}\frac{\partial}{\partial t}\left(\nabla^2 T_e\right) = \frac{1}{\alpha_E}\frac{\partial T_e}{\partial t} + \frac{1}{C_E^2}\frac{\partial^2 T_e}{\partial t^2}$$ (5.19b)

which is the *single* energy equation governing heat transport through the electron gas. Except for the change of the dependent variables from T_e to T_l, equations (5.17b) and (5.19b) have identical forms.

Several distinct features exist in equations (5.17b) and (5.19b). First, although diffusion is assumed for heat transport through the electron gas, a *wave* term (the second-order derivative with respect to time) does exist in these equations. The phonon-electron coupling factor G, evidenced by equation (5.18), dictates the equivalent thermal wave speed. Assuming an *infinite* value of G, implying either that the energy transport from electrons to phonons occurring at zero time (an *infinite* time rate) or that electrons and phonons collide with each other at an *infinite* frequency, the thermal wave speed approaches infinity ($C_E \to \infty$). Equations (5.17b) and (5.19b) reduce to the conventional diffusion equation in this case, implying that Fourier's law in heat conduction bears these assumptions in its framework. Second, under a finite value of G (and hence a finite value of C_E), the mixed-derivative term and the wave term in equations (5.17b) and (5.19b) must exist *simultaneously*. Unless the value of α_e approaches zero, implying that either thermal conductivity of the electron gas approaches zero or the heat capacity of the electron gas approaches infinity, the classical thermal wave model containing only the wave term without the mixed-derivative term in equations (5.17b) and (5.19b) *cannot* stand on its own. Similar to the case of diffusion, the *CV* wave equation inherits this assumption in its framework. Third, the coefficient (α_e/C_E^2) in front of the mixed-derivative term has a dimension of *time*, while the coefficient $(1/C_E^2)$ in front of the wave term can be expressed as the ratio of another characteristic time to the equivalent thermal diffusivity:

$$\frac{\alpha_e}{C_E^2} \equiv \tau_T, \quad \frac{1}{C_E^2} \equiv \frac{\tau_q}{\alpha_E} \tag{5.20}$$

where τ_T and τ_q are two *characteristic times* governing the phonon-electron interaction in heat transport. In terms of these characteristic times, equations (5.17b) and (5.19b) can be written in a more concise fashion:

$$\nabla^2 T + \tau_T \frac{\partial}{\partial t}\left(\nabla^2 T\right) = \frac{1}{\alpha_E}\frac{\partial T_e}{\partial t} + \frac{\tau_q}{\alpha_E}\frac{\partial^2 T_e}{\partial t^2}. \tag{5.21}$$

Equation (5.21) is in the same form as equation (2.10), employing the linearized dual-phase-lag model, equation (2.7). The two characterized times in equation (5.21), τ_T and τ_q, can thus be viewed as the phase lags of the temperature gradient and the heat flux vector in the fast-transient process in microscale. The phase lag of the temperature gradient, τ_T, captures the time delay due to the microstructural interaction effect, while the phase lag of the heat flux vector, τ_q, captures the fast-transient effect of thermal inertia.

Because equation (5.21) (the phonon-electron interaction model with emphasis on the microscale effect in space) has exactly the same form as equation

(2.10) (the dual-phase-lag model with emphasis on the resulting delayed response in time), all the special features discussed in Chapter 2 apply. In the presence of the fast-transient effect, the wave term led by τ_q in equation (5.21), the most significant effect of microstructural interaction (the mixed-derivative term led by τ_T in equation (5.21)) may be diminution of the sharp wavefront in transporting heat, as illustrated in Figure 2.8.

5.5 PHASE LAGS IN METAL FILMS

Equations (5.18) and (5.20) provide useful relations for the estimate of the equivalent thermal diffusivity (α_E), the thermal wave speed (C_E), and the phase lags of the temperature gradient and the heat flux vector (τ_T and τ_q). Based on the values of the phonon-electron coupling factor (G), heat capacities of the electron gas (C_e) and the metal lattice (C_l), copper (Cu), silver (Ag), gold (Au), and lead (Pb) are tabulated in Table 5.1 for an overview. Note, however, that these values are based on the correlations assuming *constant* microscopic properties in the two-step model. The temperature dependency of thermal conductivity and heat capacity of the electron gas are not taken into account. From Table 5.1, it is clear that the value of τ_T is greater than the value of τ_q in the short-pulse laser heating. According to the phase-lag concept, this implies the heat flux vector *precedes* the temperature gradient in the process of heat transport. The heat flux vector is the *cause* of heat flow, while the temperature gradient is the *effect*. Compared to the heat transport in superfluid liquid helium discussed in Chapter 4, this is a *reversed* sequence of precedence.

Since the microscale interaction effect can be lumped into the resulting delayed response in time, evidenced by the equivalence between equations (2.10) and (5.21) for governing heat transport and equations (5.18) and (5.20) for the perfect correlation, the thermalization process shown in Figure 5.4 must be explainable in terms of the lagging behavior. In such an attempt, however, the *constant* values of τ_T and τ_q shown in Table 5.1 need to be adjusted to account for the strong temperature dependence of thermal conductivity and heat capacity of the

Table 5.1 The equivalent thermal diffusivity (α_E), the equivalent thermal wave speed (C_E), and the phase lags of the temperature gradient and the heat flux vector (τ_T and τ_q) for typical metals ($C_e = 2.1 \times 10^4$ J/m^{3}K at room temperature; ps $\equiv$ picosecond)

	k, W/m K	C_l, J/m^3 K $\times 10^6$	G, W/m^3 K $\times 10^{16}$	α_E, m^2/s $\times 10^{-4}$	τ_T, ps	τ_q, ps	C_E, m/s $\times 10^4$
Cu	386	3.4	4.8	1.1283	70.833	0.4348	1.6109
Ag	419	2.5	2.8	1.6620	89.286	0.7438	1.4949
Au	315	2.5	2.8	1.2495	89.286	0.7438	1.2961
Pb	35	1.5	12.4	0.2301	12.097	0.1670	1.1738

electron gas during the thermalization process.

The energy equation governing the lagging behavior is equation (2.10),

$$\frac{\partial^2 T}{\partial x^2} + \tau_T \frac{\partial^3 T}{\partial x^2 \partial t} + \frac{1}{k}\left(S + \tau_q \frac{\partial S}{\partial t}\right) = \frac{1}{\alpha}\frac{\partial T}{\partial t} + \frac{\tau_q}{\alpha}\frac{\partial^2 T}{\partial t^2}, \tag{5.22}$$

with

$$S(x,t) = 0.94 J \left(\frac{1-R}{t_p \delta}\right) e^{-\frac{x}{\delta} - \frac{a\left|t - 2t_p\right|}{t_p}} \tag{5.23}$$

describing laser heating with the intensity function shown by equation (5.6a). The factor $(t - 2t_p)$ results from the shift of initial time from zero to $(-2t_p)$ in correspondence with equation (5.12) in Qiu and Tien's approach. The time-derivative of S in equation (5.22) is the *apparent* heat source resulting from the fast-transient effect of thermal inertia described by τ_q, equation (2.10).

The presence of a wave term requires *two* initial conditions:

$$T(x,0) = T_0 \quad \text{and} \quad \frac{\partial T}{\partial t}(x,0) = 0, \tag{5.24}$$

which assumes heating from a stationary state. The film thickness is assumed to be a constant, L. During the short period of laser heating, heat losses from the front and back surfaces are assumed negligible (Qiu and Tien, 1992, 1993, 1994), implying

$$\frac{\partial T}{\partial x}(0,t) = \frac{\partial T}{\partial x}(L,t) = 0. \tag{5.25}$$

The heat source term represented by equation (5.23) facilitates a direct Laplace transform solution. The transformed solution satisfying equation (5.22) and initial and boundary conditions (5.24) and (5.25) is

$$\overline{T}(x,p) = A_1 e^{Bx} + A_2 e^{-Bx} + A_3 e^{-x/\delta} \tag{5.26}$$

where

$$A_1 = \frac{(A_3/\delta)(e^{-L/\delta} - e^{-BL})}{B(e^{BL} - e^{-BL})}, \quad A_2 = A_1 - \frac{A_3}{B\delta},$$

$$A_3 = \frac{S_0(C_2 e^{-2a} - C_1 S_b)}{(1/\delta)^2 - B^2}, \quad B = \sqrt{\frac{p(1 + p\tau_q)}{\alpha(1 + p\tau_T)}},$$

$$C_1 = \frac{1 + p\tau_q}{k(1 + p\tau_T)}, \quad C_2 = \frac{\tau_q}{k(1 + p\tau_T)}, \quad S_b = t_p \left[\frac{e^{-2a} - e^{-2pt_p}}{pt_p - a} + \frac{e^{-2pt_p}}{pt_p + a} \right].$$

(5.27)

Regardless of the complicated expression, equation (5.26) is in a standard form to be implemented in the function subroutine, FUNC(P), in the Appendix for the Laplace inversion. Based on the temperature in the physical domain thus obtained, the normalized temperature change, $\Delta T(0,t)/[\Delta T(0,t)]_{max} \equiv [T(0,t) - T_0]/[T(0,t) - T_0]_{max}$ can be calculated at the front surface of the gold film at $x = 0$. The subscript "max" used here, again, refers to the maximum value of temperature change, $T(0,t) - T_0$, in the transient stage. For $L = 0.1$ μm, $\tau_T = 90$ ps, and $\tau_q = 8.5$ ps, the results thus obtained are shown in Figure 5.5 along with the experimental results obtained by Brorson et al. (1987) and Qiu et al. (1994). The dual-phase-lag model agrees very well with the experimental results, supporting the lagging behavior in the short-pulse laser heating process on metals.

Brorson et al. (1987) also obtained the reflectivity change at the front surface of a thin film of thickness 0.2 μm, twice as thick as that shown in Figure 5.5. Based on the *same* value of $\tau_T = 90$ ps and $\tau_q = 8.5$ ps, Figure 5.6 shows excellent agreement between the dual-phase-lag model and the experimental result, implying that the values of τ_T and τ_q, and hence the cumulated delayed times reflecting the microstructure in the thickness direction of the film, do not vary sensitively, as the thickness is greater than 0.1 μm.

Figure 5.7 shows the effect of τ_T and τ_q on the transient response of reflectivity change. At a fixed value of τ_T, Figure 5.7(a), increasing the value of τ_q promotes (demotes) the temperature level at earlier (later) times. At a fixed value of

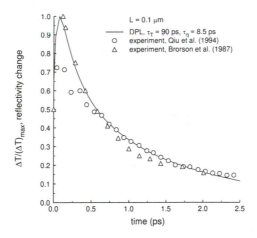

Figure 5.5 Normalized temperature change (reflectivity change) in gold film predicted by the dual-phase-lag model. Comparison with the experimental results by Brorson et al. (1987) and Qiu et al. (1994). $J = 13.4$ J/m^2, $R = 0.93$, $\delta = 15.3$ nm, $\alpha = 1.2 \times 10^{-4}$ m^2/s, and $k = 315$ W/m K.

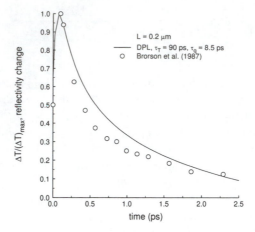

Figure 5.6 Reflectivity change at the front surface of a gold film with thickness being 0.2 μm. Comparison between the dual-phase-lag model and the experimental result (Brorson et al., 1987).

τ_q, Figure 5.7(b), decreasing the value of τ_T produces the same effect. These are useful trends in understanding the relative effect between the microscale effect in space (phonon-electron interaction) and the microscale effect in time (fast-transient effect of thermal inertia). From a microscopic point view, according to equations (5.18) and (5.20), increasing (decreasing) the value of τ_q (τ_T) is equivalent to decreasing (increasing) the value of the phonon-electron coupling factor (G). When the value of G increases, in other words, it increases the phase lag of the heat flux vector and decreases the phase lag of the temperature gradient, promoting the precedence of the temperature gradient to the heat flux vector in the process of heat transport. The use of this trend has been shown in the design of the stacking sequence preventing laser damage for multiple-layered film subjected to short-pulse

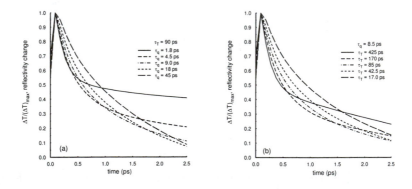

Figure 5.7 Effect of (a) τ_q and (b) τ_T on the transient response of reflectivity change. The front surface of a gold film of thickness 0.1 μm.

laser heating (Qiu and Tien, 1994; Qiu et al., 1994).

From a mathematical point of view, indeed, the lagging response depends on the respective values of τ_T and τ_q. Evidenced by the dimensionless analysis made in Chapter 2 (equation (2.25) for instance), however, the *ratio* of (τ_T/τ_q) dominates the lagging behavior in the transient stage. This is true even when a volumetric heat source (including the apparent heating) is present in the problem.

Under the same value of $\tau_T/\tau_q = 90/8.5 \cong 10.588$, Figure 5.8 shows the transient curves at various values of τ_T. Increasing the value of τ_T (and hence the value of τ_q) by more than 1 order of magnitude results in variations of the transient responses only in *longer* time. The peak reflectivity change remains at the same location, while the response curve varies in the same neighborhood. For the transient time shorter than approximately 0.5 ps, the ratio of (τ_T/τ_q) dominates the transient response.

The classical diffusion and *CV* wave models *cannot* describe the slow thermalization process shown in Figure 5.5. From a physical point of view, the *macroscopic* approach employed in these models neglects the microstructural interaction effect in the short-time transient, rendering an *over*estimated temperature in the transition response. This is shown in Figure 5.9. The solution for the diffusion model results from equations (5.26) and (5.27) for the dual-phase-lag model with $\tau_T = \tau_q = 0$, while the solution for the *CV* wave model results from $\tau_T = 0$ and $\tau_q = 8.5$ ps. In the thermalization process, the diffusion model predicts the highest temperature among all three. Compared to the experimental result, the large difference results from negligence of both the microstructural interaction effect in space and the fast-transient effect in time. The *CV* wave model accounting for the fast-transient effect in the short-time response, labeled as the "τ_q effect" in Figure 5.9, redeems the difference between the diffusion model and the experimental result. Negligence of the microstructural interaction effect in space, however, still significantly *over*estimates the transient temperature. As the dual-phase-lag model further incorporates the delay time caused by the phonon-electron interaction in

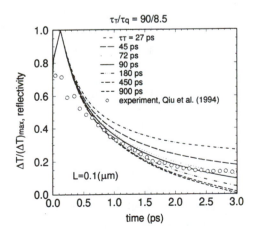

Figure 5.8 Transient response of reflectivity change under the same ratio of $\tau_T/\tau_q = 90/8.5 \cong 10.588$ in gold film.

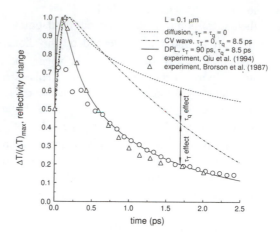

Figure 5.9 Reflectivity change at the front surface of the gold film. Comparison among the dual-phase-lag model (DPL), the diffusion model, the *CV* wave model, and the experimental results by Qiu et al. (1994). $L = 0.1$ μm.

microscale in the framework, labeled as the "τ_T effect" in Figure 5.9, the transient temperature becomes close to the experimental observation. The slow thermalization process observed in the experiment by Qiu et al. (1994), therefore, can be viewed as a lagging behavior that combines both the effects of microstructural interaction and fast-transient inertia.

5.6 EFFECT OF TEMPERATURE-DEPENDENT THERMAL PROPERTIES

Compared to the values of τ_T and τ_q listed in Table 5.1, which are estimated under constant thermal properties, $\tau_T = 89.286$ ps and $\tau_q = 0.7438$ ps, the values used in Figure 5.5, $\tau_T = 90$ ps and $\tau_q = 8.5$ ps, which reflect the effect of temperature-dependent thermal properties, are significantly different. Since the ratio of (τ_T/τ_q) dominates the transient response, this implies a decrease of the ratio (from 120.04 under constant thermal properties to 10.588, reflecting the effect of temperature-dependent thermal properties) by approximately 1 order of magnitude to fit the experimental result. However, note that the effect of temperature-dependent properties is defined in an "averaged" sense. The values of τ_T and τ_q are still held constant in equation (5.22), and hence in the solution of equations (5.26) and (5.27), but their values are adjusted according to the experimental result where the effect of temperature dependence is known to be important.

Comparing the results of the dual-phase-lag model based on two sets of values of τ_T and τ_q, therefore, reveals the lumped effect of temperature dependence of thermal properties. This is illustrated in Figure 5.10, where the results for $\tau_T = 89.286$ ps and $\tau_q = 0.7438$ ps, estimated under constant properties, and $\tau_T = 90$ ps

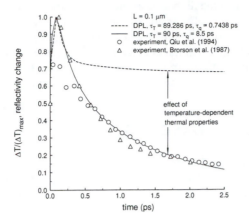

Figure 5.10 The lumped effect of temperature-dependent thermal properties on the transient response of reflectivity change at the front surface of a gold film with thickness of 0.1 μm.

and $\tau_q = 8.5$ ps, accounting for the temperature-dependent effect, are both displayed along with the experimental results by Brorson et al. (1987) and Qiu et al. (1994). Evidently, the temperature-dependent thermal properties, referring to the curve with $\tau_T = 90$ ps and $\tau_q = 8.5$ ps, render a *lower* temperature level in the thermalization process, which can be viewed as the effect of reducing the ratio of (τ_T / τ_q).

Based on the values of τ_T and τ_q estimated under constant thermal properties, shown in Table 5.1, for comparing the transient temperatures developed in the representative metals, Figure 5.11 shows the reflectivity change at the front surface for copper (Cu), silver (Ag), gold (Au), and lead (Pb). These curves,

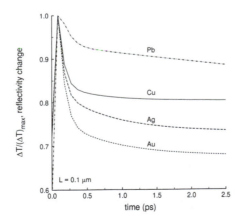

Figure 5.11 Comparison of the reflectivity change at the front surface among lead (Pb), copper (Cu), silver (Ag), and gold (Au), assuming the constant thermal properties in Table 5.1.

however, are only qualitative because the effect of the temperature-dependent thermal properties is *not* taken into account. Lead (Pb) produces the highest transient temperature change among the four, copper (Cu) is the next, silver (Ag) third, and gold (Au) fourth. Until a detailed knowledge is obtained about the way in which the raised temperature varies the thermal properties in these materials, the temperature-dependent effect remains unknown for these curves.

5.7 CUMULATIVE PHASE LAGS

At least from a mathematical point of view, equations (5.26) and (5.27) can be used to calculate the reflectivity change at the rear surface, $x = L$. For metal films with different thicknesses, these equations can also be used to study the size effect on reflectivity change. For $L = 0.1$ µm under exactly the same conditions as those in Figure 5.5, Figure 5.12 shows the normalized reflectivity change at the rear surface of the gold film. The normalized temperature change (reflectivity change), as expected, increases at a much slower rate compared to that in the front surface; see Figure 5.5. It reaches the peak value at a later instant of time, at about 1.9 ps, followed by the thermalization process with a much slower decay rate of temperature in time.

Under the same values of τ_T and τ_q, Figure 5.13 shows the effect of film thickness on the normalized reflectivity change at the front surface. The temperature level in the thermalization process is higher for the gold film with a smaller thickness. The film thickness has a more pronounced effect on the reflectivity change for L smaller than 0.1 µm. As $L \geq 0.1$ µm, the normalized reflectivity change does not sensitively vary with the film thickness.

From a physical point of view, unfortunately, Figure 5.12, describing the reflectivity change at the rear surface, and Figure 5.13, illustrating the size effect of gold films are only *qualitative*. The reflectivity change (temperature change) at the

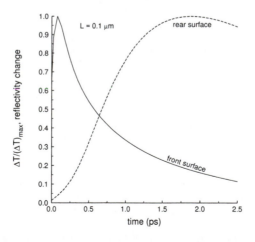

Figure 5.12 Reflectivity change at the rear surface of a gold film of thickness 0.1 µm at $x = L = 0.1$ µm.

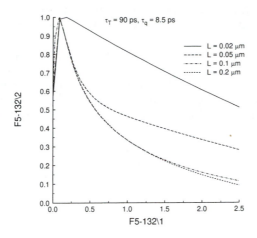

Figure 5.13 Effect of film thickness on the normalized reflectivity change at the front surface. The values of τ_T and τ_q remain constant at different film thicknesses.

rear surface is a *cumulative* result of delay times in the thickness direction. The averaged values of τ_T and τ_q used in describing the reflectivity change in the front surface, in other words, should *not* be the same as those used in the rear surface owing to the additional delay caused by the microstructures in the thickness direction. *Spatial* variations for $\tau_T(x)$ and $\tau_q(x)$ must be modeled to reflect the *nonhomogeneous* microstructures in a more precise description. The size effect on the reflectivity change encounters a similar situation.   The microstructural configuration in a thicker film is evidently different from that in a thinner film, causing different delay times in the cumulative lagging response. At this stage of development, the effect of microstructures on phase lags is still under investigation.

The dual-phase-lag model can be made more rigorous by incorporating the temperature-dependent phase lags ($\tau_T(T)$ and $\tau_q(T)$), thermal conductivity ($k(T)$), and heat capacity ($C_p(T)$) in its framework. In this case, the correlation to the two-step model with temperature-dependent heat capacity ($C_e(T_e)$) and thermal conductivity ($k(T_e)$) becomes extremely complicated. It can be shown that no correlation in algebraic form, such as equations (5.18) and (5.20), is possible, and the phase lags and equivalent heat capacity in the correlation, instead, are governed by coupled, nonlinear partial differential equations. At this stage in the development, however, our purpose is to describe the lagging behavior in the *simplest* possible situation. The *averaged* values of τ_T and τ_q used in Figure 5.5 and thereafter to describe the transient response in experiments reflect this attempt well.

5.8 CONDUCTION IN METAL LATTICE

In modeling heat transport through the metal lattice, equations (5.9) and (5.14b), note that heat conduction in the metal lattice is assumed to be negligibly small. The resulting excellent agreement between the two-step model and the experimental result, as shown in Figure 5.4, supports the validity of this assumption when used in problems involving short-pulse laser heating (small time) on metal films (small space). When the transient time lengthens or the physical domain transporting heat enlarges, however, heat conduction in the metal lattice may gradually come into the picture. To understand the potential influence, a simpler problem is designed in this section to isolate such an effect from the other factors. The thermal properties are still assumed constant, and the heat source term is removed from the energy equation.

Taking into account the effect of heat conduction in the metal lattice, the two-step energy equations become

$$C_e \frac{\partial T_e}{\partial t} = k_e \nabla^2 T_e - G(T_e - T_l) \qquad (5.28a)$$

$$C_l \frac{\partial T_l}{\partial t} = k_l \nabla^2 T_l + G(T_e - T_l), \qquad (5.28b)$$

where a subscript "e" is added to the thermal conductivity of the electron gas to distinguish it from the thermal conductivity of the metal lattice, k_l. Addition of the conduction effect in the metal lattice intrinsically alters the fundamental characteristics of the original two-step equations (5.14a) and (5.14b). This can be illustrated by eliminating the electron-gas temperature, T_e, from equations (5.28a) and (5.28b). From equation (5.28b), temperature of the electron gas can be expressed as

$$T_e = \left(\frac{C_l}{G}\right) \frac{\partial T_l}{\partial t} - \left(\frac{k_l}{G}\right) \nabla^2 T_l + T_l. \qquad (5.29)$$

Substituting equation (5.29) into (5.28a) and using the result of $G(T_l - T_e)$ from equation (5.28b) results in

$$-\left[\frac{k_e k_l}{G(k_e + k_l)}\right] \nabla^4 T_l + \nabla^2 T_l + \left[\frac{C_e k_l + C_l k_e}{G(k_e + k_l)}\right] \frac{\partial}{\partial t} \nabla^2 T_l =$$

$$\left[\frac{C_e + C_l}{k_e + k_l}\right] \frac{\partial T_l}{\partial t} + \left[\frac{C_e C_l}{G(k_e + k_l)}\right] \frac{\partial^2 T_l}{\partial t^2}. \qquad (5.30)$$

Compared to equation (5.17b), the original two-step equation without taking into account the conduction effect in the metal lattice, a *biharmonic* term is present that results from the $\nabla^2 T_e$ term in equation (5.28a) in the use of equation (5.29). In a two-dimensional situation, $\nabla^4 = \nabla^2 \nabla^2 = \partial^4/\partial x_1{}^4 + 2(\partial^4/\partial x_1{}^2 \partial x_2{}^2) + \partial^4/\partial x_2{}^4$. It appears as the *highest* order differential in equation (5.30), and hence dominates the fundamental characteristics of the solution.

The coefficient in front of the mixed-derivative term, the second bracketed quantities in equation (5.30), has a dimension of *time*, say τ_T:

$$\tau_T = \frac{C_e k_l + C_l k_e}{G(k_e + k_l)} .$$ (5.31a)

The coefficient in front of the diffusion term, the third bracketed quantities in equation (5.30), is proportional to the reciprocal of an equivalent thermal diffusivity (α):

$$\frac{1}{\alpha} = \frac{C_e + C_l}{k_e + k_l} .$$ (5.31b)

The coefficient in front of the wave term, the last bracketed quantities in equation (5.30), is thus the ratio of another characteristic time, say τ_q, to the equivalent thermal diffusivity:

$$\frac{\tau_q}{\alpha} = \frac{C_e C_l}{G(k_e + k_l)} \quad \text{with} \quad \tau_q = \frac{1}{G}\left[\frac{1}{C_l} + \frac{1}{C_e}\right]^{-1} .$$ (5.31c)

Comparing the combined results of equations (5.18) and (5.20), it becomes evident that addition of the conduction effect in the metal lattice only changes the definition of α and τ_T. The definition for τ_q remains the same. In terms of the equivalent thermal diffusivity α and the two characteristic times τ_T and τ_q, equation (5.30) can be written as

$$-\left(\alpha_p \tau_T\right)\nabla^4 T + \nabla^2 T + \tau_T \frac{\partial}{\partial t}\nabla^2 T = \frac{1}{\alpha}\frac{\partial T}{\partial t} + \frac{\tau_q}{\alpha}\frac{\partial^2 T}{\partial t^2}$$ (5.32)

with α_p being the thermal diffusivity assuming a parallel assembly between the electron gas and the metal lattice in transporting heat:

$$\frac{1}{\alpha_p} = \frac{1}{\alpha_l} + \frac{1}{\alpha_e} \quad \text{with} \quad \alpha_{e(l)} = \frac{k_{e(l)}}{C_{e(l)}} .$$ (5.33)

The subscript "*l*" has been omitted in equation (5.32) for brevity.

Equation (5.32), again, has a new appearance in heat transfer. In the case of $\alpha_p = 0$, it reduces to equation (5.21), describing the linearized lagging response, with $\alpha \equiv \alpha_F$. To characterize the fundamental solution depicted by equation (5.32), let us consider heat transport in a semi-infinite, one-dimensional medium initially at a stationary state:

$$T(x,0) = T_0, \quad \frac{\partial T}{\partial t}(x,0) = 0 .$$ (5.34)

Equation (5.32) reduces to the following form in this case:

$$-\left(\alpha_p \tau_T\right)\frac{\partial^4 T}{\partial x^4} + \tau_T \frac{\partial^3 T}{\partial x^2 \partial t} + \frac{\partial^2 T}{\partial x^2} = \frac{1}{\alpha}\frac{\partial T}{\partial t} + \frac{\tau_q}{\alpha}\frac{\partial^2 T}{\partial t^2}. \tag{5.35}$$

At the boundary of $x = 0$, the temperature suddenly raises to a constant value T_w,

$$T(0,t) = T_w h(t), \tag{5.36}$$

which produces a thermal disturbance propagating through the medium.   At a distance far away from the heated section, the thermal disturbance vanishes:

$$T(x,t) \rightarrow T_0 \quad \text{as} \quad x \rightarrow \infty. \tag{5.37}$$

Owing to the presence of the fourth-order derivative in space in equation (5.35), an additional boundary condition is needed at $x = 0$.   Assuming that both the electron gas and the metal lattice are heated to T_w, $T_e = T_l = T_w$, at the boundary of $x = 0$, equation (5.28b) gives

$$C_l \frac{\partial T_l}{\partial t} = k_l \nabla^2 T_l, \quad \text{implying} \quad C_l \frac{\partial T}{\partial t} = k_l \frac{\partial^2 T}{\partial x^2} \quad \text{at} \quad x = 0. \tag{5.38}$$

Formulation of the initial boundary problem is now furnished by equation (5.35) subject to the boundary conditions, equations (5.36) to (5.38), and the initial conditions, equation (5.34).   Introducing the following dimensionless variables,

$$\theta = \frac{T - T_0}{T_0}, \quad \beta = \frac{t}{\tau_q}, \quad \delta = \frac{x}{\sqrt{\alpha \tau_q}}, \tag{5.39}$$

equations (5.34) to (5.38) become

$$-A_e A_l \frac{\partial^4 \theta}{\partial \delta^4} + B \frac{\partial^3 \theta}{\partial \delta^2 \partial \beta} + \frac{\partial^2 \theta}{\partial \delta^2} = \frac{\partial \theta}{\partial \beta} + \frac{\partial^2 \theta}{\partial \beta^2} \tag{5.40}$$

$$\theta(\delta,0) = \theta_w, \quad \frac{\partial \theta}{\partial \beta}(\delta,0) = 0 \tag{5.41}$$

$$\theta(0,\beta) = \theta_w h(\beta), \quad \frac{\partial \theta}{\partial \beta}(0,\beta) = A_l \frac{\partial^2 \theta}{\partial \delta^2} \tag{5.42a}$$

$$\theta \rightarrow 0 \quad \text{as} \quad \delta \rightarrow \infty. \tag{5.42b}$$

where

$$A_l = \frac{\alpha_l}{\alpha}, \quad A_e = \frac{\alpha_e}{\alpha}, \quad B = \frac{\tau_T}{\tau_q}, \quad \theta_w = \frac{T_w - T_0}{T_0}. \tag{5.43}$$

From equations (5.31b) and (5.33), note that

$$A_l = \frac{1 + \left(\dfrac{C_e}{C_l}\right)}{1 + \left(\dfrac{k_e}{k_l}\right)}, \quad A_e = \frac{1 + \left(\dfrac{C_l}{C_e}\right)}{1 + \left(\dfrac{k_l}{k_e}\right)}, \tag{5.44}$$

in terms of the ratios of heat capacity and thermal conductivity.

Though complicated, the Laplace transform solution of equation (5.40) satisfying the initial and boundary conditions (5.41) to (5.43) is straightforward:

$$\bar{\theta}(\delta; p) = D_1 e^{-F\delta} + D_2 e^{-H\delta} \tag{5.45}$$

with

$$D_1 = \left[\frac{\left(\dfrac{p}{A_l}\right) - H^2}{F^2 - \left(\dfrac{p}{A_l}\right)}\right] D_2, \quad D_2 = \frac{\theta_w}{p}\left[\frac{F^2 - \left(\dfrac{p}{A_l}\right)}{F^2 - H^2}\right], \quad F = \sqrt{\frac{C_1 - \sqrt{C_1^2 - 4C_2}}{2}},$$

$$H = \sqrt{\frac{C_1 + \sqrt{C_1^2 - 4C_2}}{2}}, \quad C_1 = \frac{1 + Bp}{A_l A_e}, \quad C_2 = \frac{p(p+1)}{A_l A_e}. \tag{5.46}$$

Equation (5.45), as usual, is implemented into the function subroutine, FUNC(P), in the Appendix to obtain the inverse solution in the physical domain of time. The results are shown in Figures 5.14 to 5.16 to illustrate the effect of A_e, A_l, and B on the transient response of temperature.

When the value of A_e increases, as shown in Figure 5.14 for $A_l = 0.5$ and $B = 100$, the temperature level *decreases* in the physical domain of heat transport, implying an enhancement in transporting heat due to additional heat transport through the metal lattice. Table 5.1 and equation (5.44) are used for the estimate of the threshold values of A_e, A_l, and B. At room temperature, heat capacity of the electron gas is approximately 2 orders of magnitude smaller than that of the metal lattice, implying that $C_l/C_e \cong O(10^2)$. Heat capacity of the electron gas does increase with temperature in the transient stage, reducing the ratio of C_l/C_e to several tens. Thermal conductivity of the electron gas, on the other hand, is of the same order of

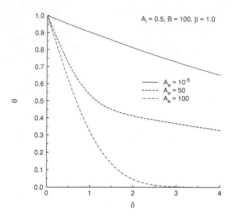

Figure 5.14 Effect of A_e (α_e/α) on the transient temperature distribution as $\beta = 1$.

magnitude as that of the metal lattice, implying that $k_l/k_e \cong O(10^0)$. The threshold values of A_l and A_e defined in equation (5.44), therefore, range from $O(10^1)$ to $O(10^2)$ (for A_e) and $O(10^{-1})$ (for A_l). The value of B (τ_T /τ_q) for gold, referring to Table 5.1, is $O(10^2)$. Temperature-dependent properties, referring to Figure 5.5, however, may reduce the ratio of B to $O(10^1)$. In studying the effect of heat conduction in the metal lattice, therefore, the threshold values of A_e, A_l, and B are selected in these domains. Although equation (5.40) reduces to the energy equation employing the dual-phase-lag model as $A_l = 0$ or $A_e = 0$, in addition, the solution represented by equations (5.45) and (5.46) does *not* retrieve the solution employing the dual-phase-lag model due to the *additional* boundary condition, equation (5.38) or the second equation in equation (5.42a), that has been *intrinsically* absorbed in the solution of (5.45). In retrieving the dual-phase-lag model neglecting the effect of heat conduction through the metal lattice, one has not only to eliminate the fourth-order derivative from equation (5.40), but also to remove equation (5.38) (and hence the second equation in (5.42a)) from the boundary conditions to yield the solution for the new system. The dual-phase-lag model, in other words, presents a *degenerated* system to equation (5.40).

    The effect of A_l on the transient temperature distribution is shown in Figure 5.15. Contrary to the A_e effect, the level of transient temperature *increases* with the value of A_l. Although coexisting with A_e in the solution, referring to C_1 and C_2 in equation (5.46), which implies the same effect on the solution when reducing the value of A_l or A_e, the *sole* contribution of A_l in the boundary condition (5.42a) is mainly responsible for this reversed behavior. It is still true that increasing the value of A_l, like the effect of A_e, improves the heat transport efficiency and hence decreases the field temperature. According to the boundary condition in equation (5.42a), however, increasing the value of A_l also increases the *time-rate* of change of temperature ($\partial\theta/\partial\beta$). The resulting *rate* effect, referring to Section 2.7, promotes the temperature level that overcomes the counterbalancing effect from the A_l effect.

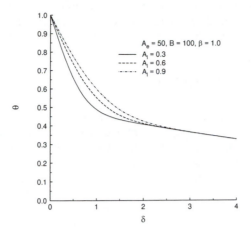

Figure 5.15 Effect of A_l (α_l/α) on the transient temperature distribution as $\beta = 1$.

Figure 5.16 shows the effect of B, the ratio of τ_T to τ_q, at constant values of $A_l = 0.5$ and $A_e = 50$. The temperature level increases with the value of B, a similar behavior to the lagging response shown in Figure 2.8.

In closing, note that the boundary condition (5.38) plays an important role for the special features shown in Figures 5.14 to 5.16. It assumes that the electron gas and the metal lattice achieve an immediate thermal equilibrium right after the boundary temperature is raised to T_w, $T_l = T_e = T_w$ at $x = 0$. This is a completely different situation from that in the short-pulse heating discussed in Sections 5.1 to 5.7. In the case of short-pulse laser heating, a heat source is imposed on the metal film and the electron gas and the metal lattice *develop* their temperatures according to the phonon-electron dynamics. In the boundary condition (5.38), on the other

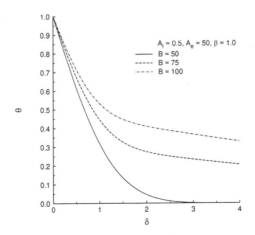

Figure 5.16 Effect of B (τ_T/τ_q) on the transient temperature distribution as $\beta = 1$.

hand, we *force* the electron gas and the metal lattice to have the same temperature at the boundary. It is indeed an idealized treatment, but it provides the simplest possible example illustrating the effect of heat conduction through the metal lattice.

5.9 MULTIPLE-LAYERED FILMS

A higher reflectivity change in a metal film refers to a more exaggerated temperature change in the electron gas. Through the phonon-electron interaction in microscale, in turn, it results in the potential buildup of lattice temperature at a later time. For applications to thermal processing of materials, therefore, the way to reduce the reflectivity change in the surface layer, and hence reduce the lattice temperature, becomes a major concern in avoiding the thermal damage caused by high-power lasers. A typical example is the use of gold film as a coating layer on the workpiece in laser processing owing to the superior reflectivity of gold, about 97%. Although the pulse duration is short, the high power delivered onto the gold film through a small area is sufficient to damage the coating layer at early times, causing total failure of thermal processing on the workpiece. According to the analysis by Qiu and Tien (1994), a temperature rise of 120°C above the ambient in a gold film may take place in about 6 ps.

One way to avoid thermal damage is to introduce a padding layer next to the surface layer that directly resists laser impingement (Tzou, 1991f; Qiu and Tien, 1994). The padding layer serves two purposes. It may function as a heat sink, taking advantage of the large heat capacity of the padding material to absorb the thermal energy transmitted through the surface layer. This is the *cooling mode* we are familiar with. From a microscopic point of view, it may also be a type of material with a large value of the phonon-electron coupling factor, taking advantage of the fast energy transfer rate through the phonon-electron interaction in microscale at short times. This provides another type of cooling that we are not so familiar with. In fact, it is a unique contribution from our evolving knowledge in microscale heat transfer.

This section is devoted to quantifying both cooling modes in view of the lagging behavior in metal films. For developing a better acquaintance with the threshold values of phase lags, τ_T and τ_q, for engineering materials, an analysis with physical dimensions is performed.

5.9.1 Mixed Formulation

The single-layered gold film with a thickness of 0.1 μm (L), the same configuration as that in Figure 5.5, is to be replaced by a double-layered film with a padding layer placed behind the gold layer, each of 0.05 μm thickness, as illustrated in Figure 5.17. When the laser energy is transmitted into the double-layered film, the energy absorption rate $S(x, t)$ is described by the same distribution, equation (5.23). Assuming no heat losses from the film surfaces in the short-time response, parallel to the previous treatment in equation (5.25),

$$q_{(1)} = 0 \quad \text{at} \quad x = 0 \quad \text{and} \quad q_{(2)} = 0 \quad \text{at} \quad x = L \tag{5.47}$$

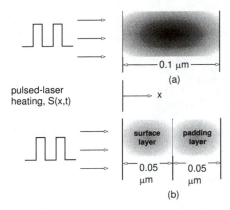

pulsed-laser
heating, S(x,t)

Figure 5.17 A padding layer made of a different material placed behind the surface layer (gold) with an equal thickness.

where the subscript "(1)" refers to the quantities in the surface layer and the subscript "(2)" refers to the quantities in the padding layer in this analysis. The thickness is denoted by L for generality. Continuity of both temperature and heat flux vector across the interface, $x = L/2$, is imposed, implying

$$T_{(1)} = T_{(2)}, \quad q_{(1)} = q_{(2)} \quad \text{at} \quad x = \frac{L}{2}. \tag{5.48}$$

Reflected by the mixed type of boundary conditions containing both heat flux and temperature in equations (5.47) and (5.48), the temperature formulation shown by equation (5.21) is not suitable for the present problem. This is because of the difficult conversion from heat flux to temperature, or vice versa, in the dual-phase-lag model, as discussed in Section 2.4. The mixed formulation shown by equations (2.65a) and (2.65b), instead, is more convenient to use, handling heat flux and temperature in a simultaneous fashion:

The surface layer, $x \in [0, L/2]$ —

$$q_{(1)} + \tau_{q(1)} \frac{\partial q_{(1)}}{\partial t} = -k_{(1)} \left[\frac{\partial T_{(1)}}{\partial x} + \tau_{T(1)} \frac{\partial^2 T_{(1)}}{\partial t \partial x} \right], \tag{5.49a}$$

$$-\frac{\partial q_{(1)}}{\partial x} + S(x,t) = C_{p(1)} \frac{\partial T_{(1)}}{\partial t}. \tag{5.49b}$$

The padding layer, $x \in [L/2, L]$ —

$$q_{(2)} + \tau_{q(2)} \frac{\partial q_{(2)}}{\partial t} = -k_{(2)} \left[\frac{\partial T_{(2)}}{\partial x} + \tau_{T(2)} \frac{\partial^2 T_{(2)}}{\partial t \partial x} \right],$$

(5.50a)

$$-\frac{\partial q_{(2)}}{\partial x} + S(x,t) = C_{p(2)} \frac{\partial T_{(2)}}{\partial t}.$$

(5.50b)

The energy absorption rate, $S(x, t)$ in equations (5.49b) and (5.50b), to repeat, in shown by equation (5.23).

5.9.2 Initial Conditions for Heat Flux

In the presence of volumetric heating, the energy absorption rate in equations (5.49b) and (5.50b), the initial conditions for heat flux, deserves special attention. According to the mixed formulation shown by equations (5.49) and (5.50), both temperature and heat flux need to be specified at $t = 0$. Assuming a disturbance from a stationary state, equation (5.24),

$$T_{(i)}(x,0) = T_0 \quad \text{and} \quad \frac{\partial T_{(i)}}{\partial t}(x,0) = 0 \quad \text{for} \quad i = 1,2,$$

(5.51)

equations (5.49b) and (5.50b) result in

$$q_{(1)}(x,0) = q_{(2)}(x,0) = -\delta S_0 e^{-\frac{x}{\delta} - 2a} \equiv q_0(x).$$

(5.52)

This is the admissible form of initial heat fluxes in both layers that is required by the energy equation. Failing to use this form as the initial heat flux renders a fluctuating solution for temperature across the film due to violation of the conservation of energy.

Equations (5.49) (for the surface layer) and (5.50) (for the padding layer) are to be solved subject to the initial conditions (5.51) and (5.52) and the boundary conditions (5.47) and (5.48).

5.9.3 Laplace Transform Solution

The governing equations (5.49) and (5.50) can be solved by the method of Laplace transform as before. Using the initial conditions (5.51) and (5.52), the temperature and heat flux in the surface and the padding layers are governed by

$$\bar{q}_{(1)} + \tau_{q(1)} \left[p\bar{q}_{(1)} - q_0(x) \right] = -k_{(1)} \left[\frac{d\bar{\theta}_{(1)}}{dx} + p\tau_{T(1)} \frac{d\bar{\theta}_{(1)}}{dx} \right],$$

(5.53a)

$$-\frac{d\bar{q}_{(1)}}{dx} + \bar{S} = pC_{p(1)}\bar{\theta}_{(1)},$$

(5.53b)

$$\bar{q}_{(2)} + \tau_{q(2)}\left[p\bar{q}_{(2)} - q_0(x)\right] = -k_{(2)}\left[\frac{d\bar{\theta}_{(2)}}{dx} + p\tau_{T(2)}\frac{d\bar{\theta}_{(2)}}{dx}\right], \qquad (5.53c)$$

$$-\frac{d\bar{q}_{(2)}}{dx} + \bar{S} = pC_{p(2)}\bar{\theta}_{(2)}, \qquad (5.53d)$$

with $\theta_{(i)} \equiv T_{(i)} - T_0$, for $i = 1, 2$, measuring the temperature rise above the ambient. The Laplace transform of the energy absorption rate is, from equation (5.23),

$$\bar{S}(x; p) = S_0 e^{-\frac{x}{\delta}}S_b, \quad \text{with} \quad S_b = t_p\left[\frac{e^{-2a} - e^{-2pt_p}}{pt_p - a} + \frac{e^{-2pt_p}}{pt_p + a}\right].$$

$$(5.54)$$

The energy equations, equations (5.53a) and (5.53b) for the surface layer and (5.53c) and (5.53d) for the padding layer, have the same form, facilitating the same type of solution. From equations (5.53a) and (5.53c),

$$\bar{q}_{(1)} = U_1\frac{d\bar{\theta}_{(1)}}{dx} + H_1 q_0, \quad \bar{q}_{(2)} = U_2\frac{d\bar{\theta}_{(2)}}{dx} + H_2 q_0, \qquad (5.55)$$

with q_0 being a spatial function defined in equation (5.52) and

$$U_i = -\frac{k_{(i)}\left[1 + p\tau_{T(i)}\right]}{1 + p\tau_{q(i)}}, \quad H_i = -\frac{\tau_{q(i)}}{1 + p\tau_{q(i)}} \quad \text{for} \quad i = 1, 2. \qquad (5.56)$$

Substituting equation (5.55) into equations (5.53b) and (5.53d) and solving for the temperature results in

$$\bar{\theta}_{(1)} = A_{11}e^{B_1 x} + A_{12}e^{-B_1 x} + A_{13}e^{-\frac{x}{\delta}} \qquad (5.57a)$$

$$\bar{\theta}_{(2)} = A_{21}e^{B_2 x} + A_{22}e^{-B_2 x} + A_{23}e^{-\frac{x}{\delta}} \qquad (5.57b)$$

where

$$B_i = \sqrt{\frac{p\left(1 + p\tau_{q(i)}\right)}{\alpha_{(i)}\left(1 + p\tau_{T(i)}\right)}} \quad \text{for} \quad i = 1, 2;$$

$$A_{13} = -\frac{S_0(C_{12}e^{-2a} - C_{11}S_b)}{(1/\delta)^2 - B_1^2}, \quad A_{23} = -\frac{S_0(C_{22}e^{-2a} - C_{21}S_b)}{(1/\delta)^2 - B_2^2};$$

$$C_{11} = \frac{1 + p\tau_{q(1)}}{k_{(1)}\left(1 + p\tau_{T(1)}\right)}, \quad C_{12} = \frac{\tau_{q(1)}}{k_{(1)}\left(1 + p\tau_{T(1)}\right)};$$

$$C_{21} = \frac{1 + p\tau_{q(2)}}{k_{(2)}\left(1 + p\tau_{T(2)}\right)}, \quad C_{22} = \frac{\tau_{q(2)}}{k_{(2)}\left(1 + p\tau_{T(2)}\right)}. \qquad (5.58)$$

The four coefficients, A_{11}, A_{12}, A_{21}, and A_{22}, are determined from the four boundary conditions in equations (5.47) and (5.48):

$$\overline{q}_{(1)} = 0 \quad \text{at} \quad x = 0 \quad \text{and} \quad \overline{q}_{(2)} = 0 \quad \text{at} \quad x = L \qquad (5.59a)$$

$$\overline{\theta}_{(1)} = \overline{\theta}_{(2)}, \quad \overline{q}_{(1)} = \overline{q}_{(2)} \quad \text{at} \quad x = \frac{L}{2}. \qquad (5.59b)$$

Though complicated, four algebraic equations solved for four unknowns are straightforward. The results are

$$A_{11} = \frac{B_2 E_2 U_2 \left(V_1 - V_2\right) + E_1 \left[U_2 W_2 - U_1 W_1 + q_0 \left(H_2 - H_1\right)\right]}{B_1 E_1 U_1 \left(e^{B_1 L/2} - e^{-B_1 L/2}\right) - B_2 E_2 U_2 \left(e^{B_1 L/2} + e^{-B_1 L/2}\right)}, \quad A_{12} = A_{11} - \frac{A_{13}}{\delta B_1};$$

$$A_{21} = \frac{2 A_{11} \sinh(B_1 L/2)}{E_1} + \frac{V_1 - V_2}{E_1}, \quad A_{22} = A_{21}e^{2B_2 L} - \frac{A_{23}}{\delta B_2}e^{\left(B_2 - \frac{1}{\delta}\right)L} \qquad (5.60)$$

where

$$E_1 = e^{(B_2 L/2)} + e^{(3B_2 L/2)}, \quad E_2 = e^{(B_2 L/2)} - e^{(3B_2 L/2)}; \qquad (5.61a)$$

$$V_1 = A_{13}\left[e^{-L/(2\delta)} - \frac{e^{-(B_1 L)/2}}{\delta B_1}\right], \quad V_2 = A_{23}\left[e^{-L/(2\delta)} - \frac{e^{-(B_2 L)/2 - (L/\delta)}}{\delta B_2}\right];$$

$$(5.61b)$$

$$W_1 = \frac{A_{13}\left[e^{-(B_1 L)/2} - e^{-L/(2\delta)}\right]}{\delta}, \quad W_2 = \frac{A_{23}\left[e^{(B_2 L)/2 - (L/\delta)} - e^{-L/(2\delta)}\right]}{\delta}.$$

$$(5.61c)$$

With all the coefficients thus determined, equations (5.57a) and (5.57b) are ready for the Laplace inversion in each layer. The Fortran code provided in the Appendix can be used equally well regardless of the complexity in the present solution. Separating the coefficients into multiple stages, in fact, is for this purpose.

5.9.4 Surface Reflectivity

The lower the reflectivity change in the surface layer, the lower the metal lattice temperature would be developed at the later stage. The temperature solution obtained in the surface layer, $T_{(1)}(x, t)$, can thus be used to calculate the reflectivity change at the front surface at $x = 0$, according to $[\theta_{(1)}(0,t)]/[\theta_{(1)}(0,t)]_{max} \equiv \Delta T_{(1)}(0,t)/[\Delta T_{(1)}(0,t)]_{max}$.

Total thickness of the double-layered film remains the same as that in the previous example of a single-layered film, $L = 0.1$ μm. The gold film is the surface layer with thickness reduced by half, 0.05 μm. Because the reflectivity change is due to the temperature change in the electron gas, the values of $k_{(1)} = 315$ W/m K, $\alpha_{(1)} = 1.2 \times 10^{-4}$ m^2/s, $\tau_{T(1)} = 90$ ps, and $\tau_{q(1)} = 8.5$ ps shown in Figure 5.5 are used for the gold film (surface layer). The base properties for the padding layer of an equal thickness (0.05 μm) are taken from chromium at room temperature, $k_{(2)} = 94$ W/m K, $\alpha_{(2)} = 2.8 \times 10^{-5}$ m^2/s, $\tau_{T(2)} = 7.86$ ps, and $\tau_{q(2)} = 0.136$ ps. They are calculated from the phonon-electron coupling factor provided by Qiu and Tien (1994). These properties in the padding layer, however, will be varied to study their effects on the reduction of the reflectivity change at the front surface of the gold film.

Figure 5.18 shows the effect of $\tau_{T(2)}$, phase lag of the temperature gradient describing the microstructural interaction effect in the padding layer, on the reflectivity change at the front surface, $x = 0$. The reflectivity change, and hence the temperature change of the electron gas in the surface (gold) layer, decreases as the value of $\tau_{T(2)}$ increases. A lower temperature change in the electron gas in the surface layer, again, would result in a lower temperature rise in the lattice at the later stage. This is desirable for preventing the surface layer from thermal damage in laser processing of materials. For reducing the potential temperature buildup in the

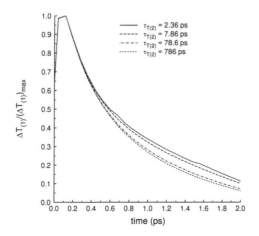

Figure 5.18 Effect of $\tau_{T(2)}$ on the surface reflectivity change ($x = 0$) of the gold film.

surface layer, therefore, it is desirable to select a padding layer with a larger value of $\tau_{T(2)}$. From a microscopic point of view, a combination of equations (5.18) and (5.20) results in

$$\alpha = \frac{k}{C_e + C_l}, \quad \tau_T = \frac{C_l}{G}, \quad \text{and} \quad \tau_q = \frac{1}{G}\left(\frac{1}{C_e} + \frac{1}{C_l}\right)^{-1}. \tag{5.62}$$

A larger value of $\tau_{T(2)}$ is equivalent to the selection of a padding layer with a larger value of heat capacity ($C_{l(2)}$) or a smaller value of the phonon-electron coupling factor (G). The use of a heat sink, a material with a large heat capacity, in microchip cooling falls into this category.

The fast-transient effect of thermal inertia described by the phase lag of the heat flux vector, $\tau_{q(2)}$, displays a reversed trend, as shown in Figure 5.19. All the thermal properties remain, except for the value of $\tau_{T(2)}$ fixed at 7.86 ps and that of $\tau_{q(2)}$ varying from 0.0136 ps to 0.408 ps. The temperature level decreases with the value of $\tau_{q(2)}$, implying that the desirable type of padding layer would have the smallest possible value of $\tau_{q(2)}$ to avoid an excessive temperature buildup in the surface layer at short times. From a microscopic point view, on the contrary, equation (5.8) suggests the use of a padding layer with the largest possible value of G. Because heat capacity of the metal lattice (C_l) is approximately 1 to 2 orders of magnitude larger than that of the electron gas (C_e), the reciprocal of heat capacity of the electron gas ($1/C_e$) dominates the phase lag of the heat flux vector (τ_q) in equation (5.62). The phonon-electron coupling factor (G) dominates the reduction of temperature in the surface layer in this case, which replaces the role of heat capacity of the metal lattice in the previous example. This implicit method of cooling emphasizing the phonon-electron coupling factor in microscale heat transport was first indicated by Qiu and Tien (1994) and Qiu et al. (1994). It can be viewed as a unique contribution of microscale heat transport to cooling technologies.

Keeping the values of τ_T and τ_q constant in the padding layer, Figure 5.20

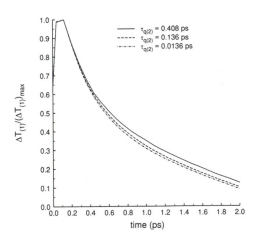

Figure 5.19 Effect of $\tau_{q(2)}$ on the surface reflectivity change ($x = 0$) of the gold film.

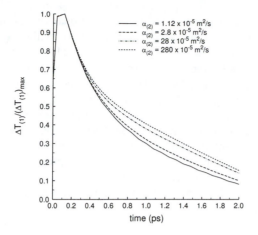

Figure 5.20 Effect of $\alpha_{(2)}$ on the surface reflectivity change ($x = 0$) of the gold film.

shows the effect of equivalent thermal diffusivity ($\alpha_{(2)}$) on the reduction of temperature. The temperature change in the surface layer decreases with the value of α, indicating the use of a padding layer with the smallest possible value of thermal conductivity ($K_{(2)}$) or the largest possible value of heat capacity ($C_{p(2)}$). The effect of thermal conductivity coincides with that for the enhancement of thermal resistance under steady state (Tzou, 1991f).

Note that the purpose of Figures 5.18 to 5.20 is to show the individual effect of τ_T, τ_q, and α in the fast-transient process in small scale. Reduction of temperature shown in these figures does *not* reveal the most pronounced effect of the padding layer owing to the complicated film properties involved in this problem. Using the trend shown in these figures, for example, Figure 5.21 displays the result

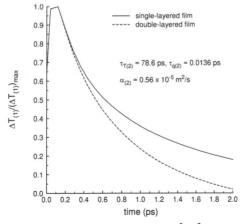

Figure 5.21 Combined effect of $\alpha_{(2)} = 0.56 \times 10^{-5}$ m²/s, $\tau_{T(2)} = 78.6$ ps, and $\tau_{q(2)} = 0.0136$ ps in the padding layer on the reduction of surface reflectivity change at $x = 0$.

by using the values of $\tau_{T(2)} = 78.6$ ps, $\tau_{q(2)} = 0.0136$ ps, and $\alpha_{(2)} = 0.56 \times 10^{-5}$ m^2/s. The effect of temperature reduction due to the additional energy absorption by the padding layer is much more pronounced than that shown in all the previous cases.

Although the temperature change in the surface layer varies with the values of τ_T and τ_q independently, the ratio of (τ_T/τ_q) dominates the transient response. Maintaining the ratio of τ_T/τ_q at 57.8 while the individual values of τ_T and τ_q vary by 2 orders of magnitude, Figure 5.22 shows only a minor change of transient temperature in the surface layer. This has been demonstrated in Section 2.5 from a general treatment in terms of dimensionless analyses. Note, however, this is only a special feature for problems involving temperature-*independent* thermal properties. Should the temperature dependence of thermal properties be taken into consideration, such as heat capacity of the electron gas at the early-time transient, the effect of the padding layer becomes even more pronounced, as shown by Qiu and Tien (1994), and the amount of temperature reduction in the surface layer will strongly depend on the individual values of τ_T and τ_q. The correlation between the microscopic two-step model and the dual-phase-lag model is no longer represented by equation (5.62) in this case.

Also, the effect of the padding layer on the reduction of temperature in the surface layer is studied in terms of the reflectivity change in this section. The reflectivity change, however, is an index for the temperature change of the electron gas rather than that of the metal lattice. Since the values of τ_T and τ_q used in this analysis are determined from the transient temperature of the electron gas, referring to the comparison shown in Figure 5.5, the cooling phenomena observed from Figures 5.18 to 5.22 are indirect in essence. Should a direct approach evaluating the temperature change in the metal lattice be intended, which is more appropriate for damage prevention in thermal processing of materials, the values of τ_T and τ_q must be determined from the transient temperature of the metal lattice. At this point in the development, these experimental data have not yet become available.

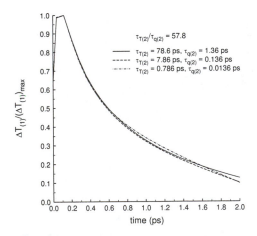

Figure 5.22 Dominance of the ratio of $(\tau_{T(2)}/\tau_{q(2)})$ in the transient response of surface reflectivity change $(x = 0)$.

NONHOMOGENEOUS LAGGING RESPONSE IN POROUS MEDIA

From the experimental results described in Chapters 4 and 5, it is clear that success of the dual-phase-lag model lies in its unique way of describing the microscale interaction effect in *space* by the resulting delayed response in *time*. For heat propagation in superfluid liquid helium, Chapter 4, the time delay between the heat flux vector and the temperature gradient is caused by the finite time required for activating the molecules at a low temperature to an energy level appropriate for conducting heat. For the short-pulse laser heating on metal films discussed in Chapter 5, alternately, the delayed response in the metal lattice is caused by the finite time required for the energy exchange between phonons and electrons. The physical mechanisms are different, but the philosophy of trading the microstructural interaction effect in space with the fast-transient effect in time is the same.

This chapter introduces another type of lagging response caused by the interstitial gas in medium blasting sand. Owing to randomly distributed pores in the medium, a straight path for heat flow to pass the solid phase does not exist. When the heat flow encounters a pore, a major portion circulates around the pore, and heat is transported through the relatively high conducting solid phase. The remainder exchanges thermal energy with the air trapped inside the pore, and heat is transported through the low conducting gaseous phase. These two types of heat conduction need *finite* time to take place. For a material volume enclosing hundreds and thousands of pores, as illustrated in Figure 2.2, the accumulated time delays become pronounced; thus detection of a sensible thermal signal across the material volume occurs at a *later* time. This chapter aims to determine the threshold values of τ_T and τ_q so that the resulting transient curves of temperature resemble the experimental results. This supports the proposed lagging behavior as a necessary condition.

The transient experiment in sand presented in this chapter was performed independently by two research groups in the United States and China: Yun-Sheng

Xu, Ying-Kui Guo, and Zeng-Yuan Guo at the Thermophysics Division, Department of Engineering Mechanics, Tsinghua University, Beijing (China), and James R. Leith at the Department of Mechanical Engineering, University of New Mexico (United States).

6.1 EXPERIMENTAL OBSERVATIONS

The casting sand used in our transient experiments has a particle size ranging from dust to 0.8 mm. The mean particle size is 0.2 mm. The sand was placed in the laboratory for stabilization for several days before the experiment. A medium-sized sand container, 280 mm × 100 mm × 140 mm, was selected to avoid the edge effect. The detailed configuration is illustrated in Figure 6.1, where the plane containing eight thermocouples is the cross section of the median. Thermal pulses were produced by a thin-film electric heater, 250 × 100 × 0.12 mm in size and with 19.5 Ω of electric resistance. It was buried in the midplane of the sandbox and connected to a 220 V, 50 Hz AC power supply. Eight copper/constantan thermocouples were buried underneath the heater at x = 0.4, 1.5 2.1, 3.6, 4.5, 5.7, 8.3, and 10.2 mm to measure the local temperature rises. The thermocouple wires were placed parallel to the electric heater to minimize heat losses. For locking the thermocouples at their fixed positions, two wires in each thermocouple were stretched tight to both sides of the sandbox. For reducing noise, one of the wires in each thermocouple was looped back. The diameter of the thermocouples is 0.1 mm. The typical response time was less than 4 ms. The frequency in data acquisition is 100 Hz. The recorded temperature rises are transmitted to a multiple-channel A/D data transformer, and a personal computer was used to receive the processed results. The thermocouples can accurately determine the starting and ending times of heating, and hence the pulse width. The error in the time measurement is less than 0.01 s, while the error

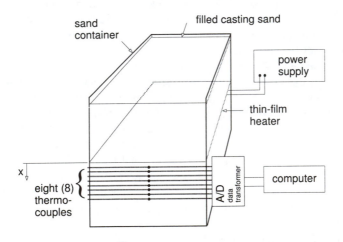

Figure 6.1 The experimental setup for measuring the transient response in casting sand.

threshold in the temperature measurement is within ±0.15°C. The pulse power produced by the heater is maintained at 5.1 W/cm². Two pulse widths, 0.14 s and 0.56 s, are emanated from the heater to disturb the sand specimen.

Figure 6.2 shows the temperature responses recorded at the eight locations (0.4, 1.5, 2.1, 3.6, 4.5, 5.7, 8.3, and 10.2 mm measured from the heater) for a pulse width of 0.56 s. The local temperature fluctuations are very small compared to the total amount of temperature rise. High accuracy achieved in our experiment is thus evident. At $x = 0.4$ mm, the location closest to the heater, the temperature starts to increase right after the pulse is activated. No delayed response is observed at this location. The time domain during which no significant temperature rise occurs increases with the distance away from the heater. Such a delayed response of temperature, however, cannot be viewed as no arrival of the thermal disturbance at these locations. The peak value of transient temperature decreases as the distance away from the heater increases, as expected.

The dark area in the bottom left corner in Figure 6.2 results from the fluctuations of temperature at the beginning and end of the thermal pulse, as shown in Figure 6.3. For the temperature response at $x = 10.2$ mm, the fluctuation starts at $t \cong 0.83$ s and ends at $t \cong 1.39$ s, giving a pulse width of 0.56 s.

6.2 MATHEMATICAL FORMULATION

The classical theories of diffusion and thermal waves are used as ad hoc models to describe the transient process of heat transport. In an approach employing diffusion, the thermal diffusivity and thermal conductivity are the primary quantities characterizing the transient response. In the thermal wave model employing the Cattaneo-Vernotte constitutive equation, in addition, the relaxation time is needed to describe the wave character in heat transport. The transient temperature shown in Figure 6.2 shall be first analyzed by these two popular models. For the casting sand under consideration, the *bulk* values of the thermal diffusivity and thermal

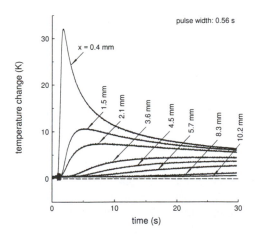

Figure 6.2 Transient temperatures at $x = 0.4$, 1.5, 2.1, 3.6, 4.5, 5.7, 8.3, and 10.2 mm measured from the heater.

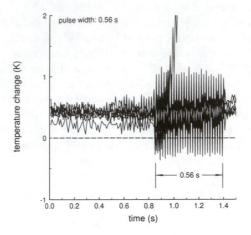

Figure 6.3 Temperature fluctuations start at the beginning $(t \cong 0.83$ s) and end $(t \cong$ 1.39 s) of the thermal pulse. Pulse width $\cong 1.39$ s $-$ 0.83 s = 0.56 s.

conductivity were determined experimentally, $\alpha = 0.3 \times 10^{-6}$ m²/s and $k = 0.29$ W/m K. The relaxation time in the *CV* wave model, on the other hand, shall be determined so that the model results come *closest* to the experimental data.

The situation shown in Figure 6.1 can be simulated as a one-dimensional problem because the thermocouples are buried in the central plane of the sand container, where the boundary effects are negligible. Heat propagation is along the direction of x. Using the *CV* wave model, the constitutive and the energy equations governing the temperature response are equations (2.3) and (2.5):

$$q + \tau \frac{\partial q}{\partial t} = -k \frac{\partial T}{\partial x} \qquad (6.1a)$$

$$-\frac{\partial q}{\partial x} = C_p \frac{\partial T}{\partial t}. \qquad (6.1b)$$

Combining equations (6.1a) and (6.1b) yields

$$\frac{\partial^2 T(x,t)}{\partial x^2} = \frac{1}{\alpha} \frac{\partial T(x,t)}{\partial t} + \frac{\tau}{\alpha} \frac{\partial^2 T(x,t)}{\partial t^2} \qquad (6.2)$$

which is the thermal wave equation with τ/α being $1/C^2$, equation (2.4). In the absence of relaxation behavior in heat propagation, $\tau = 0$, equation (6.2) reduces to the diffusion equation. The *mixed* formulation shown by equation (6.1) is more convenient to use for the present problem because it involves a heat flux emanated by the heater at $x = 0$:

$$q = \begin{cases} q_s & \text{for} \quad 0 \leq t \leq t_s \\ 0 & \text{otherwise} \end{cases} \tag{6.3a}$$

An insulated condition is assumed at the bottom of the sand container:

$$\frac{\partial T}{\partial x} = 0 \quad \text{at} \quad x = l. \tag{6.3b}$$

Along with the boundary conditions shown by equation (6.3), the initial conditions are

$$T(x,0) = T_0, \quad \frac{\partial T}{\partial t}(x,0) = 0, \quad \text{implying} \quad q(x,0) = 0. \tag{6.4}$$

Equation (6.4) reflects no disturbance existing in the sand sample before instrumentation. This is the case in our experiment because the sand sample was set for hours of stabilization between any two tests.

For comparing to the experimental data where the quantities with dimensions are present, equations (6.1), (6.3), and (6.4) are solved in their original forms with dimensions. Applying the Laplace transform to equations (6.1) and (6.3) and using the initial condition (6.4), the transformed solution is obtained:

$$\overline{T}(x;p) = D_1 e^{-Ax} + D_2 e^{Ax} \tag{6.5a}$$

where

$$D_1 = \frac{\overline{q}_b \alpha}{k} \frac{A}{p\left(1 - e^{-2Al}\right)}, \quad D_2 = D_1 e^{-2Al},$$

$$A = \sqrt{\frac{p(1 + p\tau)}{\alpha}}, \quad \overline{q}_b = \frac{q_s\left(1 - e^{-pt_s}\right)}{p}. \tag{6.5b}$$

For $\tau = 0$, equation (6.5) reduces to the transformed solution obtained by diffusion. The inversion solution can be found by the Riemann-sum approximation shown by equation (2.49):

$$T(x,t) = \frac{e^{\gamma t}}{t}\left[\frac{\overline{T}(x,\gamma)}{2} + \mathrm{Re}\sum_{n=1}^{N}(-1)^n \overline{T}\left(x;\gamma + \frac{in\pi}{t}\right)\right]. \tag{6.6}$$

Unlike equation (2.50), the quantity γ in equation (6.6) has a dimension of 1/s due to the use of real time (t) in the present analysis. The appropriate value of γ selected for a faster convergence remains the same, referring to equation (2.50),

$$\gamma\, t \cong 4.7\,. \tag{6.7}$$

The numerical code for the Laplace inversion enclosed in the Appendix can be applied, with the function subroutine FUNC(P) replaced by equation (6.5).

6.3 SHORT-TIME RESPONSES IN THE NEAR FIELD

The transient temperature predicted by diffusion is obtained by setting $\tau = 0$ in equation (6.5). For the case of a pulse width (t_s) of 0.56 s, the results are displayed in Figure 6.4. At the eight locations of $x = 0.4, 1.5, 2.1, 3.6, 4.5, 5.7, 8.3$, and 10.2 mm, the results of diffusion are displayed along with the experimental data. At distances far away from the heater, $x \geq 3.6$ mm (about twenty particles away from the heater), the diffusion model gives satisfactory results. When approaching the heater, especially for the local response at $x = 0.4$ mm, which is about twice the mean particle size in the casting sand, deviations from the experimental results start to increase. This is further illustrated by the penetration depth versus the penetration time response shown in Figure 6.5. At a certain distance measured from the heater, say x_p, the penetration time t_p is defined as the instant at which the temperature rise at x_p reaches a certain fraction (f) of the maximum temperature rise measured at the heater. Mathematically,

$$f = \frac{\Delta T}{(\Delta T)_{\text{heater}}^{\max}}\,. \tag{6.8}$$

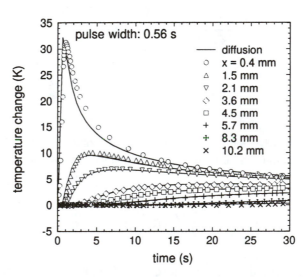

Figure 6.4 The transient response of temperature at the eight locations in the casting sand and comparison with the diffusion model. Here, $t_s = 0.56$ s, $q_s = 5.1$ W/cm^2, $\alpha = 0.3 \times 10^{-6}$ m^2/s, and $k = 0.29$ W/m K.

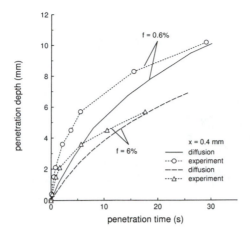

Figure 6.5 Large deviations between the diffusion model and the experimental results in the penetration depth versus penetration time response for values of f = 0.6 and 6 %.

Figure 6.5 shows the response curves at two values of f covering an order of magnitude, f = 0.6 and 6%. Despite the fact that the response curve of diffusion at x = 0.4 mm resembles that of the experimental data, large deviations exist in the penetration depth versus penetration time response. Owing to the local temperature rise occurring in a short period of time, in fact, the penetration depth versus penetration time curve seems to be a reliable index for a more detailed comparison. As shown in Figure 6.4, a slight difference in the response time results in a large difference in the response of temperature owing to the significant time-rate of change of temperature.

The *nonlinear* behavior in the penetration depth and penetration time response shown in Figure 6.5, most important, blemishes the *CV* wave model when it is used for describing the transient behavior in sand. Should a wave behavior be present in the history of heat propagation, the penetration depth and penetration time curve would display a *straight* line, and the slope would simply give the thermal wave speed. This approach was adopted by Brorson, Fujimoto, and Ippen (refer to Chapter 5) in justifying the finite transport velocity in gold films. For any physically admissible value of τ selected for the relaxation time, therefore, it is not likely that the *CV* wave model could capture the temperature response in the entire transient process.

Figure 6.6 shows the results predicted by the thermal wave model with τ = 0.5 s. The exaggerated response on arrival of the sharp wavefront (at t = x/C), unfortunately, results in excessively high peak temperatures that cannot be overcome by any value of τ. A smaller value than 0.5 s for τ results in an even more exaggerated response of peak temperature on arrival of the thermal wavefront, while a larger value than 0.5 s delays the temperature response to a later time than that

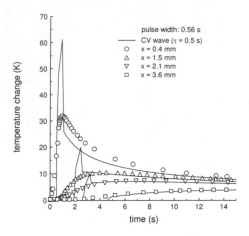

Figure 6.6 The overly exaggerated temperature response on arrival of the thermal wavefront. Comparison of the *CV* wave model with the experimental results.

observed experimentally. Reflected by the large differences shown in Figure 6.6, the *CV* wave model does not seem to describe heat propagation in sand satisfactorily.

No remarkable difference between the model and the experimental results exists for locations distant from the heater, $x \geq 3.6$ mm. The agreements at these locations demonstrate the applicability of Fourier's law in the *far* field and the reliability of the *bulk* values of the thermal diffusivity ($\alpha = 0.3 \times 10^{-6}$ m^2/s) and thermal conductivity ($k = 0.29$ W/m K) determined in our experiment. Note that these values already reflect the effect of porosity due to intrinsic pores in the sand specimen.

Deviations in the near field further increase as the thermal pulse is shortened. For $t_s = 0.14$ s, the transient temperatures predicted by diffusion are displayed in Figure 6.7. A shorter pulse advances the transient response to earlier times, rendering a more significant deviation from the experimental results when coupling with the discrete structures of sand media near the heater. Figure 6.8 compares the results of the *CV* wave model with the experimental data. For the "closest" overall resemblance, the relaxation time has been shortened to 0.1 s. The exaggerated response of temperature on arrival of the thermal wavefront, as in Figure 6.6, blemishes the suitability of using the thermal wave model for describing the transient behavior in sand.

Failure of the macroscopic models like diffusion and *CV* waves, in summary, is due to the absence of modeling the substructural effect in the fast-transient process. Especially when the transient time becomes short, as shown by Figures 6.7 and 6.8, the way in which substructures (such as the interstitial gas in the porous medium) exchange heat with the matrix material becomes extremely important. It renders a slower time-rate of increase of temperature and a lower peak temperature than those predicted by diffusion and *CV* wave models. When used for describing the short-time heat transport in a medium with internal structures, the

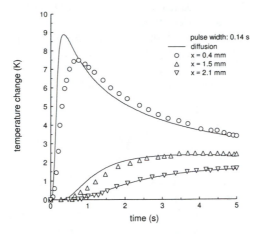

Figure 6.7 The transient response of temperature in the near field of the casting sand and comparison with the diffusion model for $t_s = 0.14$ s.

classical diffusion and *CV* wave models assuming an averaged response on macroscale must be applied with extreme care. They may overestimate or underestimate the thermal response, depending on the substructural interaction.

6.4 TWO-STEP PROCESS OF ENERGY EXCHANGE

Neither diffusion nor *CV* wave models provide sufficient details for the transient response in casting sand. Especially for the near field close to the heater where the sand media become discrete, both models employing a *macroscopic* approach result

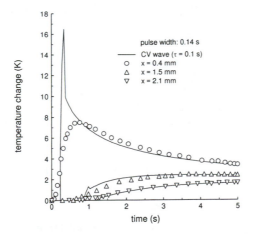

Figure 6.8 Comparison of the *CV* wave model with the experimental results for $t_s = 0.14$ s.

in large deviations. The discrepancies shown in Figures 6.5 to 6.8 reflect the need for modeling the substructural effect on heat transport.

Within the framework of a macroscopic approach, I postulate a two-step process to describe the thermal energy exchange between the solid and the gaseous phases in short times. For the solid phase, the time-rate of change of temperature, $\partial T_s/\partial t$, results from the temperature *difference* from the gaseous phase. Conservation of energy reads

$$C_s \frac{\partial T_s}{\partial t} = G(T_g - T_s),\tag{6.9}$$

where G is the *coupling factor* (in units of W/m^3 K) representing the thermal *power* required for changing the temperature of the solid phase with unit volume by 1 K. The gas temperature, T_g, in equation (6.9) is depicted by the way in which heat is transported through the gaseous phase. As a simplest possible case, let us consider a *diffusion* behavior:

$$C_g \frac{\partial T_g}{\partial t} = \nabla \bullet \left[k_g \nabla T_g\right] + G(T_s - T_g).\tag{6.10}$$

Mathematically, equations (6.9) and (6.10) display two coupled equations for two unknowns, T_s and T_g. They are parallel to the two-step equations in the microscopic phonon-electron interaction model shown in Section 1.2. Assuming constant thermal properties, elimination of the gas temperature (T_g) from equations (6.9) and (6.10) results in

$$\nabla^2 T_s + \frac{\alpha_g}{C_E^2} \frac{\partial}{\partial t}\left[\nabla^2 T_s\right] = \frac{1}{\alpha_E} \frac{\partial T_s}{\partial t} + \frac{1}{C_E^2} \frac{\partial^2 T_s}{\partial t^2}\tag{6.11}$$

where the *equivalent* thermal diffusivity, α_E, and the equivalent thermal wave speed, C_E, are defined as

$$\alpha_E = \frac{K_g}{C_g + C_s}, \quad C_E = \sqrt{\frac{K_g G}{C_g C_s}}.\tag{6.12}$$

Eliminating the temperature of the solid phase, T_s, on the other hand, the following equation is obtained:

$$\nabla^2 T_g + \frac{\alpha_g}{C_E^2} \frac{\partial}{\partial t}\left[\nabla^2 T_g\right] = \frac{1}{\alpha_E} \frac{\partial T_g}{\partial t} + \frac{1}{C_E^2} \frac{\partial^2 T_g}{\partial t^2}.\tag{6.13}$$

Like the situation in the microscopic two-step model for metals, equations (6.11) (for T_s) and (6.13) (for T_g) have an *identical* form. The complete analogy of these equations to those in the phonon-electron interaction model is taken from the parallel formulation made in equations (6.9) and (6.10).

6.5 LAGGING BEHAVIOR

Equation (6.13) derived from the macroscopic two-step process falls within the framework of the dual-phase-lag model, evidenced by the identical form of equation (2.10) without the heat source term:

$$\nabla^2 T + \tau_T \frac{\partial}{\partial t}\left[\nabla^2 T\right] = \frac{1}{\alpha}\frac{\partial T_s}{\partial t} + \frac{\tau_q}{\alpha}\frac{\partial^2 T}{\partial t^2} \tag{6.14}$$

A direct comparison of the coefficients gives

$$\alpha = \alpha_E, \quad \tau_T = \frac{\alpha_g}{C_E^2}, \quad \text{and} \quad \tau_q = \frac{\alpha_E}{C_E^2} \tag{6.15}$$

which are parallel to those given in equations (2.11) and (2.12) for the correlations to the microscopic models. Three material parameters, α, τ_T, and τ_q, involved in the dual-phase-lag model are thus related to the thermal properties of the solid and gaseous phases in the porous medium. The phase lag of the temperature gradient (τ_T) is proportional to the thermal diffusivity of the gaseous phase. This may be the clearest evidence showing that τ_T describes the *substructural* effect of heat transport in the fast-transient process.

Since the phase lags τ_T and τ_q reflect the substructural effect of heat transport in the fast-transient process, they strongly depend on the detailed configuration of the material volume. For distance measurements involving a larger material volume of many sand particles, as shown in Figure 6.9(a), the lagging

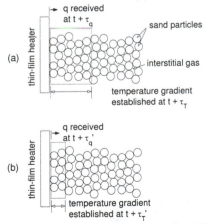

Figure 6.9 Different lagging behavior in (*a*) a larger and (*b*) a smaller material volume with different configurations of substructures.

response is characterized by the *averaged* values of τ_T and τ_q over numerous gaps occupied by the interstitial gas. For near-field measurements involving a smaller material volume containing several sand particles, as illustrated in Figure 6.9(b), the individual gaps play an important role in the energy exchange with the sand particles over short times. The sand medium becomes more discrete in a smaller material volume, resulting in different values of τ_T' and τ_q', reflecting the local response of the lagging behavior.

In passing, note that the heat transport equation (6.11), and hence equation (6.14) in the dual-phase-lag model, contains only the first-order effect for heat transport through the porous medium. Should the conductive term through the solid phase be included in equation (6.9),

$$C_s \frac{\partial T_s}{\partial t} = \nabla \bullet \left[k_s \nabla T_s\right] + G(T_g - T_s),$$ (6.16)

equation (6.14) becomes

$$\alpha\tau_q \nabla^4 T + \nabla^2 T + \tau_T \frac{\partial}{\partial t}\left[\nabla^2 T\right] = \frac{1}{\alpha}\frac{\partial T_s}{\partial t} + \frac{\tau_q}{\alpha}\frac{\partial^2 T}{\partial t^2}$$ (6.17)

where ∇^4 stands for the biharmonic operator, $\nabla^4 = \nabla^2\nabla^2$. The coefficient of $(\alpha\tau_q)$ in front of the additional term is extremely small for engineering materials. For metals, for example, it is of the order of 10^{-19} m^2. Compared to the existing Laplacian term in equation (6.17), only film structures with a thickness of the order of nanometers might show some effects. A detailed discussion of this high-order effect is postponed to Chapter 10.

6.6 NONHOMOGENEOUS PHASE LAGS

The dual-phase-lag model is used to redeem the difference from the experimental results in the neighborhood of the heater. For this purpose, the energy equation is replaced by equation (6.14), while the boundary and initial conditions remain the same, equations (6.3) and (6.4). The transformed solution, $\overline{T}(x;p)$, is given by equation (6.5) except that the coefficient A in equation (6.5b) is replaced by

$$A = \sqrt{\frac{p\left(1 + p\tau_q\right)}{\alpha\left(1 + p\tau_T\right)}}.$$ (6.18)

For $\tau_T = 0$ and $\tau_q = \tau$, equation (6.5b) is retrieved. The inverse solution is still represented by equation (6.6):

$$T(x,t|\tau_T,\tau_q) = \frac{e^{\gamma t}}{t}\left[\frac{\overline{T}(x,\gamma)}{2} + \text{Re}\sum_{n=1}^{N}(-1)^n \overline{T}\left(x;\gamma + \frac{in\pi}{t}\right)\right]_{\tau_T,\tau_q}$$ (6.19)

with the two phase lags, τ_T and τ_q, to be determined by the experimental results. The method of inverse analysis (Beck and Arnold, 1977; Özisik, 1993; Orlande et al., 1996) can be used for this purpose, but a direct method employing the contour pattern of the error estimate shall be used instead.

Let us first consider the case of a shorter pulse (t_s = 0.14 s) with all the system parameters remaining the same, namely, q_s = 5.1 W/cm^2, α = 0.3 × 10^{-6} m^2/s, and k = 0.29 W/m K. Under a pair of arbitrarily chosen values of τ_T and τ_q, equation (6.19) can be used to calculate the local temperature at (x_i, t_i) where the experimental data, denoted by $T_{\exp}(x_i, t_i)$, exist. The local temperature thus obtained is denoted by $T_{\mathrm{DPL}}(x_i, t_i; \tau_T, \tau_q)$. The error threshold defined on the basis of average deviation is then

$$E(\tau_T, \tau_q) = \frac{\sum\limits_{i=1}^{M} |T_{\mathrm{DPL}}(x_i, t_i; \tau_T, \tau_q) - T_{\exp}(x_i, t_i)|}{M}. \tag{6.20}$$

The appropriate values of τ_T and τ_q are those rendering a *minimum* value of E in the state-space of τ_T versus τ_q. For the temperature response closest to the heater, i.e., x_i = 0.4 mm for i = 1, 2, ..., M in equation (6.20) at various times of t_i recorded in the experiment, Figure 6.10 shows the contour pattern of $E(\tau_T, \tau_q)$ in the domain of $\tau_T \in$ [3.0, 5.5] and $\tau_q \in$ [8.0, 10.0]. The value of M is 33, the total number of discretized data points in Figure 6.4. A distinct minimum exists within the contour of E = 0.06. A finer scan results in $E_{\min} \cong 0.054$ at $\tau_T \cong 4.48$ s and $\tau_q \cong 8.94$ s. By using τ_T = 4.48 s and τ_q = 8.94 s, we calculate the transient temperature according to the dual-phase-lag model and compare the results with the experimental data. Figure 6.11 demonstrates the excellent agreement in the entire transient stage.

The discrete structure of the sand medium in the near field of the heater (x = 0.4, 1.5, and 2.1 mm) has definite effects on the thermal diffusivity (α) and

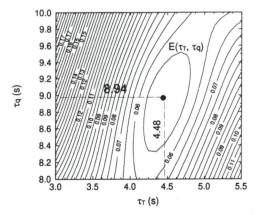

Figure 6.10 Minimum of the error threshold existing at $\tau_T \cong 4.48$ s and $\tau_q \cong 8.94$ s ($\tau_q/\tau_T \cong 2$) for t_s = 0.14 s, q_s = 5.1 W/cm^2, α = 0.3 × 10^{-6} m^2/s, and k = 0.29 W/m K.

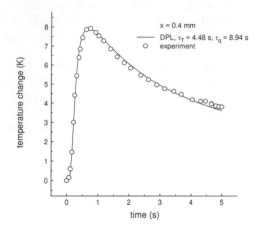

Figure 6.11 Comparison of the dual-phase-lag (DPL) model with $\tau_T = 4.48$ s and τ_q = 8.94 s with the experimental results at $x = 0.4$ mm for $t_s = 0.14$ s, $q_s = 5.1$ W/cm^2, $\alpha = 0.27 \times 10^{-6}$ m^2/s, and $k = 0.29$ W/m K.

conductivity (k). As a general trend, τ_T and τ_q dictate the short-time response, while α and k are more important to the response at longer times. In Figure 6.11 and the similar comparisons ($x = 1.5$ and 2.1 mm) to follow, we slightly adjust the values of α and k to within 10% of their bulk values so that a better fit in the tail region (longer times) of the transient curves results.

    At any given location in the sand bed, the phase lags should be independent of the pulse width. Based on the same values of τ_T (4.48 s) and τ_q (8.94 s) determined from $t_s = 0.14$ s, Figure 6.12 compares the dual-phase-lag model with the experimental results for a longer pulse, $t_s = 0.56$ s. Especially for the short-time transient within 5 s (where the lagging behavior is predominant), the model accurately captures the experimental data.

    As illustrated in Figure 6.9, a location further from the heater involves a different configuration of discrete structures in the subscale. The different configurations induce different delayed times in thermal responses, resulting in different values of the two phase lags τ_T and τ_q in the lagging behavior.

    The same procedure can be used to determine the values of τ_T and τ_q at $x =$ 1.5 mm. The results are shown in Figure 6.13 for the shorter pulse ($t_s = 0.14$ s) and Figure 6.14 for the longer pulse ($t_s = 0.56$ s). The error threshold (E_{min}) is about the same as that in Figure 6.10. The dual-phase-lag model with $\tau_T \cong 1.0$ s and $\tau_q \cong 1.36$ s agrees well with the experimental results. Compared to the previous values of (τ_T, τ_q) = (4.48 s, 8.94 s) at $x = 0.4$ mm, values of both τ_T and τ_q become smaller at a further distance of $x = 1.5$ mm. This implies a *reduction* of the lagging behavior in a larger material volume between the heater and the probed location at $x = 1.5$ mm. Also, note that the ratio of τ_T/τ_q increases from approximately 0.5 (4.48/8.94) at $x =$ 0.4 mm to 0.74 at $x = 1.5$ mm. The value of τ_T is still less than that of τ_q, implying the precedence of the temperature gradient to the heat flux vector in the history of

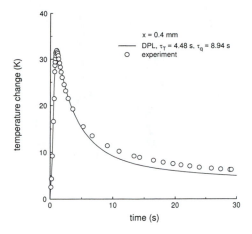

Figure 6.12 Comparison of the dual-phase-lag model (DPL) with $\tau_T = 4.48$ s and τ = 8.94 s with the experimental results at $x = 0.4$ mm for $t_s = 0.56$ s, $q_s = 5.1$ W/cm^2, $\alpha = 0.27 \times 10^{-6}$ m^2/s, and $k = 0.29$ W/m K.

heat propagation, but the amount of precedence gradually *decreases* in the direction away from the heater.

Further away from the heater at $x = 2.1$ mm, the curves of the transient temperature are shown in Figures 6.15 (for $t_s = 0.14$ s) and 6.16 ($t_s = 0.56$ s).

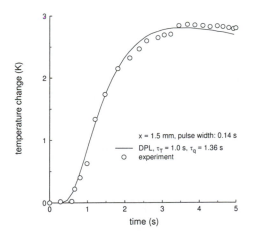

Figure 6.13 Comparison of the dual-phase-lag model (DPL) with $\tau_T = 1.0$ s and $\tau_q =$ 1.36 s with the experimental results at $x = 1.5$ mm for $t_s = 0.14$ s, $q_s = 5.1$ W/cm^2, α = 0.32×10^{-6} m^2/s, and $k = 0.28$ W/m K.

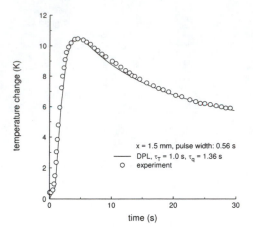

Figure 6.14 Comparison of the dual-phase-lag model (DPL) with $\tau_T = 1.0$ s and $\tau_q = 1.36$ s with the experimental results at $x = 1.5$ mm for $t_s = 0.56$ s, $q_s = 5.1$ W/cm², $\alpha = 0.27 \times 10^{-6}$ m²/s, and $k = 0.25$ W/m K.

Compared to the values of τ_T and τ_q at $x = 1.5$ mm, $(\tau_T, \tau_q) = (1.0$ s, 1.36 s) shown in Figures 6.13 and 6.14, the values of τ_T and τ_q further decrease to $(\tau_T, \tau_q) = (0.4$ s, 0.12 s). Note that at $x = 2.1$ mm, however, the ratio of $\tau_T/\tau_q = 0.4/0.12 \cong 3.3$ becomes *greater* than unity, implying that the heat flux vector *precedes* the temperature gradient in the transient process of heat flow. The transient curves predicted by the dual-phase-lag model with $(\tau_T, \tau_q) = (0.4$ s, 0.12 s), like the situation shown in Figures 6.13 and 6.14, agree well with the experimental results at different pulse widths. This is a feature necessary for all intrinsic thermal properties.

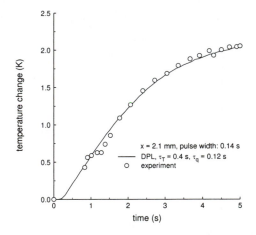

Figure 6.15 Comparison of the dual-phase-lag model (DPL) with $\tau_T = 0.4$ s and $\tau_q = 0.12$ s with the experimental results at $x = 2.1$ mm for $t_s = 0.14$ s, $q_s = 5.1$ W/cm², $\alpha = 0.32 \times 10^{-6}$ m²/s, and $k = 0.28$ W/m K.

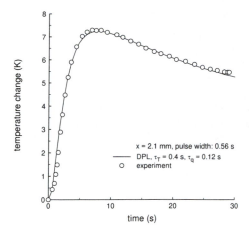

Figure 6.16 Comparison of the dual-phase-lag (DPL) model with $\tau_T = 0.4$ s and $\tau_q = 0.12$ s with the experimental results at $x = 2.1$ mm for $t_s = 0.56$ s, $q_s = 5.1$ W/cm^2, $\alpha = 0.32 \times 10^{-6}$ m^2/s, and $k = 0.28$ W/m K.

6.7 PRECEDENCE SWITCHING IN THE FAST-TRANSIENT PROCESS

The phase lags τ_T and τ_q, to reiterate, characterize the lagging behavior in the transient response. When their values are equal, $\tau_T = \tau_q$, no delay occurs between the temperature gradient and the heat flux vector, and the dual-phase-lag model reduces to diffusion. When $\tau_T < \tau_q$ ($\tau_T/\tau_q < 1$), the temperature gradient precedes the heat flux vector, implying that the temperature gradient is the cause and the heat flux vector is the effect in the transient process of heat flow. When $\tau_T > \tau_q$ ($\tau_T/\tau_q > 1$), the heat flux vector precedes the temperature gradient, implying that the heat flux vector is the cause and the temperature gradient is the effect. The precedence switch in the transient process of heat transport distinguishes the dual-phase-lag model from the others.

The values of τ_T and τ_q, as shown by Figures 6.11 to 6.16, *decrease* from $(\tau_T, \tau_q) \cong (4.48$ s, 8.94 s$)$ at $x = 0.4$ mm to $(1.0$ s, 1.36 s$)$ at $x = 1.5$ mm to $(0.4$ s, 0.12 s$)$ at $x = 2.1$ mm. The ratio between τ_T and τ_q, consequently, varies from 0.5 to 0.74 to 3.33 when departing from the heater. Somewhere between $x = 1.5$ and 2.1 mm, evidently, the ratio of τ_T/τ_q becomes greater than 1.

The exact location where the precedence switch occurs necessitates a more refined measurement including more data points between $x = 1.5$ mm and 2.1 mm. In the absence of such data, the cubic spline technique is used in Figure 6.17 to connect the data points existing at $x = 0.4, 1.5, 2.1$, and 3.6 mm. Because the curves predicted by diffusion fall onto the experimental results from $x = 3.6$ mm and thereafter, referring to Figure 6.4, the values of τ_T and τ_q are zero at $x = 3.6$ mm. According to the results using cubic splines, the curves of τ_T and τ_q intersect at $x \cong 1.68$ mm. For the material points between the heater and 1.78 mm, $\tau_T < \tau_q$ and the

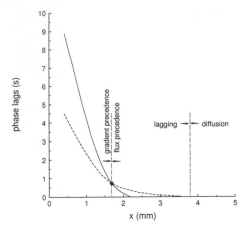

Figure 6.17   The precedence switch between the heat flux vector and the temperature gradient at $x \cong 1.78$ mm showing retrieval of the diffusion behavior after $x \geq 3.77$ mm.

temperature gradient (cause) precedes the heat flux vector (effect) in the lagging response. For the material points between 1.78 mm and 3.77 mm, $\tau_T > \tau_q$ and the heat flux vector (cause) precedes the temperature gradient (effect). From $x = 3.77$ mm and thereafter, both τ_T and τ_q reduce to zero, and the transient response is basically a diffusion phenomenon.

Moreover, Figure 6.17 shows that the phase lag of the heat flux vector, τ_q, decays at a *faster* rate than the phase lag of the temperature gradient, τ_T. For the material points located between $x \cong 2.2$ mm and 3.77 mm, clearly, the fast-transient effect already diminishes ($\tau_q = 0$), while the microstructural interaction effect on heat transport remains ($\tau_T \neq 0$). In this physical domain, therefore, the heat transport is governed by equation (6.14) with $\tau_q = 0$:

$$\frac{\partial^2 T}{\partial x^2} + \tau_T \frac{\partial^3 T}{\partial x^2 \partial t} = \frac{1}{\alpha} \frac{\partial T}{\partial t}. \tag{6.21}$$

The wave term containing the second-order derivative in time vanishes, implying the need for only one initial condition in contrast to two conditions shown in equation (6.4). Equation (6.21) also falls within the framework of the dual-phase-lag model. For the same parameters, namely, $t_s = 0.56$ s, $q_s = 5.1$ W/cm^2, $k = 0.29$ W/m K, and $\alpha = 0.3 \times 10^{-6}$ m^2/s, Figure 6.18 illustrates the effect of τ_T on the solution of equation (6.21) at $x = 3$ mm. The result of diffusion with $\tau_T = 0$ is represented by the solid line. When the value of τ_T increases, the transient curve seems to swivel around a point at $t \cong 5$ s. The temperature level for $t \in [0, 5]$ s is higher than that of diffusion and increases with the value of τ_T. These trends are reversed for the rest of the times in the transient stage. The minor difference from diffusion shown in Figure 6.18 results from the zero value of τ_q and slight perturbations imposed on τ_T from zero to

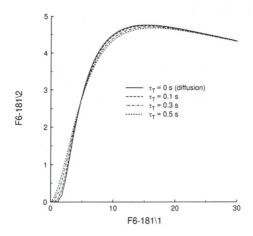

Figure 6.18 Effect of phase lag of the temperature gradient (τ_T) on the transient response represented by equation (6.21).

0.5. These values are taken from Figure 6.17 in the domain of $x \in [1.5, 3.6]$ mm. In other applications where the value of τ_T may significantly deviate from zero, larger deviations from diffusion are expected while the general features remain the same.

For transient heat transport in sand, as demonstrated in Figures 6.11 to 6.16, the dual-phase-lag model yields satisfactory results compared to the experimental data. Physically, the lagging response in this type of medium is caused by the finite time required for heat to circulate around the internal pores. The resulting phase lags strongly depend on the detailed configuration of the interstitial gas among sand particles. This is the main cause for the nonuniform distribution of phase lags in the vicinity of the heater. The lagging response induced by the presence of internal pores in heat transport, unlike the microscopic phonon scattering or phonon-electron interaction, occurs in longer times. For casting sand with a mean particle size of 0.2 mm, the phase lags characterizing the delayed response are of the order of several seconds. Evidenced by Figures 6.11 to 6.16, the dual-phase-lag model accommodating the lagging behavior is indeed capable of describing the transient behavior in discrete structures (the vicinity of the heater in the present study).

Agreement between the model and the experimental results shown in Figures 6.11 to 6.16 is necessary but, at least in this stage of development, is not sufficient. From an analytical point of view, in particular, it is difficult to show that the values of τ_T and τ_q that render transient curves closely resembling the experimental curves are *unique*. The agreement shown in Figures 6.11 to 6.16 should be viewed as a support for the *existence* of such values of τ_T and τ_q which result in transient curves closely resembling the experimental curves in the short-time transient. In spite of the great resemblance, whether these values truly represent the phase lags characterizing the delayed response in sand or not needs further exploration. Two independent tests, as previously discussed in Section 2.11, are necessary to determine the values of τ_T and τ_q in a sufficient and necessary fashion. Based on the values of τ_T and τ_q determined from the independent constant temperature gradient and constant heat flux experiments, a detailed comparison for the temperature in the *entire* transient stage is still needed. The procedure discussed in this chapter aims to fulfill this requirement.

Compared to pure metals, the lagging response in porous media should be easier to produce in the laboratory. If the transient time of primary concern is much longer than the phase lags characterizing the delayed response, however, the lagging behavior may be overcome by the diffusion behavior and cannot be observed in the experiment. The delay times depend on the energy-carrying capacity of the solid phase. Transient heat transport in sand is studied first because sand is a medium easy to obtain. The possible lagging response studied in this chapter can thus be reproduced anywhere at a minimum cost. In fact, the lagging behavior in medium blasting sand with a larger particle size of 2 mm was also studied. While all the lagging behavior remains the same, the ratio of the two phase lags, τ_q/τ_T, increases from the present value of 2 to 6.

The transient experiment for the lagging behavior in reticulated metallic structures is under way. This type of medium is used in the thermal systems of spacecraft. For aluminum-based reticulated structures, some preliminary results have shown that the phase lags range from milliseconds (for coarse phases) to sub-microseconds (for finer phases), depending on the mass fraction of the aluminum tissues. The needed response time reaches the upper end of the traditional thermocouple capability, requiring the use of fast-responding devices such as optical fibers and/or infrared detectors to capture the short-time response.

Another direction of development is the characterization of the transient performance for ceramics. Because ceramics are used in electronic devices such as microchips in computers, a detailed understanding of the way in which thermal energy is dissipated in short times is necessary to prevent the early breakdown in the switch-on operation. The experience we have obtained from the transient experiment on sand directly assists these further activities involving more generic materials.

SEVEN
THERMAL LAGGING IN AMORPHOUS MATERIALS

Amorphous materials are noncrystalline in nature. Typical examples include the roughness layer in carbon samples, random assemblies of metal spheres, silica aerogels, and silicon dioxide. Unlike continuous structures in which heat transport is dominated by the characteristic length on a macroscopic level, heat transport in amorphous materials is dictated by the averaged size of holes and clusters, called correlation length, which spans from mesoscale to microscale (from millimeters to nanometers). The continuum models are not expected to describe the refined mechanisms of transient heat transport in this type of medium, owing to the percolating networks that make the concept of macroscopic average difficult. In contrast to crystalline materials, from a microscopic point of view, amorphous materials do not possess a periodic lattice structure. The microscopic models assuming periodicity of lattices, therefore, are not suitable for use either.

Recent advancement of short-pulse lasers and their applications in material processing and diagnoses reveals the need for a detailed understanding of the way in which thermal energy dissipates through the percolating networks at short times. For an assembly of copper spheres of a mean diameter 100 μm (Fournier and Boccara, 1989), special behavior in transient heat transport, termed *anomalous diffusion*, occurs on the order of milliseconds. For silica dioxide, a popular material used in electronic devices, anomalous diffusion occurs in picoseconds (Goldman et al., 1995). Fracton and percolating theories based on random walkers in fractal geometry have been used to study the time dependence of transient temperature in fractal networks. Spectral density of states characterizing the vibrations of the percolating network has the same form as the Debye density of states for phonon vibration in lattice structures, making a consistent framework describing microscale heat transfer with different mechanisms. The review article by Majumdar (1992) provides a broad basis that summarizes the development of fractal geometry from various aspects of heat transfer.

This chapter extends the dual-phase-lag concept to characterize the fundamental transient behavior of heat transport in amorphous materials. Delayed response (in *time*) between the heat flux vector and the temperature gradient is attributed to the longer distance (in *space*) traveled by the random walker in fractal geometry.

7.1 EXPERIMENTAL OBSERVATIONS

Fournier and Boccara (1989) performed transient experiments for the surface temperature of carbon samples heated by a pulsed laser (15 ns). Their experimental setup is illustrated in Figure 7.1. Two types of carbon sample were studied, one with a polished surface, and one with an untreated rough surface. Thermal response of the sample is reflected by an elliptical mirror to an infrared detector, which is coupled to a digital oscilloscope to average the signal. On a log-log plot, the surface temperature varying as a function of time is displayed Figure 7.2. The polished sample results in a straight line with a constant slope of -1/2, an expected behavior from classical diffusion (see Section 7.2). The sample with a rough surface, on the other hand, results in a flatter curve, with a slope of -1/3 at short times, which merges onto the straight line for diffusion as the transient time lengthens ($t > 1$ ms, approximately). Physically, a smaller slope (-1/3) implies slowing down of heat transport compared to that of diffusion.

The anomalous behavior at short times, $0 < t < 1$ μs for the rough carbon sample shown in Figure 7.2, was definitely not due to the roughness effects in the equivalent layer (Fournier and Boccara, 1989) near the surface. Rather, the anomalous diffusion was attributed to the fractal behavior of heat transport through the rough surface, which was characterized by the fractal and fracton (spectral) dimensions on the rough carbon sample.

An assembly of slightly bonded copper spheres, with an averaged diameter of 100 μm, displays a similar behavior, as shown in Figure 7.3. The extremely short time response ($t < 7 \times 10^{-4}$ s) seems to have a slope of -1/2, a behavior similar to that of diffusion. The stage of anomalous diffusion follows for 7×10^{-4} s $< t < 0.1$ s, rendering a flatter curve with a slope of approximately -0.2. The classical behavior

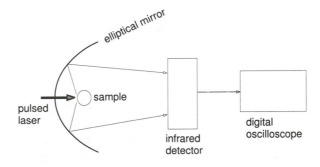

Figure 7.1 Experimental setup for measuring the surface temperature of carbon (polished and rough) samples heated by a laser pulse. Reproduced from Fournier and Boccara (1989).

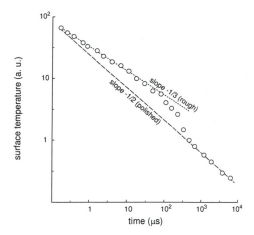

Figure 7.2 Log-log plot of surface temperature (in arbitrary units) versus time for carbon samples. Reproduced from Fournier and Boccara (1989).

dominates the long-time response ($t > 0.1$ s), as expected, where the slope of the response curve being -1/2 is retrieved.

The temperature versus time response reflects the rate of heat transport. This is important information reflecting the energy-bearing capacity of amorphous materials at short times, and is depicted by the characteristic length in heat transport which strongly depends on the internal structure of the conducting media.

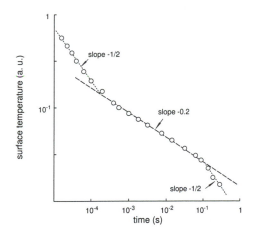

Figure 7.3 Log-log plot of surface temperature (in arbitrary units) versus time for an assembly of slightly bonded copper spheres (100 μm). Reproduced from Fournier and Boccara (1989).

7.2 $t^{1/2}$ BEHAVIOR IN DIFFUSION

A three-dimensional medium heated by an instantaneous surface source, as illustrated in Figure 7.4, is a convenient setup determining the short-time behavior of heat transport in experiments. Should the short-time response be governed by diffusion, the surface temperature will decay with time according to the $t^{-1/2}$ behavior. This is an intrinsic behavior resulting from the characteristic length of diffusion, $\lambda_D = \sqrt{\alpha t}$; see Section 1.5.

In the absence of volumetric heating in diffusion,

$$\frac{\partial^2 T}{\partial x^2} = \frac{1}{\alpha}\frac{\partial T}{\partial t} , \tag{7.1}$$

the eigenmode of diffusive temperature can be constructed as follows:

$$T(x,t) \sim e^{x^2/\lambda_D^2} \sim f(t)e^{Ax^2/\alpha t} , \tag{7.2}$$

where $f(t)$ stands for the time-dependent amplitude and the exponent, $x^2/\alpha t$, results from the ratio of the mean-square displacement traveled by the energy carriers to the square of the characteristic length in diffusion. The constant A and function $f(t)$ are to be determined. Substituting equation (7.2) into equation (7.1) results in two equations for A and $f(t)$:

$$4A^2 + A = 0, \quad \frac{df}{dt} - \left(\frac{2A}{t}\right)f = 0 . \tag{7.3}$$

The solution of $A = 0$ yields a trivial solution because it implies a uniform temperature throughout the conductor. The remaining solution for A is thus $A = -1/4$. With this value of A, the function $f(t)$ is obtained from equation (7.3) by a direct integration:

$$f(t) \sim t^{-1/2}, \quad \text{resulting in} \quad T(x,t) \sim t^{-1/2}e^{-x^2/4\alpha t} . \tag{7.4}$$

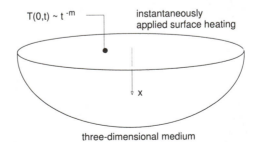

Figure 7.4 A three-dimensional medium heated by a surface source that is suddenly applied and removed.

At the heated surface, $x = 0$, the characteristic behavior of time in diffusion, $T \sim t^{-1/2}$, is thus derived without considering a special type of initial or boundary condition. This behavior can thus be viewed as *intrinsic* for heat transport in uniform media (Euclidean geometry).

In the absence of edge effects, note that equation (7.1) may represent heat diffusion in a one-dimensional conductor ($d = 1$) subjected to a point heat source ($d_s = 0$), a two-dimensional conductor (planar, $d = 2$) subjected to a line source ($d_s = 1$), or a three-dimensional medium ($d = 3$) subjected to a planar source ($d_s = 2$). In two- and three-dimensional conductors, the space variable x represents the distance measured from the heat source. Dimensionality of the heat source (d_s) is smaller than the dimensionality of the conducting medium (d) by 1, resulting in a general behavior of $t^{-(d-d_s)/2} \sim t^{-1/2}$ in Euclidean (uniform) geometry.

7.3 FRACTAL BEHAVIOR IN SPACE

Amorphous (noncrystalline) materials are noncontinuous in nature. They have a randomly connected solid phase, forming a correlation length characterizing heat transport through the percolating network. For physical domains larger than the correlation length, heat transfer occurs in Euclidean geometry and appears to be homogeneous. For physical domains smaller than the correlation length, anomalous behavior may be present, and fractal geometry has been used to describe the way in which random walkers (energy carriers) travel through the percolating structure. Typical examples, with emphasis on the transient response in fractal networks, include the works by Alexander and Orbach (1982), Gefen et al. (1983), Alexander et al. (1983), Zallen (1983), Fournier and Boccara (1989), Jagannathan et al. (1989), de Oliveira et al. (1989), Havlin and Bunde (1991), Bernasconi et al. (1992), and Goldman et al. (1995).

The concept of heated mass has been used in characterizing the transient behavior in time for amorphous structures. According to the definition of specific heat, temperature is inversely proportional to the heated mass (M), $T \sim 1/M \sim 1/\rho V$, where ρ stands for mass density and V the representative volume. The representative volume is proportional to the mean distance ($R(t)$), traveled by the random walker (the energy carrier), $V \sim R(t)$.

For heat transport in a three-dimensional ($d = 3$) Euclidean (uniform) medium heated by a planar source ($d_s = 2$), the mass density remains constant, $\rho =$ constant, while the mean distance is proportional to the square root of travel time, $R(t) \sim \sqrt{t}$. The temperature versus time relation is thus

$$T \sim \frac{1}{\rho V}\bigg|_{\rho=\text{constant}} \sim \frac{1}{V} \sim \frac{1}{R(t)} \sim \frac{1}{\sqrt{t}} \sim t^{-\frac{1}{2}} \quad \text{(Euclidean geometry)}. \quad (7.5)$$

For heat transport in three-dimensional fractal geometry with fractal and fracton (spectral) dimensions being D_f and D_n, respectively, the mean distance traveled by the random walker and the mass density in percolating networks are (Zallen, 1983; Fournier and Boccara, 1989; Goldman et al., 1995)

$$R(t) \sim t^{(D_n/2D_f)}, \quad \rho \sim R^{(D_f-3)} \qquad (7.6)$$

The temperature versus time behavior becomes

$$T \sim \frac{1}{\rho V} \sim \frac{1}{R^{(D_f-3)}R} \sim \frac{1}{R^{(D_f-2)}} \sim t^{D_n(2-D_f)/(2D_f)} \quad \text{(fractal geometry)}.$$

$$(7.7)$$

In terms of a single exponent, alternatively,

$$T \sim t^{-m}, \quad m = \frac{D_n}{2} - \frac{D_n}{D_f} \quad \text{(fractal geometry)}. \qquad (7.8)$$

The rate of heat transport, reflected by the exponent m, thus depends on both fractal and fracton dimensions in amorphous materials. The value of m is in general smaller than 1/2 (0.5), the value for heat transport in Euclidean geometry, implying a slower rate of heat transport in fractal networks. For example, the value of m is 0.154 in percolating networks, 0.33 for carbon samples with a rough surface (in sub-milliseconds, Figure 7.2), 0.2 for an assembly of weakly bonded copper spheres of mean diameter 100 μm (sub-milliseconds, Figure 7.3), 0.06 for silica aerogels (nanoseconds to picoseconds), and 0.13 for silicon dioxide (picoseconds, Goodman et al., 1995). These values of m range from zero to 0.5, with anomalous diffusion time domains depending on the correlation length reflecting different micro-structures.

Figure 7.5 shows the log-log plot of surface temperature versus time for silica aerogels predicted on the basis of the fracton model (Goldman et al., 1995). The correlation length is about 20 nm, with the basic particle size being about 5 nm. Anomalous diffusion occurs in the time domain 30 ps $< t <$ 8 ns. The value of m is 0.06, reflecting an extremely slow rate in transporting heat. The long-time response

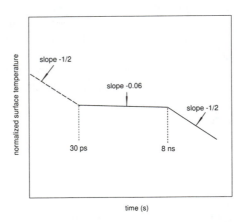

Figure 7.5 Time-dependent surface temperature of silica aerogels predicted by the fracton theory, $D_f = 2.2$ and $D_n = 1.35$. Reproduced from Goldman et al. (1995).

($t > 8$ ns) reflects the well-known behavior of diffusion with $m = 1/2$,[1] but the slope of -1/2 (identical to that of diffusion) at extremely short times is uncertain. The thermal penetration depth at extremely short times is expected to be comparable to the correlation length (20 nm). The slope of log-temperature versus log-time, consequently, is suspected to be different from -1/2 (Goldman et al., 1995).

Figure 7.6 displays the response curve of log-temperature versus log-time for silicon dioxide. It may be produced from a 0.2 ps laser pulse of intensity 6.6×10^{13} W/m². Anomalous diffusion occurs at times when the thermal penetration depth is less than the correlation length. Euclidean diffusion takes place at times greater than about 4.8 ps, where the slope of the response curve reduces to the familiar value, $m = 1/2$. Owing to the slowing down of heat transport, the effect of anomalous diffusion produces a temperature of 562 K (Goldman et al., 1995), about 78% higher than that predicted by the macroscopic diffusion model (316 K).

Thermal diffusivity can be derived from the time derivative of the mean-square distance traveled by random walkers:

$$\alpha = \frac{1}{2} \frac{d\left[R^2(t)\right]}{dt}. \qquad (7.9)$$

For heat transport in Euclidean geometry, $R^2(t) \sim t$. For heat transport in fractal geometry, $R^2(t) \sim t^{(D_n/D_f)}$. The thermal diffusivities in Euclidean and fractal geometry, therefore, are

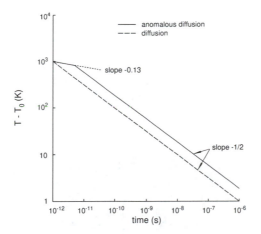

Figure 7.6 Time-dependent surface temperature of silicon dioxide predicted by the fracton theory, $D_f = 2.5$ and $D_n = 4/3$. Reproduced from Goldman et al. (1995).

---

[1] A value of $m = 1/2$ corresponds to the $t^{1/2}$ behavior, according to the definition shown in equation (7.8).

$$\alpha \sim \begin{cases} \text{constant} \quad \text{(Euclidean or uniform media)} \\ \\ t^{(D_n - D_f)/D_f} \quad \text{(fractal geometry)} \end{cases} \tag{7.10}$$

Thermal diffusivity in a fractal network decays with time in the transient process. It is no longer a constant like that in Euclidean geometry. For silica aerogels with $D_f =$ 2.2 and $D_n = 1.35$ (Bernasconi et al., 1992), $\alpha \sim t^{-0.386}$. For silicon dioxide with $D_f =$ 2.5 and $D_n = 1.333$ (Goldman et al., 1995), $\alpha \sim t^{-0.467}$.

7.4 LAGGING BEHAVIOR IN TIME

Longer conducting paths in percolating networks, from an alternate point of view, may result in delayed response in time, as illustrated in Figure 7.7. A fundamental cell, which may be on a mesoscale or microscale, is selected to cover several correlation lengths to reflect the characteristics in heat transport. The heat flux vector arriving at time $(t + \tau_q)$ results in the establishment of a temperature gradient across the same cell at a *later* instant of time $(t + \tau_T)$. The difference between the two phase lags, $\tau_q - \tau_T$, measures the *finite* time required for fractons to transport heat through the fractal network. On the basis of this observation, the dual-phase-lag model is extended to study the slowing down of heat transport in terms of lagging in *time*.

The three-dimensional medium heated by an instantaneous (as $t = 0$) surface source (at $x = 0$), Figure 7.4, is modeled by a one-dimensional problem. Equation (2.5) is reduced to

$$-\frac{\partial q}{\partial t}(x,t) + g_0 \delta(x)\delta(t) = C_p \frac{\partial T}{\partial t}(x,t) \tag{7.11}$$

with x denoting the distance measured from the heated plane at $x = 0$, $\delta(\bullet)$ denoting the Dirac delta function, and g_0 the constant heating intensity per unit area. Heat flux vector and temperature gradient are related by equation (2.7),

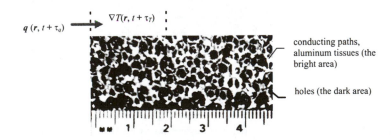

Figure 7.7 Lagging response (in *time*) resulting from heat transport in percolating networks (exemplified by a duocel aluminum foam with a relative mass density of 6%). The scale is in millimeters.

$$q(x,t) + \tau_q \frac{\partial q}{\partial t}(r,t) = -k \frac{\partial T}{\partial x}(x,t) - k\tau_T \frac{\partial^2 T}{\partial x \partial t}(x,t). \qquad (7.12)$$

Equations (7.11) and (7.12) provide a mixed formulation, which is more suitable for the present problem owing to the presence of body heating. The initial and boundary conditions for the temperature and heat flux vector are

$$T(x,0) = T_0, \quad q(x,0) = 0; \quad T(x,t) \to T_0, \quad q(x,t) \to 0 \quad \text{as} \quad x \to \infty. \qquad (7.13)$$

A dimensionless analysis is informative for the dominating parameters in thermal lagging. Introducing

$$\theta = \frac{T - T_0}{T_0}, \quad \eta = \frac{q}{T_0 C_p \sqrt{\alpha / \tau_q}}, \quad \beta = \frac{t}{\tau_q},$$

$$\xi = \frac{x}{\sqrt{\alpha \tau_q}}, \quad G_0 = \frac{g_0}{C_p T_0 \sqrt{\alpha \tau_q}}, \qquad (7.14)$$

equations (7.11) to (7.13) become

$$\eta + \frac{\partial \eta}{\partial \beta} = -\frac{\partial \theta}{\partial \xi} - Z \frac{\partial^2 \theta}{\partial \xi \partial \beta}, \quad Z = \frac{\tau_T}{\tau_q} \qquad (7.15)$$

$$-\frac{\partial \eta}{\partial \xi} + G_0 \delta(\xi)\delta(\beta) = \frac{\partial \theta}{\partial \beta} \qquad (7.16)$$

$$\theta(\xi,0) = 0, \quad \eta(\xi,0) = 0; \quad \theta(\xi,\beta) \to 0, \quad \eta(\xi,\beta) \to 0 \quad \text{as} \quad \xi \to \infty. \qquad (7.17)$$

Although the general lagging response depends on both τ_T and τ_q, the ratio of τ_T to τ_q, $Z = \tau_T/\tau_q$, appears to be the dominating parameter. This seems to be a general feature in thermal lagging, as shown by the examples in previous chapters.

Laplace transform solutions satisfying equations (7.15) to (7.17) are straightforward:

$$\bar{\theta}(\xi; p) = \frac{G_0}{2} \sqrt{\frac{1 + p}{p(1 + Zp)}} e^{-\sqrt{p(1+p)/(1+Zp)} \, \xi}, \qquad (7.18)$$

where p is the Laplace transform variable. At the surface subject to instantaneous heating, $\xi = 0$, the surface temperature in the Laplace transform domain is

$$\bar{\theta}(0; p) = \frac{G_0}{2} \sqrt{\frac{1 + p}{p(1 + Zp)}}. \qquad (7.19)$$

7.4.1 Classical Diffusion, $Z = 1$

In the case of classical diffusion assuming an instantaneous response between the heat flux vector and the temperature gradient, $\tau_T = \tau_q$, implying that $Z = 1$, equations (7.18) reduce to

$$\bar{\theta}_D(\xi; p) = \left(\frac{G_0}{2}\right)\frac{e^{-\sqrt{p}\,\xi}}{\sqrt{p}} \; . \tag{7.20}$$

where the subscript "D" refers to diffusion. The Laplace inversion for equation (7.20) is tabulated:

$$\theta_D(\xi, \beta) = L^{-1}\left[\bar{\theta}_D(\xi; p)\right] = \frac{G_0}{2}\frac{1}{\sqrt{\pi\beta}}\,e^{-\xi^2/4\beta} \; . \tag{7.21}$$

In terms of the physical variables defined in equation (7.14),

$$T_D(x, t) = T_0 + \frac{g_0}{C_p}\frac{e^{-x^2/4\alpha t}}{2\sqrt{\pi\alpha t}} \; . \tag{7.22}$$

Equation (7.22) is the well-known result for heat diffusion in a three-dimensional medium heated by an instantaneous plane source (Carslaw and Jaeger, 1959). At the heated surface at $x = 0$, the temperature is inversely proportional to $\sqrt{t}$, i.e., $T_D(x, 0) \sim 1/\sqrt{t}$, a well-known behavior depicted by equation (7.4) for classical diffusion assuming Fourier's law.

7.4.2 Partial Expansions

In the case of general lagging, $Z \neq 1$, a closed-form solution for the Laplace inversion of equation (7.18) is impossible. The asymptotic behavior of surface temperature at extremely short and long times, however, can be directly extracted from equation (7.19) by the method of partial expansions (Section 2.5). At very long times, $t \to \infty$ in correspondence with $p \to 0$,

$$\lim_{t \to \infty} \theta(0, t) \sim L^{-1}\left[\lim_{p \to 0} \bar{\theta}(0; p)\right] = L^{-1}\left[\frac{G_0}{2}\sqrt{\frac{1}{p}}\right] \sim L^{-1}\left[\frac{1}{\sqrt{p}}\right] = \left(\frac{1}{\sqrt{\pi}}\right)\frac{1}{\sqrt{t}} \; . \tag{7.23}$$

Retrieval of the $t^{1/2}$ behavior as t approaches infinity is thus clear. The long-time response of surface temperature is *independent* of the phase lags. At extremely short times, $t \to 0$ in correspondence with $p \to \infty$; on the other hand,

$$\lim_{t\to 0}\theta(0,t) \sim L^{-1}\left[\lim_{p\to\infty}\overline{\theta}(0;p)\right] = L^{-1}\left[\frac{G_0}{2}\sqrt{\frac{p}{Zp^2}}\right] \sim L^{-1}\left[\frac{1}{\sqrt{Zp}}\right] = \left(\frac{1}{\sqrt{Z\pi}}\right)\frac{1}{\sqrt{t}}.$$

(7.24)

The short-time response is shown to have *the same* $t^{-1/2}$ behavior as that in long-time diffusion, but the surface temperature depends on the ratio of τ_T to τ_q (Z). According to the lagging behavior, in summary, both the short-time and long-time responses of surface temperature are proportional to $1/\sqrt{t}$, a salient feature in the experimental result obtained by Fournier and Boccara (1989) for weakly bonded copper spheres (Figure 7.3).

7.4.3 Riemann-Sum Approximation

The lagging response of surface temperature at intermediate times relies on the Laplace inversion of equation (7.19). Analytically, however, equation (7.19) contains three branch points at $p = 0$, -1, and -1/Z, which result in improper integrals from the Bromwich contour integration (Tzou and Zhang, 1995). Numerical evaluation for the improper integrals is still necessary in the final stage. For this reason, the Riemann-sum approximation for the Laplace inversion is applied to equation (7.19) to obtain the inverse solution. The function subroutine, FUNC(P), in the Appendix is replaced by equation (7.19).

For the case of $Z > 1$, the flux-precedence type of heat flow with $\tau_T > \tau_q$, Figure 7.8 shows the logarithmic plot of surface temperature versus time, $Z = 1$ (corresponding to the case of diffusion, $\tau_T = \tau_q$), 3, 5, 10, and 50. The response curves bear great resemblance to those shown in Figures 7.2, 7.3, 7.5, and 7.6 for different types of fractal networks. The flatter portions at intermediate times on these curves correspond to the

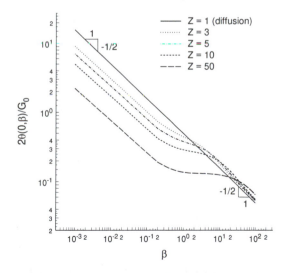

Figure 7.8 Logarithmic representation of surface temperature versus time for flux-precedence type of heat flow in amorphous materials. $Z > 1$.

region of anomalous diffusion. Heat transport rate (the slope of the response curve) in these regions decreases as the value of Z increases. The time duration in which anomalous diffusion lasts, on the other hand, increases with the value of Z. At extremely short times (smaller values of β) and relatively long times (larger values of β), the response curves of surface temperature are straight lines with slopes being *exactly* one-half ($m = 1/2$). The important analytical features obtained from the partial expansion technique, equations (7.23) and (7.24), are thus preserved in the Riemann-sum approximation. The slope in the anomalous region is reduced by approximately half as the value of Z reaches 5. The slope further reduces to almost zero as the value of Z increases to 50. Lagging temperature ($Z \neq 1$) may be higher or lower than the diffusion temperature ($Z = 1$).

At extremely short times, before the anomalous region, the lagging temperature is lower than the diffusion temperature because of the thermalization process. The solid phase in the amorphous material tries to reach thermodynamic equilibrium with the gaseous phase during this period of time. The stage of anomalous diffusion follows where heat transport rate becomes slow, reflected by a smaller slope than one-half ($m < 1/2$). At relatively long times, after the anomalous region, the lagging temperature becomes higher than the diffusion temperature, which is the feature shown in Figures 7.2, 7.3, 7.5, and 7.6 described by fracton and fractal models.

For a more precise comparison to the experimental results (Figures 7.2 and 7.3) and fracton heat transport models (Figures 7.5 and 7.6), the lagging response at special values of Z is displayed individually to describe the corresponding fractal behavior in different timescales. Figure 7.9 displays the response curve of surface temperature versus time extracted from Figure 7.8 for $Z = 10$. The timescale is reduced to $2 < \beta < 10^4$ to describe the similar behavior shown in Figure 7.2 for the rough carbon sample. The dual-phase-lag model, referring to Chapter 2, retrieves the classical diffusion model as transient time approaches infinity ($\beta \to \infty$). As a result, as shown in Figure 7.9, the lagging temperature asymptotically approaches

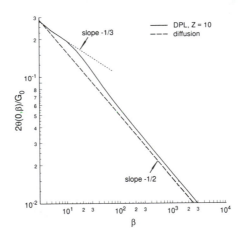

Figure 7.9 Response curve of thermal lagging with $Z = 10$ describing the equivalent fractal behavior shown in Figure 7.2 for the rough carbon sample. DPL, dual-phase-lag model.

the diffusion temperature (calculated by using $Z = 1$ in equation 7.21) after the stage of anomalous diffusion ($m = 1/3$). The diffusion-like response with a slope of $m = 1/2$, however, is achieved shortly after the anomalous region ($\beta \cong 100$). This is the situation shown in Figure 7.2.

Figure 7.10 extracts the response curve of $Z = 5$ in Figure 7.3 for the copper sphere assembly, in the reduced timescale $10^{-2} < \beta < 10^2$. The slope in the anomalous region is $m = 0.18$, close to the experimental value of 0.2 shown in Figure 7.3. The slope $m = 0.5$ prior to the anomalous region coincides with the analytical feature described by equation (7.24), but the temperature level is significantly lower than the diffusion temperature as shown in Figure 7.8.

Figure 7.11 prepares the transient response curve for $Z = 50$ in the time-scale (β) from 10^{-2} to 10^2 to resemble the response curve shown in Figure 7.5 for silica aerogels. From the phase-lag concept, a smaller slope (slower rate of heat transport) in the anomalous region results from a larger ratio of τ_T to τ_q (Z). Should the value of Z further increase, the slope in the anomalous region approaches zero, and the time domain for anomalous diffusion lengthens. It is thus clear that the ratio of τ_T to τ_q, $Z = \tau_T/\tau_q$, is a dominant parameter for anomalous diffusion. The rate of heat transport (slope m) in the anomalous region decreases as the value of Z increases for the present flux-precedence type of heat flow ($Z > 1$).

Figure 7.12 displays the transient response of surface temperature for $Z = 5$ in the timescale $2 < \beta < 10^2$ to resemble the response shown in Figure 7.6 for silicon dioxide. The response of the surface temperature has a slope ($m = 0.13$) close to that shown in Figure 7.6, but the temperature "overshoots" slightly before it retrieves the response of classical diffusion ($m = 1/2$). This behavior is similar to the experimental result shown in Figure 7.2 for the rough carbon sample. After the "overshooting" stage, the response curve of thermal lagging is then parallel to the response curve of classical diffusion in the rest of the process. This is the feature shown in Figure 7.6 for silicon dioxide.

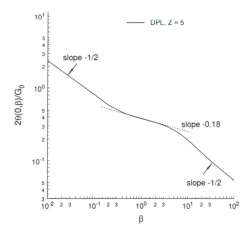

Figure 7.10 Response curve of thermal lagging with $Z = 5$ describing the equivalent fractal behavior shown in Figure 7.3 for the copper-sphere assembly. DPL, dual-phase-lag model.

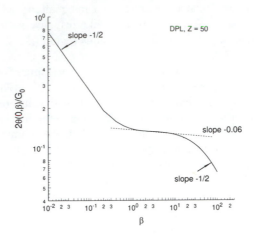

Figure 7.11 Response curve of thermal lagging with $Z = 50$ describing the equivalent fractal behavior shown in Figure 7.5 for silica aerogels. DPL, dual-phase-lag model.

7.4.4 Physical Responses

The dual-phase-lag model, with the flux-precedence type of heat flow ($Z > 1$), seems to preserve well the fractal behavior of heat transport in amorphous materials, evidenced by the similar responses in Figures 7.2 and 7.9 (for rough carbon samples), Figures 7.3 and 7.10 (copper-sphere assembly), Figures 7.5 and 7.11 (silica aerogels), and Figures 7.6 and 7.12 (silicon dioxide).

The two-side $t^{-1/2}$ behavior across the region of anomalous diffusion deserves further discussion: How does the short-time diffusion-*like* behavior transit

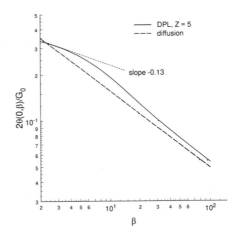

Figure 7.12 Response curve of thermal lagging with $Z = 5$ describing the equivalent fractal behavior shown in Figure 7.6 for silicon dioxide. DPL, dual-phase-lag model.

to the long-time *exact*-diffusion behavior through the region of anomalous diffusion? Figure 7.13 recovers the logarithmic scales used in Figure 7.8 to the physical scales and represents the same response curves for $Z = 1$ (classical diffusion), 3, 5, 10, and 50. The curve of $Z = 1$ can be viewed as the reference along which $\log(T) \sim (-1/2) \log(t)$ $(T \sim t^{-1/2}, m = 1/2)$.   At extremely short times, the lagging temperature is significantly lower than the temperature of diffusion owing to the thermalization process between the solid and the gaseous phases in amorphous materials.[2] The temperature-versus-time curve in this time domain is parallel to that of diffusion, resulting in the same $t^{-1/2}$ type of response. Also, the time duration for the thermalization process shortens as the value of Z (τ_T/τ_q) increases. This behavior becomes more evident for $Z = 10$ and 50.

Anomalous diffusion follows after the initial stage of thermalization. The temperature-time curve flattens, implying a slower time-rate of change of temperature and, consequently, a slower rate of heat transport. The temperature level remains lower than that of diffusion, until the stage of crossover to diffusion. The lagging temperature then exceeds the temperature of diffusion after the crossover, with the $t^{-1/2}$ behavior gradually recovered.

Although the temporal behavior of temperature at extremely short times is the same as that in the classical theory of diffusion, $T(0, t) \sim t^{-1/2}$, it should not be confused with the exact behavior of diffusion.   The lagging (thermalization) temperature at extremely short times, as shown in Figure 7.13, can be many times smaller than the temperature of classical diffusion.

The dual-phase-lag model has been applied to describe the transient behavior of heat transport in amorphous materials. First, it shows that the fractal behavior in *space* may be described in terms of the lagging response in *time*. The feasibility is supported by the great resemblance between Figures 7.2 and 7.9 for the

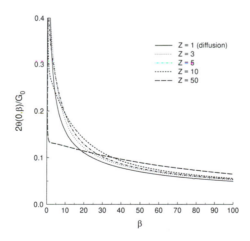

Figure 7.13 Lagging responses of surface temperature versus time in the physical domain.

---

[2] This is a phenomenon consistent with the thermalization temperature in microscopic phonon-electron interactions, see Figures 5.7 and 5.9 in Chapter 5.

rough carbon sample, Figures 7.3 and 7.10 for the copper-sphere assembly, Figures 7.5 and 7.11 for silica aerogels, and Figures 7.6 and 7.12 for silicon dioxide. Second, while the fracton model employing fractal geometry provides useful information for the understanding of anomalous diffusion, the slope m in particular, an energy equation describing the full range of the short-time transient is still absent. Should the intrinsic time behavior in amorphous materials be $T \sim t^{-m}$, in parallel with the derivation of equation (7.4) for the diffusion equation (7.1)? What is the appropriate form of the energy equation that can describe the short-time behavior of temperature in amorphous materials? If this form of the energy equation is known, it sheds light on the full understanding of the fractal behavior in both *space* and time. Supported by the perseverance of several unique features of anomalous diffusion in Figure 7.8 at different values of Z, and hence different values of τ_T and τ_q, the dual-phase-lag model might be a possible choice toward this end.

EIGHT
MATERIAL DEFECTS IN THERMAL PROCESSING

Voids or cracks may be formed in thermal processing of materials owing to thermal expansion. When such defects are initiated in the workpiece, the thermal energy in the neighborhood of the defects may be amplified, resulting in severe material damage and, consequently, total failure of the thermal processing. A detailed understanding of the way in which the local defects dissipate the thermal energy is thus necessary not only to avoid the damage but also to improve the efficiency of thermal processing. This chapter studies the physical mechanisms of energy accumulation around the local defects in both the steady and transient stages. The steady-state analysis quantifies the local damage in terms of the energy concentration or intensification factor. Because no transient effect is involved, the generalized lagging behavior reduces to Fourier's law in heat conduction. The transient analysis emphasizes the transitional behavior of suddenly formed cracks. The effect of thermal inertia and the additional delayed time induced by the microstructural interaction may reduce the degree of damage in some cases. A quasi-stationary analysis will be provided for identifying the governing physical parameters.

8.1 ENERGY ACCUMULATION AROUND DEFECTS

Geometric curvature of the local defect is the main reason for the accumulation of thermal energy. A circular hole in an infinite plate subjected to the impingement of a remote heat flux q_0 may be the simplest example illustrating the physical concept. Without loss of generality, let us consider a circular hole insulated around its circumference as shown in Figure 8.1. The radius of the hole is a and a steady-state response shall be examined first for a better focus on the concept of energy accumulation. The energy equation governing the temperature distribution is

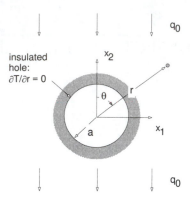

Figure 8.1   The local defect, a circular hole with radius a, subject to the impingement of the heat flux q_0.

$$\frac{\partial^2 T}{\partial r^2} + \frac{1}{r}\frac{\partial T}{\partial r} + \frac{1}{r^2}\frac{\partial^2 T}{\partial \theta^2} = 0 \qquad (8.1)$$

with r and θ being the polar coordinates centered at the hole.   The boundary conditions are

$$k\frac{\partial T}{\partial x_2} = q_0 \quad \text{as} \quad r \to \infty \quad \text{and} \quad \frac{\partial T}{\partial r} = 0 \quad \text{at} \quad r = a . \qquad (8.2)$$

Equation (8.1) displays a differential equation of the equidimensional type in r, allowing for a product solution of the form

$$T(r,\theta) = r^\lambda \cos\theta . \qquad (8.3)$$

The cosine function in equation (8.3) reflects symmetry of the problem with respect to θ. Should the heat flux q_0 at the bottom of Figure 8.1 be downward, the same direction as that at the top, an antisymmetric situation results, and a sine function should be selected instead. In determining the value of λ, a direct substitution of equation (8.3) into (8.1) gives $\lambda = \pm 1$. The temperature solution satisfying equation (8.1) is thus

$$T(r,\theta) = (C_1 r + \frac{C_2}{r})\cos\theta \qquad (8.4)$$

where the coefficients C_1 and C_2 are to be determined from the boundary conditions in equation (8.2). Noticing that the factor $r\cos(\theta)$ in equation (8.4) is indeed x_2 according to the coordinate systems shown in Figure 8.1, direct differentiation and substitutions result:

$$C_1 = \frac{q_0}{k}, \quad C_2 = \left(\frac{q_0}{k}\right)a^2.$$ (8.5)

The temperature distribution and the heat flux vector, from Fourier's law in steady-state, are thus determined:

$$T(r,\theta) = \frac{q_0}{k}\left(r + \frac{a^2}{r}\right)\cos\theta$$ (8.6)

$$q_r = -q_0\left[1 - \left(\frac{a}{r}\right)^2\right]\cos\theta, \quad q_\theta = q_0\left[1 + \left(\frac{a}{r}\right)^2\right]\sin\theta.$$ (8.7)

At the circumference of $r = a$, $q_r = 0$, which is derived from the insulated boundary condition, equation (8.2). When approaching the circumference from the solid side, on the other hand,

$$\lim_{r \to a^+}\left(\frac{q_\theta}{q_0}\right) = 2\sin\theta.$$ (8.8)

The maximum value of q_θ occurs at $\theta = \pi/2$ (the east side) and $3\pi/2$ (the west side), both being downward. The local magnitude of the heat flux vector is *twice* the remotely applied heat flux vector q_0, and for the sake of convenience, the intensity factor of heat flux (IFHF) measuring the local intensity is defined as 2.

A value of IFHF greater than 1 implies local *amplification* of the heat flux vector, and hence a more pronounced accumulation of thermal energy. Multi-dimensional heat fluxes may further increase its value, depending on the directions of impingement. The biaxial fluxes shown in Figure 8.2 are planed for illustrating this effect. Since the problem is linear, the principle of superposition is applicable, implying that

$$q_\theta = q_\theta^{(I)} + q_\theta^{(II)}.$$ (8.9)

While the expression for $q_\theta^{(I)}$ has already been obtained in equation (8.7), the expression for $q_\theta^{(II)}$ results from replacing θ in equation (8.7) by $(\theta + \pi/2)$ owing to the rotational symmetry of the problem. Mathematically,

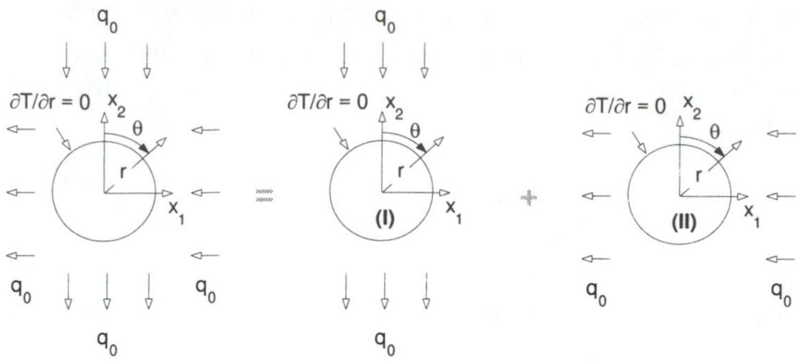

Figure 8.2   The effect of biaxial fluxes on the intensity factor of heat flux, illustrating the principle of superposition.

$$q_\theta^{(II)}(r,\theta) = q_\theta^{(I)}(r,\theta + \pi/2) = q_0\left[1+\left(\frac{a}{r}\right)^2\right]\cos\theta. \qquad (8.10)$$

Superimposing equations (8.7) and (8.10) at $r = a$ then gives

$$\lim_{r\to a^+}\left(\frac{q_\theta}{q_0}\right) = 2(\sin\theta + \cos\theta) \quad \text{for biaxial fluxes.} \qquad (8.11)$$

The locations possessing local maxima shift to $\pi/4$ (the northeast side) and $5\pi/4$ (the southwest side) with the value of IFHF amplified to $2\sqrt{2} \cong 2.8284$. The presence of a local defect, therefore, amplifies the local magnitude of the heat flux vector.

8.1.1 The Crack Damage

The local defect with an abrupt change of geometric curvature significantly increases the IFHF. For a line crack where the geometric curvature dramatically changes from a positive to a negative large value at the crack tip, the value of IFHF approaches *infinity*, and the crack damage in energy transport is more appropriately measured by the *intensity factor of the temperature gradient*, IFTG (Tzou, 1991d to f). The concept of IFTG is illustrated by a central cracked panel shown in Figure 8.3, where the crack with a half-length a is impacted by an incoming heat flux q_0. The polar coordinate system centered at the crack tip, (r, θ) with θ measured positive counterclockwise, is used to describe the temperature gradient in the vicinity of the crack tip. It is equivalent to the heat flux vector in a steady state.

Under a steady state, the energy equation in the near-tip region is given by equation (8.1):

$$\nabla^2 T(r,\theta) = \frac{\partial^2 T}{\partial r^2} + \frac{1}{r}\frac{\partial T}{\partial r} + \frac{1}{r^2}\frac{\partial^2 T}{\partial\theta^2} = 0. \qquad (8.1')$$

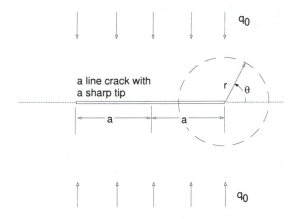

Figure 8.3   A crack defect impacted by a heat flux vector q_0 and the polar coordinates centered at the crack tip measuring the thermal energy intensification.

The equidimensionality in r suggests a product solution

$$T(r,\theta) = r^{\lambda+1} F(\theta) . \tag{8.12}$$

Comparing to equation (8.3) in the previous case of a circular hole, a general function, $F(\theta)$, is assumed here to account for the boundary conditions at the top and bottom surfaces of the crack. The crack surfaces are assumed to be isothermal during the heating process:

$$T = 0 \quad \text{at} \quad \theta = \pm\pi . \tag{8.13}$$

Employing both the ideal and van der Waal's behavior for the gas arrested in the crack closure, equation (8.13) has been discussed quantitatively (Tzou, 1991e). It will be shown shortly that the problem thus formulated renders an eigenvalue problem to be solved for the eigenvalue λ and the eigenfunction $F(\theta)$. The constant temperature maintained at the crack surface is thus immaterial, which has been assumed zero in equation (8.13) without loss of generality.

Substitution of equation (8.12) into (8.1) yields the eigenequation governing the eigenfunction $F(\theta)$:

$$\frac{d^2 F}{d\theta^2} + (\lambda + 1)^2 F = 0 . \tag{8.14}$$

It has a general solution

$$F(\theta) = C_1 \cos[(\lambda + 1)\theta] + C_2 \sin[(\lambda + 1)\theta] . \tag{8.15}$$

The eigenvalue λ is determined from the boundary conditions in equation (8.13). A direct substitution yields

$$\sin[2(\lambda+1)\pi] = 0 \quad \text{or} \quad \lambda_n = \frac{n}{2} - 1 \quad \text{for} \quad n = 0, 1, 2, \ldots \qquad (8.16)$$

The total temperature then results from the superposition of all the fundamental modes:

$$T(r,\theta) = \sum_{n=0}^{\infty} r^{n/2} \left[C_{1n} \cos\left(\frac{n\theta}{2}\right) + C_{2n} \sin\left(\frac{n\theta}{2}\right) \right]. \qquad (8.17)$$

The r component of the temperature gradient, consequently, is

$$\frac{\partial T}{\partial r} = \frac{C_{11}}{2\sqrt{r}} \cos\left(\frac{\theta}{2}\right) + r^0 \left[C_{22} \sin\theta \right]$$

$$+ r\left[\frac{3}{2} C_{13} \cos\left(\frac{3\theta}{2}\right) \right] + o\left(r^{n/2}\right) \quad \text{for} \quad n \geq 3. \qquad (8.18)$$

In the near-tip region with r approaching zero, clearly, the first term in equation (8.18) approaches *infinity*, while the other terms approach either a finite constant (the C_{22} term) or zero. The near-tip behavior, therefore, is dominated by

$$\frac{\partial T}{\partial r} \cong \frac{C_{11}}{2\sqrt{r}} \cos\left(\frac{\theta}{2}\right) \quad \text{as} \quad r \to 0 \qquad (8.19)$$

A $1/\sqrt{r}$ type of *singularity* exists in the r component of the temperature gradient, so does the same component of the heat flux vector due to the Fourier's law in a steady state. The asymptotic (eigenvalue) analysis, to the extent possible, determines the thermal field within an arbitrary constant C_{11}. For the special case shown in Figure 8.3, however, the value of C_{11} may be further attempted by the consideration of dimensional consistency. The quantity of $C_{11}/\sqrt{r}$ in equation (8.19), first, must have a dimension of temperature gradient (K/m), implying that the coefficient C_{11} possesses a dimension of $K/\sqrt{m}$. The coefficient C_{11} appears as the amplitude of the heat flux vector in the near-tip region. It must be a compound function of thermal loading (q_0), geometry (a), and material properties (the thermal conductivity k in a steady state). This observation suggests that

$$[C_{11}] = [q_0]^b [k]^d [a]^f \quad \text{or} \quad \frac{K}{\sqrt{m}} = \left(\frac{W}{m^2}\right)^b \left(\frac{W}{mK}\right)^d (m)^f \qquad (8.20)$$

where the brackets denote the dimensions of the enclosed quantities. A direct comparison of the dimensions gives $b = 1$, $d = -1$ and $f = 1/2$, implying that

$$C_{11} = \frac{q_0 \sqrt{a}}{k}.$$ (8.21)

The near-tip temperature gradient, $\partial T/\partial r$, and the heat flux vector, q_r, are thus

$$\frac{\partial T}{\partial r} \cong \left(\frac{q_0}{2k}\right)\sqrt{\frac{a}{r}}\cos\left(\frac{\theta}{2}\right) \quad \text{and} \quad q_r = -\left(\frac{q_0}{2}\right)\sqrt{\frac{a}{r}}\cos\left(\frac{\theta}{2}\right) \quad as \quad r \to 0.$$ (8.22)

The near-tip temperature has a regular behavior of the $\sqrt{r}$ type:

$$T = \left(\frac{q_0 \sqrt{a}}{k}\right)\sqrt{r}\cos\left(\frac{\theta}{2}\right).$$ (8.23)

The θ component of the heat flux vector is hence

$$q_\theta = -k\left(\frac{1}{r}\frac{\partial T}{\partial \theta}\right) = \left(\frac{q_0}{2}\right)\sqrt{\frac{a}{r}}\sin\left(\frac{\theta}{2}\right) \quad as \quad r \to 0$$ (8.24)

which possesses the same $1/\sqrt{r}$ type of singularity as q_r. Note that determination of the coefficient C_{11} according to dimensional consistency is very limited. It excludes any constant, such as π popularly existing in crack problems. For problems involving multiple-layered structures, in addition, it cannot incorporate the ratio of thermal properties such as k_1/k_2 (the ratio of thermal conductivity of two adjacent media). For problems in these categories, the method of dual integral equations (Tzou, 1985, for example) or the complex function analysis is required to fully determine the thermal field surrounding the crack tip.

From the examples given above, Figures 8.1 and 8.2 containing a circular hole and Figure 8.3 containing a crack with sharp tips, it is informative that local defects tend to cumulate the thermal energy in heat transport. In thermal processing of materials, the energy localization thus formed creates hot spots in the workpiece, initiating thermal damage under extreme conditions. The IFTG is introduced to measure the macrocrack damage (Tzou, 1991d to f). According to equation (8.22),

$$\text{IFTG} = \lim_{r \to 0}\left[2\sqrt{r}\frac{\partial T}{\partial r}\right]_{\max} = \frac{q_0 \sqrt{a}}{k}$$ (8.25)

It is the maximum value occurring at $\theta = 0$ in front of the crack tip. In the near-tip region with r approaching zero, the IFTG measures the *finite* value of infinity ($\partial T/\partial r$) multiplied by zero ($\sqrt{r}$). It represents a combined effect of thermal loading (q_0), crack geometry (a), and thermal conductivity (k) of the medium. A larger value of IFTG implies a more pronounced thermal energy accumulation, and hence more severe crack damage, in the vicinity of the crack tip. This corresponds to the case of a larger value of q_0, a smaller value of k, or a longer crack length (a) in the sample.

For microcracks with small *a*, however, the intensity factor of the temperature gradient does not provide a reliable index for measuring the crack damage. The microcrack density measuring the total number of microcracks per unit volume should be used instead (Tzou, 1991g, h, 1994, 1995d; Tzou and Chen, 1990; Tzou and Li, 1994a, 1995).

In passing, note that either the IFHF (Figures 8.1 and 8.2) or the IFTG (Figure 8.3) is the result of *steady-state* heat transport. Formation of local defects, including voids and cracks, requires a finite period of time to take place in reality. In a fast-transient process such as femtosecond laser heating on metallic films, whether the same type of damage could be developed during the *short*-time response or not requires further studies.

8.2 ENERGY TRANSPORT AROUND A SUDDENLY FORMED CRACK

In the process of thermal treatment, including laser irradiation, local defects such as cracks are created when the local strain energy density induced by the thermal expansion exceeds a critical value. Such a critical value is called the surface energy density, a material constant. For brittle solids such as glass, the value of the surface energy density is 10 J/m^2. For more ductile media such as aluminum, its value is larger, 20 J/m^2. Criteria governing crack initiation/extension is one of the main areas in fracture mechanics. A detailed discussion for the applicability of the existing stress-, strain- and energy-based criteria can be found in the work by Sih and Tzou (1985).

After the local energy level is accumulated to a critical value, a crack is formed and suddenly propagates along the direction of the minimum strain energy density (Sih, 1973; Sih and Tzou, 1985; Tzou and Sih, 1985). A typical situation is illustrated in Figure 8.4. A laser beam impinges upon the front surface of a workpiece, introducing a tensile stress due to the thermal expansion. If the laser

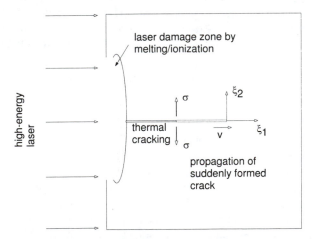

Figure 8.4 Lengthening of a suddenly formed crack during the transient process of laser irradiation.

energy is sufficiently high, the tensile stress may dislocate the metal lattice and open up free surfaces along the weakest links. The grain boundary is a typical example. The crack induced by the excessive stress drives the crack to grow along the grain boundary, until the local stress in the vicinity of the crack tip becomes balanced. The crack formation in material processing thus involves a moving crack, and the way in which thermal energy is dissipated from the propagating crack tip is of major concern.

Energy dissipation in the near-tip region has two characteristics: First, owing to the high crack speed, which can reach terminal velocities from 1000 to 1500 m/s in metals (Bluhm, 1969; Tzou, 1990a, b), the thermal field in the vicinity of the crack tip undergoes a rapid change in *time*. Second, the physical domain may be on a microscale when describing the state of the affairs near the crack tip. Not only the fast-transient (in *time*) but also the small-scale (in *space*) effects are thus necessary for modeling energy dissipation in the near-tip region. In addition, a straight path describing the crack trajectory is a mathematical idealization. For a suddenly formed crack propagating along the grain boundary in microscale, obviously, a *curved* trajectory provides a more realistic simulation. Based on these observations, let us consider a crack propagating along a smooth but curved path in a solid as shown in Figure 8.5. The linear and angular velocities of the crack are denoted by v and ω, respectively. The instantaneous radius of curvature is R. The (ξ_1, ξ_2) coordinates are the moving coordinates convecting with the crack tip, while the (x_1, x_2) coordinates are the fixed coordinates in space. In terms of the polar coordinates (r, θ), the material coordinates ξ_1 and ξ_2 are

$$\xi_1 = r\cos\theta, \quad \xi_2 = r\sin\theta. \tag{8.26}$$

The dual-phase-lag model is used to describe the fast-transient heat transport in the vicinity of the fast-running crack tip:

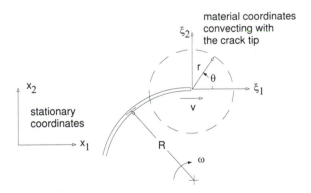

Figure 8.5 A propagating crack along a curved trajectory with a linear velocity v and an angular velocity ω. The moving coordinates (ξ_1, ξ_2) are convecting with the crack tip, and the stationary coordinates (x_1, x_2) are fixed in space.

$$\bar{q} + \tau_q \frac{\partial \bar{q}}{\partial t} = -k\nabla T - k\tau_T \frac{\partial}{\partial t}\nabla T \qquad (8.27a)$$

$$\nabla^2 T + \tau_T \frac{\partial}{\partial t}\left[\nabla^2 T\right] = \frac{1}{\alpha}\frac{\partial T}{\partial t} + \frac{\tau_q}{\alpha}\frac{\partial^2 T}{\partial t^2}. \qquad (8.27b)$$

Its suitability for the present problem is further justified because it is desirable to approach as close as possible to the crack tip, hence entering the microscale, in measuring the crack damage. Should the phonon-electron interaction dominate heat transport in the near-tip region,

$$\alpha = \frac{K}{C_e + C_l}, \quad \tau_T = \frac{C_l}{G}, \quad \tau_q = \frac{1}{G}\left[\frac{1}{C_e} + \frac{1}{C_l}\right]^{-1}. \qquad (8.28)$$

This mode of heat transport occurs at early times in crack formation, on the order from picoseconds to femtoseconds. Should the phonon scattering from mutual collision or grain boundaries dominate, on the other hand,

$$\alpha = \frac{\tau_R c^2}{3}, \quad \tau_T = \frac{9\tau_N}{5}, \quad \tau_q = \tau_R. \qquad (8.29)$$

This mode of heat transport follows the early stage of heat transport by phonon-electron interaction and occurs from nanoseconds to submicroseconds. These correlations were discussed in Chapter 2 in great detail, equations (2.11) and (2.12).

For a crack penetrating into a grain (transgranular cracking) or along the grain boundaries (intergranular cracking), the fundamental characteristics of the near-tip temperature are governed by equation (8.27b). The mixed-derivative term led by τ_T describes the delay time (phase lag) caused by the microstructural interaction such as phonon-electron interaction or phonon scattering. The wave term led by τ_q, with $\tau_q/\alpha = 1/C^2$, C denoting the thermal wave speed, describes the relaxation time due to the fast-transient effect of inertia.

8.3 THERMAL SHOCK FORMATION — FAST-TRANSIENT EFFECT

When the time-rate of change of temperature satisfies the relation

$$\nabla^2 \dot{T} << \left(\frac{\tau_q}{\alpha \tau_T}\right)\frac{\partial \dot{T}}{\partial t}, \qquad (8.30)$$

which includes the flux-precedence case of $\tau_q << \tau_T$, the fast-transient effect of thermal inertia overcomes the microstructural interaction effect. Equation (8.27b) in this case becomes

$$\nabla^2 T = \frac{1}{\alpha}\frac{\partial T}{\partial t} + \frac{\tau_q}{\alpha}\frac{\partial^2 T}{\partial t^2} = \frac{1}{\alpha}\frac{\partial T}{\partial t} + \frac{1}{C^2}\frac{\partial^2 T}{\partial t^2}.$$ (8.31)

As recognized, it is the thermal wave equation employing Cattaneo-Vernotte's constitutive behavior. It describes a heat transport process macroscopic in *space* but microscopic in *time*.

For describing the temperature field in the vicinity of the crack tip, it is more convenient to employ the material coordinate system, (ξ_1, ξ_2) in Figure 8.5, that propagates with the crack tip. With the assistance of equation (8.26), application of the chain rule results in

$$\nabla^2 \rightarrow \frac{\partial^2 T}{\partial \xi_1^2} + \frac{\partial^2 T}{\partial \xi_2^2},$$ (8.32a)

$$\frac{\partial T}{\partial t} \rightarrow \frac{\partial T}{\partial t} - v\frac{\partial T}{\partial \xi_1} - \omega\left(\xi_1\frac{\partial T}{\partial \xi_2} - \xi_2\frac{\partial T}{\partial \xi_1}\right),$$ (8.32b)

$$\frac{\partial^2 T}{\partial t^2} \rightarrow \frac{\partial^2 T}{\partial t^2} - 2v\frac{\partial^2 T}{\partial t \partial \xi_1} + v^2\frac{\partial^2 T}{\partial \xi_1^2} - \omega\xi_1\left[\frac{\partial^2 T}{\partial t \partial \xi_2} - (v - \omega\xi_2)\frac{\partial^2 T}{\partial \xi_1 \partial \xi_2}\right.$$

$$+ \omega\frac{\partial T}{\partial \xi_1} - \omega\xi_1\frac{\partial^2 T}{\partial \xi_2^2}\left] + \omega\xi_2\left[\frac{\partial T}{\partial t \partial \xi_1} - (v - \omega\xi_2)\frac{\partial^2 T}{\partial \xi_1^2} - v\frac{\partial T}{\partial \xi_2} - v\xi_1\frac{\partial^2 T}{\partial \xi_1 \partial \xi_2}\right]$$

$$+ \omega\left[\xi_2\frac{\partial T}{\partial t \partial \xi_1} - \xi_1\frac{\partial T}{\partial t \partial \xi_2}\right] + v\omega\left[\xi_1\frac{\partial T}{\partial \xi_1 \partial \xi_2} + \frac{\partial T}{\partial \xi_2} - \xi_2\frac{\partial^2 T}{\partial \xi_1^2}\right].$$
(8.32c)

If the effect of Jaumann rate resulting from the angular motion of crack vanishes, $\omega = 0$, equation (8.32) reduces to the material derivatives for a crack propagating along a straight path (Tzou, 1990a, b). Substituting equation (8.32) into (8.31) results in the energy (thermal wave) equation describing the *unsteady* temperature field observed from the moving coordinates:

$$\alpha\nabla^2 T = \alpha M^2\frac{\partial^2 T}{\partial \xi_1^2} + \left[\left(\frac{\alpha}{C^2}\right)\frac{\partial^2 T}{\partial t^2} - \left(\frac{M^2}{c}\right)\frac{\partial^2 T}{\partial t \partial \xi_1}\right]$$

$$+ \left[\frac{\partial T}{\partial t} - v\frac{\partial T}{\partial \xi_1}\right] - \omega\left[\xi_1\frac{\partial T}{\partial \xi_2} - \xi_2\frac{\partial T}{\partial \xi_1}\right] + \frac{M^2}{2cR}\left\{\xi_2\left[\frac{\partial^2 T}{\partial t \partial \xi_1} - (v - \omega\xi_2)\frac{\partial^2 T}{\partial \xi_1^2}\right.\right.$$

$$-\omega\frac{\partial T}{\partial\xi_2}-\omega\xi_1\frac{\partial^2 T}{\partial\xi_1\partial\xi_2}\Bigg]-\xi_1\Bigg[\frac{\partial^2 T}{\partial t\partial\xi_2}-(v-\omega\xi_2)\frac{\partial^2 T}{\partial\xi_1\partial\xi_2}+\omega\frac{\partial T}{\partial\xi_1}-\omega\xi_1\frac{\partial^2 T}{\partial\xi_2^2}\Bigg]$$

$$-\left(\xi_1\frac{\partial^2 T}{\partial t\partial\xi_2}-\xi_2\frac{\partial^2 T}{\partial t\partial\xi_1}\right)+v\left(\frac{\partial T}{\partial\xi_2}+\xi_1\frac{\partial^2 T}{\partial\xi_1\partial\xi_2}-\xi_2\frac{\partial^2 T}{\partial\xi_1^2}\right)\Bigg\}$$

(8.33 Cont.)

where M is the *thermal Mach number* defining the linear crack velocity (v) relative to the thermal wave speed (C), $M = v/C$. The quantity c is defined as $v/2\alpha$.

8.3.1 Asymptotic Analysis

In the vicinity of the crack tip, which is of primary concern in studying crack damage, not every term in equation (8.33) carries the same weight. To demonstrate this important *asymptotic* behavior, assuming a product form of temperature (Williams, 1952; Achenbach and Bazant, 1975; Tzou, 1990a, b):

$$T(r,\theta,t) = \Gamma(t)r^\lambda H(\theta), \tag{8.34}$$

where $\Gamma(t)$ measures the time-varying amplitude of temperature and the polar coordinates (r, θ) centered at the crack tip relate to the material coordinates (ξ_1, ξ_2) by equation (8.26). Applying the chain rule for the spatial derivatives,

$$\frac{\partial T}{\partial\xi_1} = \cos\theta\frac{\partial T}{\partial r}-\frac{\sin\theta}{r}\frac{\partial T}{\partial\theta}, \quad \frac{\partial T}{\partial\xi_2} = \sin\theta\frac{\partial T}{\partial r}+\frac{\cos\theta}{r}\frac{\partial T}{\partial\theta}, \quad \text{etc.,} \tag{8.35}$$

equation (8.33) can be categorized into three groups in different orders of r:

$$\nabla^2 T,\frac{\partial^2 T}{\partial\xi_1^2}\sim r^{\lambda-2}, \tag{8.36a}$$

$$\frac{\partial T}{\partial\xi_1},\frac{\partial T}{\partial\xi_2},\frac{\partial^2 T}{\partial t\partial\xi_1},\xi_2\frac{\partial^2 T}{\partial\xi_1^2},\xi_1\frac{\partial^2 T}{\partial\xi_1\partial\xi_2}\sim r^{\lambda-1}, \tag{8.36b}$$

$$\frac{\partial T}{\partial t},\xi_1\frac{\partial T}{\partial\xi_1},\xi_2\frac{\partial T}{\partial\xi_2},\xi_1\frac{\partial T}{\partial\xi_2},\xi_2\frac{\partial T}{\partial\xi_1},\frac{\partial^2 T}{\partial t^2},\xi_2\frac{\partial^2 T}{\partial t\partial\xi_1},\xi_2^2\frac{\partial^2 T}{\partial\xi_1^2},$$

$$\xi_1\xi_2\frac{\partial^2 T}{\partial\xi_1\partial\xi_2},\xi_1\frac{\partial^2 T}{\partial t\partial\xi_2},\xi_2\frac{\partial^2 T}{\partial t\partial\xi_1}\quad\text{and}\quad\xi_1^2\frac{\partial^2 T}{\partial\xi_2^2}\sim r^\lambda. \tag{8.36c}$$

In the near-tip region with r approaching zero, the terms in equation (8.36b) (proportional to $r^{\lambda-1}$) and (8.36c) (proportional to r^{λ}) approach zero at a *faster* rate than those in equation (8.36a) (proportional to $r^{\lambda-2}$), implying that

$$\nabla^2 T \cong M^2 \frac{\partial^2 T}{\partial \xi_1^2} \quad \text{as} \quad r \to 0 \text{ (the near - tip region)} \tag{8.37}$$

is the remainder of equation (8.33). Equation (8.37) suggests that, in the near-tip region,

1. the effect of transient terms containing derivatives with respect to time is a high-order effect, supporting the fact that the singular behavior (see point 2 below) remains *the same* for both steady-state and transient problems; and
2. the effect of angular motion reflected by the terms containing ω is a high-order effect, supporting the fact that the singular behavior remains *the same* for a crack propagating along a straight or a curved path. The crack turning along a grain boundary in microscale, therefore, is a high-order effect.

In fact, the asymptotic analysis made here bears the same merit as the boundary layer analysis by Prandtl for fluid motion in the neighborhood of a geometric boundary (Schlichting, 1960). Such an approach, for a closer resemblance, had been successfully extended to the study of singular behavior for both forced and free convection in the corner area of a container (Tzou, 1992h).

In the absence of the microstructural interaction effect, equation (8.37) supports the fact that the conduction ($\nabla^2 T$) and thermal inertia terms ($\partial^2 T/\partial \xi_1^2$) dominate heat transport in the near-tip region. In terms of the (ξ_1, ξ_2) coordinates, it can be written as

$$(1 - M^2)\frac{\partial^2 T}{\partial \xi_1^2} + \frac{\partial^2 T}{\partial \xi_2^2} = 0 \quad \text{with} \quad r = \sqrt{\xi_1^2 + \xi_2^2} \to 0. \tag{8.38}$$

In terms of the polar coordinates (r, θ) centered at the crack tip,

$$\frac{\partial^2 T}{\partial \xi_1^2} = \left[\cos^2 \theta\right]\frac{\partial^2 T}{\partial r^2} - \left[\frac{\sin(2\theta)}{r}\right]\frac{\partial^2 T}{\partial r \partial \theta} + \left[\frac{\sin \theta}{r}\right]^2 \frac{\partial^2 T}{\partial \theta^2} + \left[\frac{\sin^2 \theta}{r}\right]\frac{\partial T}{\partial r} + \left[\frac{\sin(2\theta)}{r^2}\right]\frac{\partial T}{\partial \theta}$$

$$\tag{8.39a}$$

$$\frac{\partial^2 T}{\partial \xi_2^2} = \left[\sin^2 \theta\right]\frac{\partial^2 T}{\partial r^2} + \left[\frac{\sin(2\theta)}{r}\right]\frac{\partial^2 T}{\partial r \partial \theta} + \left[\frac{\cos \theta}{r}\right]^2 \frac{\partial^2 T}{\partial \theta^2} + \left[\frac{\cos^2 \theta}{r}\right]\frac{\partial T}{\partial r} - \left[\frac{\sin(2\theta)}{r^2}\right]\frac{\partial T}{\partial \theta}$$

$$\tag{8.39b}$$

Alternatively,

$$-M^2\left\{\left[\cos^2\theta\right]\frac{\partial^2 T}{\partial r^2}-\left[\frac{\sin(2\theta)}{r}\right]\frac{\partial^2 T}{\partial r\partial\theta}+\left[\frac{\sin\theta}{r}\right]^2\frac{\partial^2 T}{\partial\theta^2}\right.$$

$$\left.+\left[\frac{\sin^2\theta}{r}\right]\frac{\partial T}{\partial r}+\left[\frac{\sin(2\theta)}{r^2}\right]\frac{\partial T}{\partial\theta}\right\}+\left\{\frac{\partial^2 T}{\partial r^2}+\frac{1}{r}\frac{\partial T}{\partial r}+\frac{1}{r^2}\frac{\partial^2 T}{\partial\theta^2}\right\}=0. \tag{8.40}$$

Equation (8.40) is *equi*dimensional in r. A direct substitution of equation (8.34) into (8.40) results in

$$\left(1-M^2\sin\theta\right)\frac{d^2 H}{d\theta^2}-\left[M^2(1-\lambda)\sin 2\theta\right]\frac{dH}{d\theta}$$

$$+\lambda\left\{\lambda+M^2\left[(2-\lambda)\cos^2\theta-1\right]\right\}H=0, \tag{8.41}$$

which, consistent with the result of the asymptotic analysis, is valid for any function of $\Gamma(t)$ in the transient response. Equation (8.41) reveals an important characteristic for the temperature response in the near-tip region. It depends *only* upon the thermal Mach number $M\ (=v/C)$. The effect of finite speed of heat propagation (C) resulting from the phase lag of the heat flux vector ($C=\sqrt{\alpha/\tau_q}$) is thus intrinsic to the near-tip response.

The asymptotic analysis results in an *eigenvalue* problem described by equation (8.41). It is to be solved for the eigenvalue λ associated with the angular distribution of the near-tip temperature, $H(\theta)$, called the eigenfunction, surrounding the rapidly propagating crack tip. For illustrating the fundamental properties, let us specify a zero temperature at the crack surfaces:

$$T=0 \quad \text{at} \quad \theta=\pm\pi \quad \text{or} \quad H(\pm\pi)=0 \tag{8.42}$$

according to equation (8.34). From a mathematical point of view, we are to determine the fundamental value of λ that renders a *nontrivial* solution for equation (8.41). An analytical solution for the eigenvalue and eigenfunction is most desirable owing to the singularity present at the crack tip. Obtaining an analytical solution satisfying equations (8.41) and (8.42), however, is nontrivial because equation (8.41) involves variable coefficients in the second-order differential equation. We introduce the successive variable transformations proposed by Tzou (1990a, b):

$$H(\theta)=\sqrt{\left[1-\left(M\sin\theta\right)^2\right]^\lambda}\,\Phi(\theta). \tag{8.43}$$

Equation (8.43) transforms equation (8.41) into the following form:

$$(g_1 f)\frac{d^2\Phi}{d\theta^2}+\left(fM^2\sin 2\theta\right)\frac{d\Phi}{d\theta}+\left(g_1\frac{d^2 f}{d\theta^2}+g_2\frac{df}{d\theta}+g_3 f\right)\Phi=0, \tag{8.44a}$$

where

$$f = \sqrt{g_1^\lambda}, \quad g_1 = 1 - (M \sin\theta)^2, \quad g_2 = -M^2(1-\lambda)\sin(2\theta),$$

$$g_3 = \lambda\left\{\lambda + M^2\left[(2-\lambda)\cos^2\theta - 1\right]\right\}.$$

(8.44b)

The major function of equation (8.43) is to transform the independent variable from H to Φ. Introducing the second transformation that transforms the independent variable from θ to γ,

$$\left(g_1 f\right)\frac{d^2\gamma}{d\theta^2} - \left(fM^2 \sin 2\theta\right)\frac{d\gamma}{d\theta} = 0, \tag{8.45}$$

equation (8.44a) becomes

$$\frac{d^2\Phi}{d\gamma^2} + \left[\frac{g_1\dfrac{d^2f}{d\theta^2} + g_2\dfrac{df}{d\theta} + g_3 f}{g_1 f\left(\dfrac{d\gamma}{d\theta}\right)^2}\right]\Phi = 0, \tag{8.46}$$

which eliminates the first-order derivative and sheds light on obtaining a *closed-form* solution. Unlike the dependent-variable transformation, equation (8.43), the independent-variable transformation is determined from the first-order ordinary differential equation (8.45). Its solution depends on the thermal Mach number, M, and should be determined in the respective regimes of M.

8.3.2 Subsonic Regime With $M < 1$

For a crack propagating at a speed of v less than C, the thermal wave speed, $M < 1$, equation (8.45) can be integrated to give

$$\gamma(\theta) = \tan^{-1}\left[\sqrt{1-M^2}\,\tan\theta\right], \quad \text{consequently,} \quad \frac{d\gamma}{d\theta} = \frac{\sqrt{1-M^2}}{1-(M\sin\theta)^2}. \tag{8.47}$$

Substituting equation (8.47) into (8.46) gives

$$\frac{d^2\Phi}{d\gamma^2} + \lambda^2\Phi = 0, \tag{8.48}$$

allowing for a closed-form solution:

$$\Phi(\gamma) = A\cos(\lambda\gamma) + B\sin(\lambda\gamma). \tag{8.49}$$

The boundary condition, equation (8.42), however, imposes that

$$\Phi = 0 \quad \text{at} \quad \gamma = \pm\pi, \tag{8.50}$$

which requires $B = 0$ and, consequently,

$$\Phi(\gamma) = A\cos(\lambda\gamma). \tag{8.51}$$

In equation (8.50), according to equation (8.43), $\Phi = 0$ in correspondence with $H = 0$ is obvious. The branch of $\gamma = \pi$, however, is adopted in correspondence with $\theta = \pi$ in equation (8.47). Combining equations (8.50) with (8.51), for a nontrivial solution of the near-tip temperature requires

$$\cos(\lambda\pi) = 0 \quad \text{or} \quad \lambda = \frac{(2n+1)}{2} \quad \text{for} \quad n = 0, 1, 2, \dots, \quad \text{etc}. \tag{8.52}$$

In the spectrum of r dependency, therefore,

$$T \sim r^{\frac{1}{2}}, r^{\frac{3}{2}}, r^{\frac{5}{2}}, \dots \quad \text{for} \quad r \to 0. \tag{8.53}$$

In the near-tip region with $r \to 0$, obviously, the high-order terms lead by $r^{3/2}$, $r^{5/2}$, ..., etc., approach zero *faster* than the leading term led by $r^{1/2}$, implying that

$$T \sim r^{\frac{1}{2}} \quad \text{in correspondence with } \lambda = \frac{1}{2} \quad \text{for} \quad r \to 0 \tag{8.54}$$

dominates the fundamental characteristic of the near-tip solution of temperature. In terms of the physical variables H and θ, with $\lambda = 1/2$ and the use of equations (8.47) and (8.51) in equation (8.43),

$$H(\theta) = \frac{A}{\sqrt{2}}\sqrt{\sqrt{1 - (M\sin\theta)^2} + \cos\theta} \quad \text{for} \quad r \to 0, \quad M < 1. \tag{8.55}$$

In the subsonic regime with $M < 1$, the r dependency, $\lambda = 1/2$, is *independent* of the crack speed. The crack speed only affects the angular distribution of the near-tip temperature. For the case of C approaching infinity, the thermal wave model reduces to the diffusion theory. The thermal Mach number approaches zero ($M \to 0$ as $C \to \infty$) in this case, resulting in

$$H_D(\theta) = \frac{A}{\sqrt{2}}\sqrt{1 + \cos\theta} \quad \text{for} \quad r \to 0 \quad \text{(diffusion model)} \tag{8.56}$$

Figure 8.6 shows the angular distribution of the near-tip temperature, $H(\theta)$, represented by equation (8.55), with the result of diffusion, equation (8.56) appearing as a special case of $M = 0$. All the distributions are symmetric with respect to $\theta = 0$, as expected. For all cases, the temperature reaches a maximum at the leading edge of the crack tip at $\theta = 0°$. The temperature level decreases as the thermal Mach number increases, implying either increase of the crack speed at a constant thermal wave speed or decrease of the thermal wave speed at a constant crack speed. This is expected from a physical point view. When the crack speed increases at a constant thermal wave speed, the material points in the vicinity of the crack tip do not have sufficient time to respond to the appropriate temperature level before the crack tip advances. The faster the crack speed, the shorter the response time would be. Consequently, the temperature established at a fixed observation point relative to the crack tip is expected to be lower as the crack speed increases.

The near-tip temperature displays a $\sqrt{r}$ type of behavior in the near-tip region, as shown by equation (8.54). This behavior is *the same* as that for a stationary crack (Tzou, 1991d to f) and is *independent* of the crack speed. From classical Fourier's law in heat conduction, the resulting heat flux vector is

$$q_r = -k\frac{\partial T}{\partial r} \sim \frac{1}{\sqrt{r}} \quad \text{(Fourier's law, } M = 0\text{)} . \tag{8.57}$$

It displays a $1/\sqrt{r}$-type of *singularity* at the crack tip ($r \to 0$), a behavior similar to that discussed in equation (8.19), indicating an intensified accumulation of thermal energy in the vicinity of the crack tip. Geometrically, this is due to the abrupt change of geometric curvature at the crack tip.

When the response time comes into the picture through the relative speed of crack propagation, however, a dramatic change in the energy accumulation results. According to equation (8.27a), the unsteady heat flux field observed from the moving crack tip is governed by

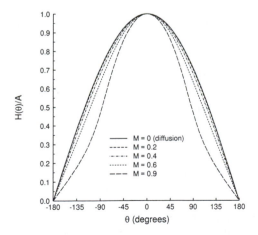

Figure 8.6 Angular distribution of the near-tip temperature, the eigenfunction $H(\theta)$ shown by equation (8.55).

$$\tau_q\left[\frac{\partial\bar{q}}{\partial t}-v\frac{\partial\bar{q}}{\partial\xi_1}-\omega\left(\xi_1\frac{\partial\bar{q}}{\partial\xi_2}-\xi_2\frac{\partial\bar{q}}{\partial\xi_1}\right)\right]+\bar{q}=-k\nabla T\,. \tag{8.58}$$

Assuming a product form for the heat flux components,

$$\bar{q}=\left[q_r,q_\theta\right]=\Lambda(t)r^s[Q_r(\theta),Q_\theta(\theta)]\,, \tag{8.59}$$

similarly, the asymptotic analysis categorizes the left side of equation (8.58) into two groups:

$$\frac{\partial\bar{q}}{\partial t},\xi_1\frac{\partial\bar{q}}{\partial\xi_2},\xi_2\frac{\partial\bar{q}}{\partial\xi_1},\bar{q}\sim r^s;\quad\frac{\partial\bar{q}}{\partial\xi_1}\sim r^{s-1} \tag{8.60}$$

The term of order r^s approaches zero faster than the term of order r^{s-1} in the near-tip region with $r\to 0$, implying that

$$\frac{-v\alpha}{C^2}\frac{\partial\bar{q}}{\partial\xi_1}\cong-k\nabla T\quad\text{or}\quad\frac{\partial\bar{q}}{\partial\xi_1}\cong\frac{2ck}{M^2}\nabla T\quad\text{for}\quad r\to 0\quad\text{(near-tip region)} \tag{8.61}$$

Again, the transient term and angular motion of the crack appear as high-order effects for heat transport in the vicinity of the moving crack tip. Applying the chain rule (equation (8.35)) to equation (8.61), clearly, equation (8.61) requires that $s = \lambda$, implying that the r dependency of the heat flux vector must be *identical* to that of the temperature. Since the case of $\lambda = 1/2$ describes the fundamental characteristic of the eigenfunction for the near-tip temperature, referring to equation (8.54), the case of $s = 1/2$ characterizes the fundamental solution of the heat flux vector. We conclude, therefore,

$$T\sim r^{\frac{1}{2}}\quad\text{and}\quad\bar{q}\sim r^{\frac{1}{2}}\quad\text{for}\quad r\to 0\quad\text{and}\quad M<1\quad\text{(subsonic regime)} \tag{8.62}$$

Contrary to the result of diffusion ($M = 0$), where a $1/\sqrt{r}$ type of singularity is present in the heat flux vector at the crack tip, equation (8.62) results in a $\sqrt{r}$ type of behavior for the heat flux vector that is *bounded* at the *moving* crack tip. The intensified energy accumulation in the near-tip region, in other words, *diminishes* when the thermal inertia effect (the finite speed of heat propagation) becomes pronounced. From a physical point of view, accumulation of thermal energy needs a finite period of time to take place. When the crack speed is high, though in the subsonic regime, thermal energy does not have sufficient time to accumulate before the crack tip advances to another location. The bounded heat flux vector at the crack tip shown in equation (8.62) fully reflects this behavior. From this result, it can be clearly seen that a rapidly propagating crack in solids *does not* cause as much damage as a stationary crack in transporting heat. In fact, crack motion *decreases* the amount of energy accumulation at the crack tip.

8.3.3 Supersonic Regime With $M > 1$

For crack propagation in *pure* metals at room temperature, the result in the subsonic regime is sufficient because the terminal velocity (10^3 m/s) is about 2 orders of magnitude smaller than the thermal wave speed (10^5 m/s). Consequently, the thermal Mach number is small, of the order of 10^{-2}. Angular distribution of the near-tip temperature, as shown by Figure 8.6, will be close to that predicted by diffusion, but the singular behavior of the heat flux vector reflecting the intensified energy accumulation at the crack tip vanishes, as shown by equation (8.62).

Thermal wave speed strongly depends on temperature. At extremely low temperatures such as superfluid liquid helium, its value can decrease to as small as 18 m/s (Peshkov, 1944). For possible applications to the cryogenic system, therefore, it is desirable to develop a detailed understanding of the thermal inertia effect in the *supersonic* regime with $M > 1$. For $M > 1$, implying that the crack propagates at a speed faster than the thermal wave speed in solids, $v > C$, equation (8.43) can still be used, but the physical domain for θ is restricted by

$$0 \le \theta \le \sin^{-1}\left(\frac{1}{M}\right) \quad \text{for a crack tip moving to the left,} \tag{8.63a}$$

$$\pi - \sin^{-1}\left(\frac{1}{M}\right) \le \theta \le \pi \quad \text{for a crack tip moving to the right.} \tag{8.63b}$$

Owing to the symmetry of the problem, only the physical domain in the upper-half plane is shown in equation (8.63). Note that the eigenvalue λ in equation (8.43) is no longer 1/2. It has to be redetermined in the present supersonic regime with $M > 1$. Equation (8.63) defines the physical domain of the *heat-affected zone* surrounding a moving crack tip or a heat source (Tzou, 1989a, b, 1990c, 1991a; Tzou and Li, 1993a, b). Owing to rapid motion of the crack, the material point outside of the domain specified by equation (8.63) cannot even sense the presence of the crack tip. As a result, the temperature field remains at the reference value in such a *thermally undisturbed zone*. This situation resembles high-speed aerodynamics, including the *thermal Mach angle* defined as $\theta_M = \sin^{-1}(1/M)$. Evidently, a thermal shock wave is located at θ_M, which separates the heat-affected zone from the thermally undisturbed zone, as illustrated in Figure 8.7. Since the temperature remains undisturbed (stays at the reference value) in the thermally undisturbed zone, our effort will be on the determination of the temperature field in the heat-affected zone.

For $M > 1$, equation (8.45) possesses a branched solution,

$$\gamma(\theta) = \frac{1}{2}\ln\left[\frac{\sqrt{M^2-1}\tan\theta+1}{\sqrt{M^2-1}\tan\theta-1}\right], \quad \frac{d\gamma}{d\theta} = \frac{\sqrt{M^2-1}}{\left[1-\left(M\sin\theta\right)^2\right]} \quad \text{for} \quad M > 1.$$
$$\tag{8.64}$$

Substituting equation (8.64) into (8.46), in parallel to equation (8.48) gives

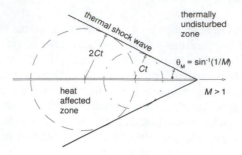

Figure 8.7 Thermal shock wave, thermal shock angle (θ_M), heat-affected zone ($0 \leq \theta \leq \theta_M$ measuring from the *trailing* edge of the crack) and thermally undisturbed zone ($\theta_M \leq \theta \leq \pi$) surrounding a rapidly propagating crack tip to the right. Supersonic regime with $M > 1$.

$$\frac{d^2\Phi}{d\gamma^2} - \lambda^2\Phi = 0, \quad M > 1. \tag{8.65}$$

For a better illustration of the eigenstructure in the supersonic regime, we introduce another transformation from γ to ζ:

$$\gamma = \frac{1}{2}\ln\zeta \pm \frac{\pi}{2}i, \quad i = \sqrt{-1}. \tag{8.66}$$

Mathematically, this is equivalent to the special selection of $\zeta(\theta)$ from equation (8.64), so that

$$\zeta(\theta) = \frac{1 + \sqrt{M^2 - 1}\tan\theta}{1 - \sqrt{M^2 - 1}\tan\theta}, \quad M > 1. \tag{8.67}$$

Note that the physical domain of the heat-affected zone defined by equation (8.63) can be expressed alternately:

$$0 \leq \theta \leq \tan^{-1}\left(\frac{1}{\sqrt{M^2 - 1}}\right) \quad \text{for a crack tip moving to the left}, \tag{8.68a}$$

$$\pi - \tan^{-1}\left(\frac{1}{\sqrt{M^2 - 1}}\right) \leq \theta \leq \pi \quad \text{for a crack tip moving to the right}. \tag{8.68b}$$

The value of $\zeta(\theta)$ defined in equation (8.67) is thus positive definite in the heat-affected zone. The solution of equation (8.65) can thus be written in terms of ζ:

$$\Phi(\zeta(\theta)) = \cos\left(\frac{\lambda\pi}{2}\right)\left[Ae^{\frac{\lambda\ln\zeta}{2}} + Be^{-\frac{\lambda\ln\zeta}{2}}\right] + i\sin\left(\frac{\lambda\pi}{2}\right)\left[Ae^{\frac{\lambda\ln\zeta}{2}} + Be^{-\frac{\lambda\ln\zeta}{2}}\right].$$

(8.69)

Since Φ describes the angular distribution of the near-tip temperature, referring to equation (8.43), it must be *real*. This implies, from equation (8.69), that

$$\sin\frac{\lambda\pi}{2} = 0 \quad \text{or} \quad \lambda = 2n, \quad n = 1, 2, 3, \ldots, \quad \text{etc.,}$$

(8.70)

for a nontrivial solution. In the supersonic regime with $M > 1$, therefore, the *smallest* eigenvalue characterizing the fundamental solution is $\lambda = 2$. Mathematically,

$$T \sim r^2 \quad \text{for} \quad r \to 0, \quad M > 1 \quad \text{(supersonic regime)}.$$

(8.71)

Referring to the fundamental behavior shown in equation (8.62) for the subsonic regime, the r dependency of temperature changes from $1/2$ to 2 in transition of the thermal Mach number from subsonic ($M < 1$) to supersonic ($M > 1$) regimes. With $\lambda = 2$, equation (8.69) thus becomes

$$\Phi(\zeta(\theta)) = -\left(A\zeta + \frac{B}{\zeta}\right),$$

(8.72)

with A and B determined from the boundary conditions at the crack surfaces. At $\theta = \pm\pi$, according to equation (8.67), the top and bottom surfaces of the crack collapse onto the same mathematical boundary of $\zeta = 1$. The boundary condition in the ζ domain thus becomes

$$\Phi = 0 \quad \text{at} \quad \zeta = 1.$$

(8.73)

This yields $A = -B$ according to equation (8.72), resulting in

$$\Phi(\zeta) = B\left(\zeta - \frac{1}{\zeta}\right),$$

(8.74)

with B being the amplitude of the eigenfunctions. From equation (8.43) with $\lambda = 2$ (for converting Φ to H) and equation (8.67) (for converting ζ to θ), the angular distribution of the near-tip temperature in the supersonic regime is obtained:

$$H(\theta) = 2B\sqrt{M^2 - 1}\, \sin(2\theta), \quad M > 1 \quad \text{(supersonic regime)}. \quad (8.75)$$

Equation (8.75) is applicable only in the heat-affected zone defined by equation (8.63b) or (8.68b) for a crack tip propagating to the right. Equation (8.63a) or (8.68a) is used for a crack tip moving to the left, a conjugate situation to the present problem with the same distributions of temperature in the heat-affected zone. The temperature in the thermally undisturbed zone remains at the reference value, which is assumed to be zero without loss of generality. The result is shown in Figure 8.8 at various values of M in the supersonic regime. The thermal shock surface is located at $\theta_M = \pi - \sin^{-1}(1/M)$, 150° for $M = 2$, 165.5° for $M = 4$, 170.4° for $M = 6$, 172.8° for $M = 8$, and 174.3° for $M = 10$. A finite jump of temperature exists in transition from the thermally undisturbed zone $(0 \le \theta \le \pi - \theta_M)$ to the heat-affected zone $(\pi - \theta_M \le \theta \le \pi)$. The finite jump is, according to equation (8.75),

$$\lim_{\theta \to \theta_M^+} \frac{H(\theta)}{B} = \frac{4(M^2 - 1)}{M^2}, \quad M > 1 \qquad (8.76)$$

with θ_M^+ denoting the limit approaching from the heat-affected zone. When the thermal Mach number approaches infinity $(M \to \infty)$, the temperature jump reaches an ultimate value of 4. In addition, at a constant value of the thermal wave speed, Figure 8.8 shows that the temperature level in the heat-affected zone *increases* with the crack speed. This is the reverse behavior to that observed in Figure 8.6 in the subsonic regime. It is still true that the response time for temperature rise is shortened as the crack speed increases (the argument made in Figure 8.6). The heat-affected zone, however, becomes *narrower* at a higher crack speed, resulting in a closer distance between the thermal shock wave and the observation point (for the temperature rise) in the heat-affected zone. The thermal shock surface carries a higher temperature that also increases with the thermal Mach number (the crack

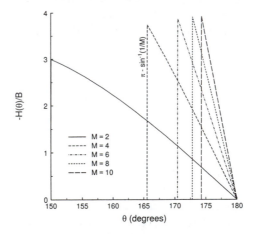

Figure 8.8 Angular distribution of the near-tip temperature in the supersonic regime with $M > 1$, equation (8.75).

speed) as shown by equation (8.76). When the heat-affected zone closes up as a result of increase of the crack speed, the thermal shock waves serve as additional *heating* to the heat-affected zone, resulting in a higher temperature level at a higher crack speed. This is called the *swinging* phenomenon of temperature in transition from the subsonic to the supersonic regimes (Tzou, 1989a, b, 1990a, b).

8.3.4 Transonic Stage With M = 1

In reality, it is rare for a crack to reach *exactly* the thermal wave speed in the history of crack propagation. In heating problems with a large temperature gradient present in the workpiece, however, the thermal wave speed may be *nonhomogeneous* as a result of the nonhomogeneous temperature distribution. As the crack penetrates through physical domains with different thermal wave speeds, therefore, a transonic stage may exist for an extremely short period of time should the crack speed pass the *local* thermal wave speed. Unlike crack propagation in the subsonic and supersonic regimes, the temperature response at the transonic stage is a more localized behavior.

At the transonic stage with $M < 1$, equation (8.44b) reduces to

$$f = \cos^\lambda \theta, \quad g_1 = \cos^2 \theta, \quad g_2 = (\lambda - 1)\sin(2\theta), \quad g_3 = \lambda \left[\lambda \sin^2 \theta + \cos(2\theta) \right],$$

$$(8.77)$$

rendering a simple equation in correspondence with equation (8.46):

$$\frac{d^2\Phi}{d\gamma^2} = 0, \quad M = 1 \quad \text{(transonic stage)}. \tag{8.78}$$

A direct integration gives

$$\Phi = A\gamma + B . \tag{8.79}$$

With the assistance of equation (8.77), the solution of equation (8.45) describing the relation between γ and θ is easily obtained:

$$\gamma(\theta) = \tan\theta . \tag{8.80}$$

The top and bottom surfaces of the crack, $\theta = \pm\pi$, again collapse onto the same branch of $\gamma = 0$. The boundary condition in correspondence with equation (8.73) is thus

$$\Phi = 0 \quad \text{at} \quad \gamma = 0 , \tag{8.81}$$

resulting in $B = 0$ in equation (8.79). Combining equations (8.79) (with $B = 0$) and (8.80) and substituting the result into equation (8.43) (with $M = 1$) gives

$$H(\theta) = A \cos^\lambda \theta \tan \theta = A \frac{\sin \theta}{\cos^{1-\lambda} \theta} . \tag{8.82}$$

Determination of the eigenvalue λ in the present case lies in the argument of a finite temperature across the thermal shock wave. When approaching the transonic stage from the supersonic side, i.e., as $M \to 1^+$, according to equation (8.63b), the thermal shock wave should be located at $\theta = \pi/2$ and the heat-affected zone ranges from $\pi/2$ to π, as illustrated in Figure 8.9. Because the temperature jump across the thermal shock surface is finite, evidenced by all the cases in Figure 8.8 and equation (8.76), it is suggested that equation (8.82) for the transonic stage also possesses a *finite* jump in temperature in transition from the thermally undisturbed zone to the heat-affected zone. Mathematically, this condition is expressed as

$$\lim_{\theta \to (\pi/2)^+} H(\theta) \quad \text{remains bounded} . \tag{8.83}$$

A careful inspection on equation (8.82) reveals that the *smallest* eigenvalue characterizing the fundamental solution at the transonic stage is $\lambda = 1$. For the case of $\lambda < 1$, the denominator, $\cos^{1-\lambda}(\pi/2)$ in equation (8.82), approaches zero, rendering an infinite temperature response at the thermal shock surface ($\theta_M = \pi/2$). For the case of $\lambda > 1$, on the other hand, the resulting factor $\cos^{\lambda-1}(\pi/2)$ in equation (8.82) approaches zero, which only renders a trivial solution. With $\lambda = 1$, therefore, equation (8.82) becomes

$$H(\theta) = A \sin \theta, \quad M = 1 \quad \text{(transonic stage)} . \tag{8.84}$$

The basic sine curve describing the angular distribution of the near-tip temperature at the transonic stage is shown in Figure 8.10. The temperature reaches a maximum at the thermal shock surface ($\theta = \pi/2$) and then tapers off in the direction toward the trailing edge of the crack tip ($\theta = \pi$). This is a behavior similar to that in the supersonic regime.

In summary, the fast-transient effect of thermal inertia has an intrinsic influence on heat transport. Due to insufficient response time in the short-time

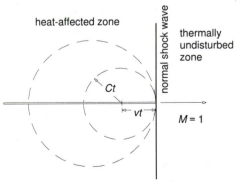

Figure 8.9 Normal shock formed at the transonic stage.

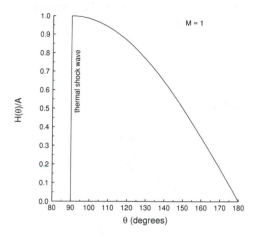

Figure 8.10 Angular distribution of the near-tip temperature at the transonic stage with $M = 1$, equation (8.84).

transient, most important, it *reduces* the amount of thermal energy accumulated in the vicinity of a moving crack tip. The r dependency of the heat flux vector is introduced to measure the crack damage, as summarized in Table 8.1. In transition from the subsonic to the transonic to the supersonic regimes, the r dependency of the heat flux vector varies from $r^{1/2}$, r, to r^2, with r denoting the radial distance measured from the crack tip. In the near-tip region with $r \to 0$, the heat flux vector *vanishes* at the crack tip, implying *diminution* of energy accumulation at the moving crack tip. Unlike the classical theory of diffusion, which always predicts a $1/\sqrt{r}$ type of *singularity* for *both* the temperature gradient and the heat flux vector regardless of the crack motion (Tzou, 1991d to f), the fast-transient effect in the short-time response results in different r-dependencies for the temperature gradient and the heat flux vector. The r dependency of the heat flux vector is the same as that of the temperature, rather than the temperature gradient. The r dependency of the heat flux vector is always greater than that of the temperature gradient by 1.

      The fast-transient effect of thermal inertia also introduces several unique features that cannot be depicted by diffusion. They include the thermal shock formation and swinging phenomenon of temperature in the transition of the thermal

Table 8.1 **Transition of the r dependencies of temperature gradient and heat flux vector in the vicinity of a moving crack tip ($r \equiv$ radial distance measured from the crack tip).**

	Temperature gradient	Heat flux vector
Subsonic regime ($M < 1$)	$r^{-1/2}$	$r^{1/2}$
Transonic stage ($M = 1$)	r^0	r
Supersonic regime ($M > 1$)	r	r^-

Mach number. Although energy accumulation at the crack tip is weakened by crack motion, the thermal shock formation and temperature swinging deserve special attention in rapid thermal processing of materials.

8.4 DIMINUTION OF DAMAGE — MICROSCALE INTERACTION EFFECT

When the delay time due to microstructural interactions (including phonon-electron interaction in metals and phonon scattering in dielectric films or semiconductors) becomes comparable to the relaxation time due to the fast-transient effect of thermal inertia, the condition shown in equation (8.30) is no longer valid. The lagging behavior describing both the microstructural interaction and the fast-transient effects is depicted by equation (8.27b). It results from the dual-phase-lag model, equation (8.27a). In Section 8.3 we demonstrated the dominance of thermal wave behavior due to the fast-transient effect in the near-tip region of a moving crack. This section demonstrates that the additional delay due to phonon-electron interaction and phonon scattering in microscale tends to further level off the energy accumulation around the crack tip. The r^2 type of behavior of the heat flux vector obtained in the supersonic regime, referring to Table 8.1, seems to be an *ultimate* response when both effects are taken into account.

The asymptotic analysis can be made in the same fashion. In the presence of a time delay due to the microstructural interaction effect, mathematically, equation (8.27b) replaces equation (8.31) in describing heat transport in the near-tip region. In terms of the material coordinates propagating with the crack tip, the additional mixed-derivative term in equation (8.27b) can be written as

$$\tau_T \frac{\partial}{\partial t} \nabla^2 T \rightarrow \tau_T \frac{\partial}{\partial t}\left(\frac{\partial^2 T}{\partial \xi_1^2} + \frac{\partial^2 T}{\partial \xi_2^2}\right) - v\tau_T\left(\frac{\partial^2 T}{\partial \xi_1^3} + \frac{\partial^3 T}{\partial \xi_1 \partial \xi_2^2}\right). \tag{8.85}$$

Equation (8.33), containing all the conduction, linear and angular convection, and fast-transient effect in time, remains, except for the addition of equation (8.85) into the left side of the equation:

$$\alpha\tau_T \frac{\partial}{\partial t}\left(\frac{\partial^2 T}{\partial \xi_1^2} + \frac{\partial^2 T}{\partial \xi_2^2}\right) - \alpha v\tau_T\left(\frac{\partial^2 T}{\partial \xi_1^3} + \frac{\partial^3 T}{\partial \xi_1 \partial \xi_2^2}\right) + \alpha\nabla^2 T = \alpha M^2 \frac{\partial^2 T}{\partial \xi_1^2} + \cdots$$

$$\tag{8.86}$$

The rest of terms in equation (8.86) are identical to those in equation (8.33). Applying the same procedure, equations (8.34) and (8.35), the additional terms in equation (8.86) possess the following r dependencies:

$$\frac{\partial^3 T}{\partial t \partial \xi_1^2}, \frac{\partial^3 T}{\partial t \partial \xi_2^2} \sim r^{\lambda-2}, \quad \frac{\partial^3 T}{\partial \xi_1^3}, \frac{\partial^3 T}{\partial \xi_1 \partial \xi_2^2} \sim r^{\lambda-3} \tag{8.87}$$

The same expressions for the conduction and fast-transient effects, as shown by equation (8.36a), are

$$\nabla^2 T, \frac{\partial^2 T}{\partial \xi_1^2} \sim r^{\lambda-2}. \tag{8.36a'}$$

In the near-tip region with $r \to 0$, obviously, the fast-transient effect represented by equation (8.36a) vanishes *faster than* the microstructural interaction effect represented by equation (8.87), implying that the microstructural interaction effect even *dominates* over the fast-transient effect in the vicinity of the moving crack tip. The remainder of equation (8.86) is thus

$$\frac{\partial^2 T}{\partial \xi_1^3} + \frac{\partial^3 T}{\partial \xi_1 \partial \xi_2^2} \cong 0 \quad \text{for} \quad r \to 0 \quad \text{(near-tip region). (8.88)}$$

Again, crack curving and unsteady motion are high-order effects compared to the microstructural interaction effect in the near-tip region. In comparison with equation (8.38), a *second*-order differential equation describing the fast-transient effect in the near-tip region, equation (8.88), describing the microstructural interaction effect, is a *third*-order differential equation with a completely different mathematical structure. A distinct fundamental structure of the near-tip temperature is thus expected.

Applying the chain rule, equations (8.35) and (8.39), to equation (8.88) and substituting equation (8.34) into the resulting equation, the equation governing the angular distribution of the near-tip temperature can be nicely arranged into the following form:

$$\sin\theta\left(\frac{d^3H}{d\theta^3} + \lambda^2 \frac{dH}{d\theta}\right) + (2-\lambda)\cos\theta\left(\frac{d^2H}{d\theta^2} + \lambda^2 H\right) = 0. \tag{8.89}$$

Its counterpart in the previous case is equation (8.41). Equation (8.89) displays a third-order differential equation, necessitating *three* boundary conditions to determine the eigenvalues and eigenfunctions. Crack curving is a high-order effect, implying that the eigenstructure of equation (8.89) remains the same for a straight or curved crack trajectory. For a crack propagating along a straight path, referring to Figure 8.5 with R (the instantaneous radius of curvature) approaching infinity, the temperature distribution in the near-tip region must be symmetric with respect to the axis of $\theta = 0$. This characteristic is shown in Figure 8.6. Mathematically, such a symmetric condition can be expressed as

$$\frac{dH}{d\theta} = 0 \quad \text{at} \quad \theta = 0. \tag{8.90a}$$

Equation (8.89) is a *third*-order differential equation, allowing for boundary conditions containing the *second*-order derivatives to the extent possible. At the top surface of the crack, therefore,

$$a_1 H + a_2 \frac{dH}{d\theta} + a_3 \frac{d^2 H}{d\theta^2} = 0 \quad \text{at} \quad \theta = \pi, \tag{8.90b}$$

$$b_1 H + b_2 \frac{dH}{d\theta} + b_3 \frac{d^2 H}{d\theta^2} = 0 \quad \text{at} \quad \theta = \pi \tag{8.90c}$$

are the most general boundary conditions from a mathematical point of view, with a and b denoting the thermal moduli of the conducting medium. The coefficients a_1 and b_1, for example, correspond to the thermal conductivity, and a_2 and b_2 may be the heat transfer coefficients for heat convection into the aerial closure between the crack surfaces. When $a_2 = a_3 = 0$, equation (8.90b) reduces to the temperature-specified boundary condition considered in the previous section. Note that Fourier's law in heat conduction no longer applies in the presence of lagging behavior. The second-order derivative in equation (8.90b) or (8.90c) describes the additional effect in the most general situation.

Equation (8.89) can be solved analytically by introducing the following transformation:

$$p = \frac{\partial^2 H}{\partial^2 \theta} + \lambda^2 H . \tag{8.91}$$

Equation (8.89) thus reduces to a first-order differential equation,

$$\frac{dp}{d\theta} + \left[(2 - \lambda) \cot \theta \right] p = 0 , \tag{8.92}$$

with singularities existing at $\theta = 0$ and π. Integrating equation (8.92) and using the result of equation (8.91) gives

$$p = \frac{\partial^2 H}{\partial^2 \theta} + \lambda^2 H = D_3 (\sin \theta)^{\lambda - 2} \tag{8.93}$$

which is a second-order, nonhomogeneous ordinary differential equation. Application of the method of variation of parameter (Hildebrand, 1976) then yields

$$H(\theta) = D_1 \cos(\lambda \theta) + D_2 \sin(\lambda \theta) + D_3 f(\theta; \lambda) , \tag{8.94}$$

with

$$f(\theta;\lambda) = \int_a^\theta \frac{\sin[\lambda(\theta - z)](\sin z)^{\lambda-2}}{\lambda} dz \qquad (8.95a)$$

and the lower bound a appearing as another arbitrary constant to be determined. Equation (8.95a) results in

$$f'(\theta;\lambda) = \int_a^\theta \cos[\lambda(\theta - z)](\sin z)^{\lambda-2} dz,$$

$$f''(\theta;\lambda) = (\sin\theta)^{\lambda-2} - \lambda \int_a^\theta \sin[\lambda(\theta - z)](\sin z)^{\lambda-2} dz. \qquad (8.95b)$$

The eigenvalues (λ) and eigenfunctions (H) are to be determined from the boundary condition (8.90). Substituting equation (8.94) into (8.90), three algebraic equations result for the determination of a nontrivial solution:

$$\lambda D_2 + f'(0;\lambda)D_3 = 0, \qquad (8.96a)$$

$$\left[a_1 \cos(\lambda\pi) - \lambda a_2 \sin(\lambda\pi) - \lambda^2 a_3 \cos(\lambda\pi)\right]D_1$$

$$+ \left[a_1 \sin(\lambda\pi) + \lambda a_2 \cos(\lambda\pi) - \lambda^2 a_3 \sin(\lambda\pi)\right]D_2$$

$$+ \left[a_1 f(\pi;\lambda) + a_2 f'\cos(\pi;\lambda) + a_3 f''(\pi;\lambda)\right]D_3 = 0, \qquad (8.96b)$$

$$\left[b_1 \cos(\lambda\pi) - \lambda b_2 \sin(\lambda\pi) - \lambda^2 b_3 \cos(\lambda\pi)\right]D_1$$

$$+ \left[b_1 \sin(\lambda\pi) + \lambda b_2 \cos(\lambda\pi) - \lambda^2 b_3 \sin(\lambda\pi)\right]D_2$$

$$+ \left[b_1 f(\pi;\lambda) + b_2 f'\cos(\pi;\lambda) + b_3 f''(\pi;\lambda)\right]D_3 = 0. \qquad (8.96c)$$

with prime denoting differentiation with respect to θ, $f' \equiv df/d\theta$. For obtaining non-trivial solutions of D, the determinant of coefficients in equation (8.96) must vanish. A careful arrangement results in

$$\left[(a_1 b_2 - b_1 a_2) + \lambda^2(a_2 b_3 - b_2 a_3)\right]\left\{\int_0^a \cos(\lambda z)(\sin z)^{\lambda-2} dz\right.$$

$$\left. + \int_a^\pi \cos(\lambda z)(\sin z)^{\lambda-2} dz\right\} = 0. \qquad (8.97)$$

The first integral results from $f'(0;\lambda)$ in equation (8.96a), while the second integral results from the combination of $f(\pi;\lambda)$, $f'(\pi;\lambda)$, and $f''(\pi;\lambda)$ in equations (8.96b) and (8.96c). This intermediate step explicitly shows that the lower bound a selected in equation (8.95a) for f is indeed *immaterial*. It can be arbitrarily chosen without affecting the solution. Since the coefficients a and b are arbitrary in nature, equation (8.97) implies

$$\int_0^\pi \cos(\lambda z)(\sin z)^{\lambda-2}\, dz = 0\,, \tag{8.98}$$

which is the eigenequation for the present problem, with λ being the eigenvalue to be determined.

8.4.1 Eigenvalues

Due to *singularities* of $\sin(z)$ existing at 0 and π in the case of $\lambda < 2$, first, the eigenvalue λ must be greater than or equal to 2 ($\lambda \geq 2$). For $\lambda < 2$, in fact, $f'(0;\lambda), f(\pi;\lambda), f'(\pi;\lambda)$, and $f''(\pi;\lambda)$ in equations (8.96a) to (8.96c) all approach infinity, implying that D_3 must be equal to zero ($D_3 = 0$). Consequently, equations (8.96a) to (8.96c) yield $D_1 = D_2 = 0$, resulting in a *trivial* solution for $H(\theta)$ defined in equation (8.94). For a nontrivial solution to exist, therefore, the eigenvalue λ must be greater than or equal to 2.

For $\lambda \geq 2$, equation (8.98) can be directly integrated to give

$$(\lambda - 1)\Gamma\!\left(\frac{\lambda - 1}{2}\right) - 2\Gamma\!\left(\frac{\lambda + 1}{2}\right) = 0\,, \tag{8.99}$$

with $\Gamma(z)$ being the gamma function defined as

$$\Gamma(z) = \int_0^\infty e^{-y} y^{z-1} dy\,. \tag{8.100}$$

According to the recurrence relation of gamma functions, $\Gamma(1+z) = z\,\Gamma(z)$, however,

$$\Gamma\!\left(\frac{\lambda + 1}{2}\right) = \left(\frac{\lambda - 1}{2}\right)\Gamma\!\left(\frac{\lambda - 1}{2}\right). \tag{8.101}$$

Substituting equation (8.101) into (8.99) results in a *permanent identity*, which is satisfied by *all* values of $\lambda \geq 2$. The special eigenstructure shown by equation (8.98) thus results in a *continuous* spectrum of eigenvalues, $\lambda \geq 2$. According to equation (8.34), therefore, the temperature field in the near-tip region is

$$T(r,\theta,t) = \Gamma(t) r^2 H(\theta; \lambda = 2), \quad \text{implying} \quad \frac{\partial T}{\partial r} \sim r \quad \text{for} \quad r \to 0\,. \tag{8.102}$$

The temperature gradient is *bounded* at the crack tip. The near-tip behavior proportional to r is *identical* to that in the supersonic regime with $M > 1$ shown in Table 8.1. When additional delay caused by the microstructural interaction comes into the picture, evidently, the temperature gradient at the crack tip remains finite, implying *diminution* of thermal energy accumulation in the near-tip region. Equation (8.88) governing the near-tip temperature in the presence of the microstructural interaction effect is *independent* of the thermal Mach number (and hence the crack speed). Continuing the near-tip behavior in the supersonic regime, it seems that the r type behavior of the temperature gradient is *ultimate* regardless of the crack speed.

8.4.2 Eigenfunctions

In the near-tip region with $r \to 0$, the eigenfunctions in correspondence with the eigenvalues greater than 2 ($\lambda > 2$) approach zero *faster* than the eigenfunction with the *lowest* eigenvalue, $\lambda = 2$. The eigenfunction corresponding to $\lambda = 2$ thus characterizes the fundamental behavior in the near-tip region.

For $\lambda = 2$, equation (8.95a) can be integrated directly to give

$$f(\theta; \lambda = 2) = \frac{1}{4} - \left[\frac{\cos(2a)}{4} \right] \cos(2\theta) - \left[\frac{\sin(2a)}{4} \right] \sin(2\theta). \qquad (8.103)$$

Although the arbitrary constant a returns, it does not affect the eigenstructure because it only appears in the *amplitudes* of the fundamental eigenfunctions. In an eigenvalue problem, recall that the modal shape (fundamental eigenfunction) can be determined within an undetermined coefficient, the amplitude, which depends on the strength of the external excitation. Combining with equation (8.94) results in the angular distribution of temperature in the near-tip region:

$$H(\theta) = C_1 \cos(\lambda\theta) + C_2 \sin(\lambda\theta) + C_3 f(\theta; \lambda), \qquad (8.104)$$

with the old coefficients D included in the new coefficients C. The boundary condition (8.90a) requires $C_2 = 0$. The crack-surface conditions, equations (8.90b) and (8.90c), on the other hand, imply

$$C_3 = \left[4\left(\frac{a_3}{a_1}\right) - 1 \right] C_1 \quad \text{and} \quad C_3 = \left[4\left(\frac{b_3}{b_1}\right) - 1 \right] C_1. \qquad (8.105)$$

Substituting equation (8.105) into (8.104),

$$H(\theta) = C_1 \left\{ \cos(2\theta) + \left[4\left(\frac{a_3}{a_1}\right) - 1 \right] \right\} \quad \text{and} \quad H(\theta) = C_1 \left\{ \cos(2\theta) + \left[4\left(\frac{b_3}{b_1}\right) - 1 \right] \right\}.$$

$$(8.106)$$

Two eigenfunctions, in other words, correspond to the same eigenvalue $\lambda = 2$, resulting a situation of *degeneracy*. Along with the continuous spectrum of eigenvalues, this is another special feature of the lagging behavior in the near-tip region. The two eigenfunctions shown in equation (8.106) become identical when

$$\frac{a_3}{a_1} = \frac{b_3}{b_1}. \tag{8.107}$$

Dividing a_1 ($\neq 0$) through equation (8.90b) and b_1 ($\neq 0$) through equation (8.90c), equation (8.107) implies an equal coefficient of the second-order derivative of $H(\theta)$ in the boundary conditions. Including the case of $a_2 = a_3 = 0$ in equation (8.90b) (a temperature-specified crack-surface condition) and $b_1 = b_3 = 0$ in equation (8.90c) (a gradient-specified crack-surface condition), degeneracy of eigenfunctions disappears when equation (8.107) is satisfied. By examining the eigenstructure shown in equation (8.106), in addition, notice that the second eigenfunction only differs from the first by a constant. Mathematically, eigenfunctions of this type can actually be generated by the Gram-Schmidt orthogonalization procedure (Arfken, 1970) based on any one of the two eigenfunctions, implying the existence of a weak degeneracy in this problem. Consideration of one eigenfunction in equation (8.106) is thus representative.

Figure 8.11 displays the angular distribution of the near-tip temperature, $H(\theta)$, for various values of a_3/a_1. They are basically the cosine curves with different bases depending on the value of $4(a_3/a_1) - 1$. At a certain value of a_3/a_1, the other curves can be viewed as the degenerated eigenfunction in equation (8.106) with a different value of b_3/b_1. The weak degeneracy results in a rigid shift of the eigenfunctions in this sense.

Generally speaking, degeneracy of eigenfunctions results from three boundary conditions considered at the crack surface. This unusual situation results from the *third*-order differential equation governing the angular distribution of the

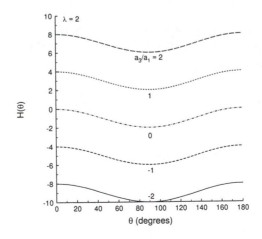

Figure 8.11 Angular distribution of the near-tip temperature and degenerated eigenfunction shown in equation (8.106).

near-tip temperature, equation (8.89), which reflects solely the microstructural interaction effect for heat transport in the near-tip region, equation (8.88). The microstructural interaction effect completely alters the eigenstructure of the temperature in the near-tip region, but the lowest eigenvalue, $\lambda = 2$, remains exactly the same as that obtained earlier in the supersonic regime with $M > 1$. Referring to equation (8.34), in other words, the near-tip temperature behaves like r^2 in the presence of the microstructural interaction effect, resulting in a bounded heat flux vector at the crack tip as $r \to 0$.

For a stationary crack under steady state, the dual-phase-lag model reduces to Fourier's law in heat conduction. The abrupt change of geometric curvature at the crack tip results in a $1/\sqrt{r}$ type of singularity in both the temperature gradient and the r component of the heat flux vector. Intensified thermal energy accumulation exists in this case, and the thermal damage around the crack tip is measured by the value of IFTG. Once the crack starts to grow, either transgranularly or intergranularly, both the thermal wave model (macroscopic in space but microscopic in time) and the dual-phase-lag model (microscopic in both space and time) result in a *bounded* heat flux vector and the temperature gradient at the crack tip. A bounded behavior of the heat flux vector at the crack tip indicates diminution of thermal energy intensification at a high crack speed or in small scale, supporting the fact that a stationary crack in a workpiece deserves more attention than a suddenly formed crack. This conclusion, of course, is restricted to a stably growing crack prior to catastrophic crack propagation. When the crack length reaches the critical value for global crack instability, although no significant energy accumulation exists in the near-tip region, the dynamically propagating crack will mechanically destroy the entire workpiece. A thorough study for the thermomechanical interaction in the entire stage from stable crack growth to catastrophic crack propagation can be found in an earlier work by Tzou (1987).

8.5 HIGH HEAT FLUX AROUND A MICROVOID

Compared with the steady state and propagation stages, the *transient* stage of heat transport around a *stationary* defect presents the most critical situation in energy accumulation. Even for a spherical void with a relatively smoother change of geometrical curvature than a crack tip, localized high heat fluxes will form in the early part of a fast-transient stage. The flux localization is a combined effect of microstructural interaction, fast-transient behavior of thermal inertia, and sudden change of geometric curvature on the surface of defects. Viewing the geometric effect alone, the two-dimensional plate containing a circular hole discussed in Section 8.1 provides an example. The flux localization occurs at the west and east sides of the hole, with a local amplification factor of 2.

A spherical microvoid in a three-dimensional body is considered here for more general treatment. The mathematical model focuses on the energy accumulation around a vertex defect at the interaction among grain boundaries, as illustrated in Figure 8.12. Since the characteristic dimension of the microvoid is on the subgrain level, impingement of the incoming heat flux on the microvoid can be simulated in an infinite domain without edge effects. Depending on the type of

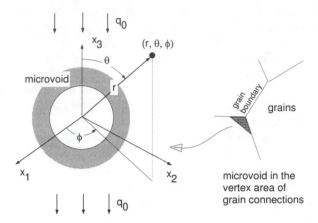

Figure 8.12 Idealization of the vertex microvoid on the grain boundary and the spherical coordinate system describing the lagging behavior in the short-time transient.

material in which the microvoid exists, heat transport in the local area of the microvoid can be either of the phonon-electron interaction or the phonon-scattering type. Owing to the presence of directional grain boundaries in the neighborhood of the microvoid, local scattering of phonons may consequently become directional. For developing a fundamental understanding of the flux localization around a microvoid, however, the effect of anisotropic scattering of phonons will be neglected.

At the short-time transient where the microstructural interaction effect on the delayed response is present, the energy equation describing the lagging behavior is equation (8.27b):

$$\nabla^2 T(\vec{r},t) + \tau_T \frac{\partial}{\partial t} \nabla^2 T(\vec{r},t) = \frac{1}{\alpha}\frac{\partial T}{\partial t} + \frac{\tau_q}{\alpha}\frac{\partial^2 T}{\partial t^2}. \qquad (8.27b')$$

For microscale heat transport dominated by phonon-electron interaction,

$$\alpha = \frac{K}{C_e + C_l}, \quad \tau_T = \frac{C_l}{G}, \quad \tau_q = \frac{1}{G}\left(\frac{1}{C_e} + \frac{1}{C_l}\right)^{-1}, \qquad (8.108)$$

with K being the thermal conductivity of the electron gas, C_e and C_l the heat capacity of the electron gas and the metal lattice, respectively, and G the phonon-electron coupling factor. For the phonon-scattering-dominated mechanism, on the other hand,

$$\alpha = \frac{\tau_R c^2}{3}, \quad \tau_T = \frac{9}{5}\tau_N, \quad \tau_q = \tau_R, \qquad (8.109)$$

with τ_R and τ_N referring to the relaxation times characterizing the "umklapp" (momentum lost) and normal (momentum conserved) processes of phonon collision, respectively. They are the correlations derived in Chapter 2.

8.5.1 Mathematical Formulation

Owing to the symmetry of the spherical geometry and impingement of heat flow, the temperature distribution around the microvoid is independent of the azimuthal angle ϕ. The temperature distribution, in other words, is only a function of r and θ. The Laplacian operator in equation (8.27b) is thus

$$\nabla^2 T = \frac{\partial^2 T}{\partial r^2} + \frac{2}{r}\frac{\partial T}{\partial r} + \frac{1}{r^2 \sin\theta}\frac{\partial}{\partial\theta}\left(\sin\theta\frac{\partial T}{\partial\theta}\right). \tag{8.110}$$

With equation (8.110), equation (8.27b) describing the lagging response in small scale becomes a third-order partial differential equation with variable coefficients. It must be solved under specified initial and boundary conditions. The microvoid is assumed to be disturbed from a stationary state, implying a uniform initial temperature and a zero time-rate of change of temperature as $t = 0$:

$$T = T_0, \quad \frac{\partial T}{\partial t} = 0 \quad \text{as} \quad t = 0. \tag{8.111}$$

Heat loss from the microvoid surface is assumed negligible in the short-time transient, implying

$$q_r = 0 \quad \text{at} \quad r = a. \tag{8.112}$$

Physically, equation (8.112) assumes that heat convection into the aerial closure of the microvoid requires longer times to become effective than that for small-scale heat transport to take place. This is equivalent to the negligence of heat loss from the microvoid surface in the short-time response. Owing to negligence of heat flow across the microvoid surface, the resulting intensity factor of the heat flux factor (IFHF, to be discussed below) will be the *highest* among all the other cases with energy release. At a distance far from the microvoid, the heat flux is equal to the incoming heat flux,

$$q_3 = -q_0 \quad \text{as} \quad x_3 \to \infty, \quad x_3 = r\cos\theta. \tag{8.113}$$

Before making an effort to determine the solution satisfying equations (8.27b) and (8.111) to (8.113), it should be noted that a spherical geometry for describing the microvoid shape is a mathematical idealization. In reality, microvoids not only have irregular shapes, but may also randomly vary from one sample to another. The solution involving a simple geometry such as a sphere, however, serves as the fundamental solution for the more refined stochastic analysis averaging over the statistical ensemble space (Tzou, 1988, 1989e).

Equation (8.27b) subject to the initial condition (8.111) and boundary conditions (8.112) and (8.113) is to be solved by semi-analytical means. For identifying the physical parameter characterizing the lagging response around the microvoid, a dimensionless analysis is always desirable. Introducing

$$\Theta = \frac{T - T_0}{T_0}, \quad \beta = \frac{t}{\tau_q}, \quad [\delta, \xi, A] = \frac{[r, x_3, a]}{\sqrt{\alpha \tau_q}},$$

$$[\eta_r, \eta_\theta, \eta_3, \eta_0] = \frac{[q_r, q_\theta, q_3, q_0]}{kT_0 / \sqrt{\alpha \tau_q}}, \quad B = \frac{\tau_T}{\tau_q}, \tag{8.114}$$

equations (8.27b) and (8.111) to (8.113) become

$$\left(1 + B\frac{\partial}{\partial \beta}\right)\nabla_s^2 \Theta = \left(1 + B\frac{\partial}{\partial \beta}\right)\left[\frac{\partial^2 \Theta}{\partial \delta^2} + \frac{2}{\delta}\frac{\partial \Theta}{\partial \delta} + \frac{1}{\delta^2 \sin\theta}\frac{\partial}{\partial \theta}\left(\sin\theta \frac{\partial \Theta}{\partial \theta}\right)\right] = \frac{\partial \Theta}{\partial \beta} + \frac{\partial^2 \Theta}{\partial \beta^2},$$

$$\tag{8.115}$$

$$\Theta = 0, \quad \frac{\partial \Theta}{\partial \beta} = 0 \quad \text{as} \quad \beta = 0, \tag{8.116}$$

$$\eta_r = 0 \quad \text{at} \quad \delta = A, \tag{8.117}$$

$$\eta_3 = -\eta_0 \quad \text{as} \quad \xi \to \infty, \quad \xi = \delta\cos\theta. \tag{8.118}$$

The quantity $\nabla_s^2 \Theta$ in equation (8.115) refers to the Laplacian operator in the dimensionless spherical coordinates. Since the problem thus formulated involves heat-flux-specified boundary conditions, the dimensionless form of equation (2.7) describing the lagging behavior will be needed:

$$[q_r, q_\theta, q_3] + \tau_q \frac{\partial}{\partial t}[q_r, q_\theta, q_3] = -k\left\{\left[\frac{\partial T}{\partial r}, \frac{1}{r}\frac{\partial T}{\partial \theta}, \frac{\partial T}{\partial x_3}\right] + \tau_T \frac{\partial}{\partial t}\left[\frac{\partial T}{\partial r}, \frac{1}{r}\frac{\partial T}{\partial \theta}, \frac{\partial T}{\partial x_3}\right]\right\},$$

$$\tag{8.119}$$

implying

$$[\eta_r, \eta_\theta, \eta_3] + \frac{\partial}{\partial \beta}[\eta_r, \eta_\theta, \eta_3] = -\left[\frac{\partial \Theta}{\partial \delta}, \frac{1}{\delta}\frac{\partial \Theta}{\partial \theta}, \frac{\partial \Theta}{\partial \xi}\right]$$

$$- B\left[\frac{\partial^2 \Theta}{\partial \beta \partial \delta}, \frac{1}{\delta}\frac{\partial^2 \Theta}{\partial \beta \partial \theta}, \frac{\partial^2 \Theta}{\partial \beta \partial \xi}\right] \tag{8.120}$$

according to equation (8.114). Note that a mixed use of cylindrical and spherical coordinate systems is made here. Derivatives of the ϕ component in the spherical coordinate system vanish due to azimuthal symmetry, while the x_3 component (ξ component) in the cylindrical coordinate system is related to the (r, θ) coordinates in the spherical coordinate system by

$$x_3 = r\cos\theta \quad \text{or} \quad \xi = \delta\cos\theta, \tag{8.121}$$

as used in equation (8.118). When x_3 approaches infinity, in terms of the spherical coordinates (r, θ) with azimuthal symmetry, it is equivalent to

$$\xi \to \infty \quad \Rightarrow \quad \delta \to \infty \quad \text{and} \quad \theta \to 0. \tag{8.122}$$

Equation (8.122) will be used below in the determination of the far-field solution satisfying the remote condition (8.118).

The parameter B, ratio of τ_T to τ_q weighing the relative delay times due to the microstructural interaction and fast-transient effects, dominates the lagging behavior in the short-time response. In the case of B approaching 1 ($B \to 1$), $\tau_T = \tau_q$, equation (8.115) reduces to

$$\left[\nabla_s^2\Theta - \frac{\partial\Theta}{\partial\beta}\right] + \frac{\partial}{\partial\beta}\left[\nabla_s^2\Theta - \frac{\partial\Theta}{\partial\beta}\right] = 0, \tag{8.123}$$

rendering a diffusion equation, the bracketed quantity being zero, as a particular solution. It is thus evident that the classical diffusion behavior is retrieved as $B = 1$. Equation (8.119) reduces to Fourier's law in this case, and the response between and the heat flux vector becomes *instantaneous*, with τ_T or τ_q being a trivial shift in the timescale (see Chapter 2). Also, note that the phase lag of the heat flux vector, τ_q, does not exist in diffusion. The characteristic time used in the dimensionless time in equation (8.114) can be arbitrarily chosen (such as the diffusion time), while the remaining expressions for diffusion stay the same. In the case of $B \to 0$, $\tau_T = 0$, the term led by B vanishes, and equation (8.115) reduces to the classical CV wave equation. The second-order time derivative, $(\partial^2\Theta/\partial\beta^2)$ on the right side of the equation, reflects the familiar wave behavior.

The other cases of B absorb the microscopic phonon-electron interaction model (parabolic) and the phonon scattering model. In terms of the corresponding microscopic properties,

$$B = \begin{cases} 1 + \left(\dfrac{C_l}{C_e}\right), & \text{phonon-electron interaction model} \\ \left(\dfrac{9}{5}\right)\left(\dfrac{\tau_N}{\tau_R}\right), & \text{phonon scattering model.} \end{cases} \tag{8.124}$$

In the correlation to the phonon-electron interaction model, the ratio B depends on the ratio of heat capacities of the electron gas and the metal lattice. The phonon-electron coupling factor G does not appear as an explicit dominating parameter.

8.5.2 Linear Decomposition

Since the problem is linear, the temperature distribution, Θ in equation (8.115), can be decomposed into the steady-state component, $\Theta^{(s)}(\delta, \theta)$, and the fast-transient component, $\Theta^{(t)}(\delta, \theta, \beta)$:

$$\Theta(\delta,\theta,\beta) = \Theta^{(s)}(\delta,\theta) + \Theta^{(t)}(\delta,\theta,\beta) . \tag{8.125}$$

Substituting equation (8.125) into equations (8.115) to (8.118) and (8.120), the governing systems for the steady-state and the fast-transient solutions are obtained:

Steady-State System —

$$\frac{\partial^2 \Theta^{(s)}}{\partial \delta^2} + \frac{2}{\delta} \frac{\partial \Theta^{(s)}}{\partial \delta} + \frac{1}{\delta^2 \sin\theta} \frac{\partial}{\partial \theta}\left(\sin\theta \frac{\partial \Theta^{(s)}}{\partial \theta} \right) = 0 , \tag{8.126a}$$

$$\eta_3^{(s)} = -\eta_0 \quad \text{as} \quad \xi \to \infty, \quad \xi = \delta \cos\theta , \tag{8.126b}$$

$$\left[\eta_r^{(s)}, \eta_\theta^{(s)}, \eta_3^{(s)} \right] = -\left[\frac{\partial \Theta^{(s)}}{\partial \delta}, \frac{1}{\delta} \frac{\partial \Theta^{(s)}}{\partial \theta}, \frac{\partial \Theta^{(s)}}{\partial \xi} \right] . \tag{8.126c}$$

Fast-Transient System —

$$\left(1 + B\frac{\partial}{\partial \beta} \right)\left[\frac{\partial^2 \Theta^{(t)}}{\partial \delta^2} + \frac{2}{\delta} \frac{\partial \Theta^{(t)}}{\partial \delta} + \frac{1}{\delta^2 \sin\theta} \frac{\partial}{\partial \theta}\left(\sin\theta \frac{\partial \Theta^{(t)}}{\partial \theta} \right) \right] = \frac{\partial \Theta^{(t)}}{\partial \beta} + \frac{\partial^2 \Theta^{(t)}}{\partial \beta^2}$$

$$\tag{8.127a}$$

$$\Theta^{(t)} = \Theta_0, \quad \frac{\partial \Theta^{(t)}}{\partial \beta} = 0 \quad \text{as} \quad \beta = 0, \tag{8.127b}$$

$$\eta_3^{(t)} = 0 \quad \text{as} \quad \xi \to \infty, \quad \xi = \delta \cos\theta , \tag{8.127c}$$

$$\left[\eta_r^{(t)}, \eta_\theta^{(t)}, \eta_3^{(t)} \right] + \frac{\partial}{\partial \beta}\left[\eta_r^{(t)}, \eta_\theta^{(t)}, \eta_3^{(t)} \right] = -\left[\frac{\partial \Theta^{(t)}}{\partial \delta}, -\frac{1}{\delta} \frac{\partial \Theta^{(t)}}{\partial \theta}, \frac{\partial \Theta^{(t)}}{\partial \xi} \right]$$

$$-B\left[\frac{\partial^2\Theta^{(t)}}{\partial\beta\partial\delta},\frac{1}{\delta}\frac{\partial^2\Theta^{(t)}}{\partial\beta\partial\theta},\frac{\partial^2\Theta^{(t)}}{\partial\beta\partial\xi}\right].$$ (8.127d Cont.)

The steady-state and the fast-transient systems are assembled in a particular way. They are to satisfy the remote boundary condition as ξ approaches infinity ($\xi \to \infty$), referring to equation (8.118) and the combination of equations (8.126b) and (8.127c). Because the steady-state solution is not a function of time (β), all the initial conditions are satisfied as well, referring to equations (8.116) and (8.127b). The boundary condition at the microvoid surface, equation (8.117), is left to a later stage until $\Theta^{(s)}(\delta)$ and $\Theta^{(t)}(\delta,\beta)$ are obtained:

$$\eta_r^{(s)} + \eta_r^{(t)} = 0 \quad \text{at} \quad \delta = A .$$ (8.128)

8.5.3 Steady-State Solution

Before extracting the remote component of $\Theta^{(s)}$ from the steady-state solution, let us first study the full steady-state solution *including* the boundary condition at $r = a$:

$$q_r^{(s)} = 0, \quad \text{at} \quad r = a \quad \text{or} \quad \eta_r^{(s)} = 0, \quad \text{at} \quad \delta = A .$$ (8.129)

The governing system describing the *full* steady-state solution around a microvoid, in other words, is composed of equations (8.126a), (8.126b), and (8.129). At steady state, note that the heat flux vector relates to the temperature gradient by Fourier's law, equation (8.126c). The boundary condition (8.129) can thus be expressed as

$$\frac{\partial\Theta}{\partial\delta} = 0 \quad \text{at} \quad \delta = A .$$ (8.130)

The governing system remains the same as that in Section 8.1 for a circular hole, except to change from a two-dimensional to a three-dimensional problem with azimuthal symmetry. Equation (8.126a) allows the same type of product solution, equation (8.3), with $\lambda =1$ and -2 in the present case (Tzou, 1991g and h). This may not be obvious at first glance at equation (8.126a) owing to the variable coefficient of the last term. However, a direct substitution of equation (8.3) into (8.126a) provides a simple proof. Using the two boundary conditions, equations (8.126b) and (8.130), the steady-state temperature is obtained as

$$\Theta^{(s)}(\delta,\theta) = \eta_0\left[\delta+\frac{A^3}{2\delta^2}\right]\cos\theta .$$ (8.131)

The heat flux vector is given by Fourier's law, equation (8.126c), rendering

At the microvoid surface, $\delta = A$, the r component of the heat flux vector vanishes ($\eta_r = 0$) as required by the boundary condition (8.129), while the θ component possesses a *maximum* value of 3/2 at $\theta = \pm\pi/2$:

$$\text{IFHF} = \left(\frac{\eta_\theta}{\eta_0}\right)_{max} = \left(\frac{q_\theta}{q_0}\right)_{max} = \frac{3}{2} \quad \text{at} \quad \theta = \pm\frac{\pi}{2}. \tag{8.133}$$

The minus sign at $\theta = -\pi/2$ is omitted in equation (8.133) because it simply refers to the direction of η_θ flowing toward the south pole with reference to the orientations defined in Figure 8.12. Local heat fluxes at the east ($\pi/2$) and the west ($-\pi/2$) sides of the microvoid are amplified. The value of IFHF is 1.5 (3/2) for a spherical microvoid, *smaller* than the value of 2 for a circular hole (equation (8.8)) in the corresponding two-dimensional problem.

In real operations, the IFHF around defects should be minimized by all means. Local amplification of the heat flux vector implies enhancement of heat flow in the local area surrounding the defect. In most situations, it produces a large temperature *gradient* in the local area, which is the very reason for the hot-spot formation and, consequently, the thermal cracking if sufficiently severe. The need for the transient value of IFHF thus becomes evident. In the presence of small-scale effects in transporting heat, how do the microstructural interaction effect (small-scale effect in space) and the fast-transient effect of thermal inertia (small-scale effect in time) affect the value of IFHF? Should the transient value be smaller than the steady-state value, paying attention to the steady-state response would be sufficient for damage prevention in material processing. Should the transient value be greater than the steady-state value, on the other hand, damage prevention should focus on the short-time response.

In passing to the transient response, note that the solution shown by equation (8.131) is to illustrate the steady-state value of IFHF. For the latter combination with the transient solution to satisfy equation (8.128) at the microvoid surface, only the component satisfying the remote boundary condition (8.126b) is needed. The steady-state solution satisfying equations (8.126a) and (8.126b) is thus

$$\Theta^{(s)}(\delta,\theta) = \eta_0\delta\cos\theta \quad \text{as} \quad \xi \to \infty \quad (\delta \to \infty, \quad \theta = 0). \tag{8.134}$$

8.5.4 Fast-Transient Component

The transient response is governed by equations (8.127a) to (8.127d). Note that only the remote boundary condition appears in the governing system. It describes a heat flux *vanishing* at infinity because the incoming heat flux has been absorbed in the steady-state component, equation (8.134). Applying the Laplace transform to equation (8.127a) and using the initial conditions in equation (8.127b), the transformed energy equation reads as

$$\frac{\partial^2 \overline{\Theta}^{(t)}}{\partial \delta^2} + \frac{2}{\delta}\frac{\partial \overline{\Theta}^{(t)}}{\partial \delta} + \frac{1}{\delta^2 \sin\theta}\frac{\partial}{\partial\theta}\left(\sin\theta\frac{\partial \overline{\Theta}^{(t)}}{\partial\theta}\right) = D^2\overline{\Theta}^{(t)}, \quad \text{with} \quad D = \sqrt{\frac{p(p+1)}{1+Bp}}\,.$$

$$(8.135)$$

The Laplace transform method brings in another advantage to the present problem. The resulting equation (8.135) greatly resembles the steady-state equation (8.126a), allowing the same type of product solution as that in the steady state:

$$\overline{\Theta}^{(t)}(\delta,\theta;p) = F(\delta;p)\cos\theta\,.$$

$$(8.136)$$

Substituting equation (8.136) into (8.135) yields a Bessel type of differential equation governing $F(\delta;p)$:

$$\delta^2\frac{d^2F}{d\delta^2} + (2\delta)\frac{dF}{d\delta} - \left[2+(D\delta)^2\right]F = 0\,.$$

$$(8.137)$$

It falls into the category of generalized Bessel equations (Hildebrand, 1976), permitting a solution of the form

$$F(\delta;p) = \frac{1}{\sqrt{\delta}}\left[C_1 I_{3/2}(D\delta) + C_2 I_{-3/2}(D\delta)\right]\,.$$

$$(8.138)$$

with $I_n(\bullet)$ denoting the modified Bessel function of the first kind of order n. With the assistance of the identities,

$$I_{3/2}(D\delta) = \sqrt{\frac{2}{\pi D\delta}}\left[\cosh(D\delta) - \frac{\sinh(D\delta)}{D\delta}\right]\,,$$

$$(8.139a)$$

$$I_{-3/2}(D\delta) = \sqrt{\frac{2}{\pi D\delta}}\left[\sinh(D\delta) - \frac{\cosh(D\delta)}{D\delta}\right]\,,$$

$$(8.139b)$$

moreover, equation (8.138) can be expressed in terms of the familiar hyperbolic functions,

$$F(\delta;p) = \sqrt{\frac{2}{\pi D}}\frac{1}{\delta}\left\{C_1\left[\cosh(D\delta) - \frac{\sinh(D\delta)}{D\delta}\right] + C_2\left[\sinh(D\delta) - \frac{\cosh(D\delta)}{D\delta}\right]\right\}\,.$$

$$(8.140)$$

The transient temperature is thus

$$\overline{\Theta}^{(t)}(\delta,\theta;p) = \sqrt{\frac{2}{\pi D}}\frac{1}{\delta}\left\{C_1\left[\cosh(D\delta) - \frac{\sinh(D\delta)}{D\delta}\right]\right.$$

$$\left. + C_2\left[\sinh(D\delta) - \frac{\cosh(D\delta)}{D\delta}\right]\right\}\cos\theta, \tag{8.141}$$

with C_1 and C_2 being constants to be determined from the boundary conditions. The transformed boundary condition, equation (8.127c), is

$$\overline{\eta}_3^{(t)} = 0 \quad \text{as} \quad \xi \to \infty, \quad \xi = \delta\cos\theta. \tag{8.142}$$

It must be expressed in terms of temperature for the determination of C_1 and C_2. The transformed heat flux vector and the transformed temperature are related by the Laplace transform of equation (8.127d):

$$\left[\overline{\eta}_r^{(t)},\overline{\eta}_\theta^{(t)},\overline{\eta}_3^{(t)}\right] + p\left[\overline{\eta}_r^{(t)},\overline{\eta}_\theta^{(t)},\overline{\eta}_3^{(t)}\right] = -\left[\frac{\partial\overline{\Theta}^{(t)}}{\partial\delta}, -\frac{1}{\delta}\frac{\partial\overline{\Theta}^{(t)}}{\partial\theta}, \frac{\partial\overline{\Theta}^{(t)}}{\partial\xi}\right]$$

$$- Bp\left[\frac{\partial\overline{\Theta}^{(t)}}{\partial\delta}, \frac{1}{\delta}\frac{\partial\overline{\Theta}^{(t)}}{\partial\theta}, \frac{\partial\overline{\Theta}^{(t)}}{\partial\xi}\right]. \tag{8.143}$$

The ξ (x_3) component, therefore, is

$$\overline{\eta}_3^{(t)} = -\left(\frac{1+Bp}{1+p}\right)\frac{\partial\overline{\Theta}^{(t)}}{\partial\xi}, \tag{8.144}$$

implying an equivalent condition to equation (8.142),

$$\frac{\partial\overline{\Theta}^{(t)}}{\partial\xi} = 0 \quad \text{as} \quad \xi \to \infty, \quad \xi = \delta\cos\theta. \tag{8.145}$$

Taking the derivative of equation (8.141) with respect to ξ ($= \delta\cos\theta$) and noting that both hyperbolic sine and hyperbolic cosine functions behave the same at a large value of δ,

$$\lim_{\delta\to\infty}\left\{\frac{\sinh(D\delta)}{\cosh(D\delta)}\right\} = \lim_{\delta\to\infty}e^{D\delta}, \tag{8.146}$$

the temperature gradient in the ξ direction can be obtained as

$$\lim_{\xi \to \infty} \frac{\partial \overline{\Theta}^{(t)}}{\partial \xi} = \lim_{\substack{\delta \to \infty \\ \theta \to 0}} \frac{\partial \overline{\Theta}^{(t)}}{\partial \delta(\cos \theta)} = \sqrt{\frac{2}{\pi D}} \left[\frac{(D\delta)^2 - 2(D\delta) + 2}{D\delta^3} \right] \left(C_1 + C_2 \right) e^{D\delta}.$$

$$(8.147)$$

Since the exponential function approaches infinity as δ approaches infinity, the only choice to satisfy equation (8.145) is $C_2 = -C_1$. The transformed transient temperature is thus

$$\overline{\Theta}^{(t)}(\delta, \theta; p) = C_1 \sqrt{\frac{2}{\pi D}} \frac{1 + D\delta}{D\delta^2} \left[\cosh(D\delta) - \sinh(D\delta) \right] \cos \theta.   (8.148)$$

The remaining unknown C_1 is now ready to be determined from the Laplace transform of the boundary condition (8.128):

$$\overline{\eta}_r^{(s)} + \overline{\eta}_r^{(t)} = 0 \quad \text{at} \quad \delta = A,   (8.149)$$

or, in terms of the transformed temperature gradients by the use of equations (8.126c) and (8.144),

$$\frac{\partial \overline{\Theta}^{(s)}}{\partial \delta} + \left(\frac{1 + Bp}{1 + p} \right) \frac{\partial \overline{\Theta}^{(t)}}{\partial \delta} = 0 \quad \text{at} \quad \delta = A,   (8.150)$$

with $\overline{\Theta}^{(s)}(\delta, \theta) = (\eta_0 \delta \cos \theta)/p$ from equation (8.134). Substituting equation (8.148) into (8.150) and solving for C_1 gives

$$C_1 = -\frac{\eta_0}{p} \left(\frac{1 + p}{1 + Bp} \right) \sqrt{\frac{\pi D}{2}} \frac{DA^3}{[(DA)^2 + 2(DA) + 2]} \times \frac{1}{[\sinh(DA) - \cosh(DA)]}.$$

$$(8.151)$$

This furnishes the analysis for temperature in the Laplace transform domain.

8.5.5 Flux Intensification

The θ component of the heat flux vector is related to the temperature gradient by equation (8.143):

$$\overline{\eta}_\theta = -\left(\frac{1 + Bp}{1 + p} \right) \frac{1}{\delta} \frac{\partial \overline{\Theta}}{\partial \theta}, \quad \overline{\Theta} = \overline{\Theta}^{(s)} + \overline{\Theta}^{(t)}.   (8.152)$$

Substituting the results of equations (8.148) and (8.151),

$$\frac{\overline{\eta}_\theta}{\eta_0} = -\frac{1+Bp}{p(1+p)}\sin\theta\left\{\frac{DA^3}{[(DA)^2 + 2(DA) + 2][\sinh(DA) - \cosh(DA)]}\right.$$

$$\left. \times \frac{(1+D\delta)[\cosh(D\delta) - \sinh(D\delta)]}{D\delta^3}\left(\frac{1+p}{1+Bp}\right) - 1\right\}. \qquad (8.153)$$

Compared to the steady-state response, equation (8.131), the transient effect does *not* alter the structure in the θ direction. The maximum heat flux occurs at the microvoid surface at $\delta = A$ and $\theta = \pm\pi/2$. Equation (8.153) gives

$$\left(\frac{\overline{\eta}_\theta(p)}{\eta_0}\right)_{max} = \left(\frac{\overline{q}_\theta(p)}{q_0}\right)_{max} = \frac{1+Bp}{p(1+p)}\left[\left(\frac{1+p}{1+Bp}\right)\frac{1+DA}{(DA)^2 + 2(DA) + 2} + 1\right]$$

$$(8.154)$$

where $D = [p(p+1)/(1+Bp)]^{1/2}$. From equation (8.154), evidently, the transient value of IFHF depends not only on the ratio of two phase lags ($B = \tau_T/\tau_q$) but also on the dimension of the microvoid (A).

Equation (8.154) can be inverted in the same manner by the Riemann-sum approximation:

$$IFHF = \left(\frac{q_\theta(t)}{q_0}\right)_{max} = \frac{e^{4.7}}{t}\left[\frac{1}{2}\left(\frac{\overline{q}_\theta(4.7/t)}{q_0}\right)_{max}\right.$$

$$\left. + Re\sum_{n=1}^{N}(-1)^n\left(\frac{\overline{q}_\theta[(4.7 + in\pi)/t]}{q_0}\right)_{max}\right] \qquad (8.155)$$

The expression in equation (8.154) can thus be programmed into the function subroutine FUNC(P) in the FORTRAN code in the Appendix.

Note that an arbitrary characteristic time (such as diffusion time) is used in the case of diffusion for making the physical time (t) dimensionless. When forming the ratio of $q_\theta(t)$ to q_0 from that of $\eta_\theta(t)$ to η_0, this arbitrarily chosen characteristic time is canceled, demonstrating that such an arbitrarily chosen time is indeed immaterial in the final result of IFHF.

The numerical results of transient IFHF for diffusion ($B = 1$) and *CV* wave ($B = 0$) are first shown in Figure 8.13. An intrinsic difference exists between the two cases. Because the diffusion model employing Fourier's law assumes an *immediate* response, implying that the remotely applied heat flux q_0 arrives at the microvoid surface at zero time, the IFHF starts at a value of 1 ($q_\theta = q_0$) as $\beta(t) = 0$. The *CV* wave model accounting for the finite speed of heat propagation, on the other hand, starts at a value of zero for the IFHF, implying the absence of any thermal signal ($q_\theta = 0$) as $\beta = 0$. Both models approach the steady-state value, 3/2 or 1.5 as shown by equation (8.133), as transient time lengthens. Due to the assumption of an

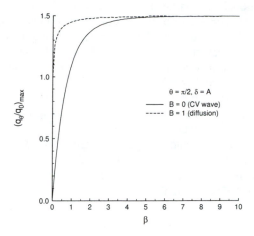

Figure 8.13   Transient response of the intensity factor of heat flux for classical diffusion ($B = 1$) and CV wave ($B = 0$) models with $A = 1$.

immediate response, however, the diffusion model approaches the steady-state value faster than the CV wave model.

At the same size of microvoid, $A = 1$, Figure 8.14 shows the effect B, the ratio of two phase lags τ_T/τ_q, on the transient value of IFHF. It shows that the transient value of IFHF significantly increases with the value of B, implying a more intensified localization of heat flux as either the delay time due to the microstructural interaction effect (τ_T) *increases* or the delay time due to the fast-transient effect of thermal inertia (τ_q) *decreases*. The transient curve for the classical CV wave model is included, showing that the results predicted by the dual-phase-lag model with different values of B approach the same steady-state value (3/2 or 1.5) as the transient time lengthens.

The difference between the dual-phase-lag model and the classical

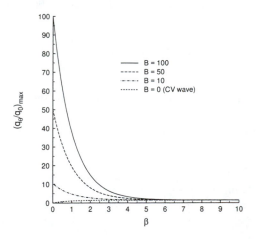

Figure 8.14   Transient response of the intensity factor of heat flux predicted by the dual-phase-lag model for $A = 1$, $B = 0$ (CV wave), 10, 50, and 100.

diffusion and wave models is significant. For the case of $B = 100$, the typical value in the microscopic phonon-electron interaction model (Tzou, 1995a to c), the difference may exceed for 2 orders of magnitude. Owing to negligence of the lagging behavior in the microstructural interactions, the classical diffusion and CV wave models cannot even preserve the qualitative trend in the short-time response. While the macroscopic models (diffusion and CV wave) were shown to fail in describing the heating and thermalization processes in the fast-transient heating of the electron gas (Qiu and Tien, 1992, 1993, 1994), Figure 8.14 reveals their limitations in terms of the transient response of IFHF.

The high heat flux around a microvoid may explain the large difference between the prediction and the experimental result for the spectrally dependent reflectivity change in metal films; see the Introduction section. Since the thermal energy at short times becomes highly localized in the vicinity of the microvoid, shown by the large value of IFHF in Figure 8.14, the electron gas in the neighborhood of the film surfaces receives less thermal energy from the laser beam. The reflectivity change attributed to the temperature change of the electron gas (Qiu and Tien, 1992, 1993, 1994), consequently, will be *lower* than the analytical result, assuming a perfect (defect free) film. Since the transient value of IFHF is large, the perfect-film assumption is expected to produce significantly higher reflectivity change than the experimental result due to the possible effect from microvoids. A rough estimate by Tzou (1995a to c) indicates that the value of τ_q is around 8.5 picoseconds for a gold film. This is approximately the scale of transient times shown in Figures 8.13 and 8.14.

A closer inspection of Figures 8.13 and 8.14 shows that the value of IFHF at $\beta = 0$ follows *exactly* the value of B. This helpful clue can be justified by the limiting theorem in the Laplace transform:

$$\lim_{p \to \infty} p \left(\frac{\overline{q}_\theta(p)}{q_0} \right) = \lim_{t \to 0^+} \left(\frac{q_\theta(t)}{q_0} \right). \tag{8.156}$$

Noting that the parameter D approaches $\sqrt{p/B}$ and the ratio $(1+Bp)/(1+p)$ approaches B as the value of p approaches infinity, equation (8.154) results in

$$\lim_{t \to 0^+} \left(\frac{q_\theta(t)}{q_0} \right) = \lim_{p \to \infty} B \left[\left(\frac{1}{A\sqrt{B}} \right) \frac{1}{\sqrt{p}} + 1 \right] = B. \tag{8.157}$$

The initial value of IFHF, which is exactly identical to the value of B, is thus verified.

The long-time response of IFHF can be confirmed by another limiting theorem,

$$\lim_{p \to 0} p \left(\frac{\overline{q}_\theta(p)}{q_0} \right) = \lim_{t \to \infty} \left(\frac{q_\theta(t)}{q_0} \right). \tag{8.158}$$

Noting that the parameter D approaches zero and the ratio $(1+Bp)/(1+p)$ approaches 1 as the value of p approaches zero, equation (8.154) results in

$$\lim_{t\to\infty}\left(\frac{q_\theta(t)}{q_0}\right) = \frac{3}{2}, \qquad (8.159)$$

the steady-state value shown by equation (8.133).

The intensified flux localization induces highly elevated temperatures in the vicinity of the microvoid at short times. Figure 8.15 displays the distributions of temperature in the direction away from the microvoid surface at various times. The local temperature rise develops to about 20 times higher than the ambient temperature at $\beta = 1$. It gradually tapers off as the transient time increases to 5. This highly localized temperature is noteworthy in thermal processing of materials because it may be the major cause of hot-spot formation.

At least in appearance, referring to equation (8.154), the transient value of IFHF should depend on the size of microvoid absorbed in parameter A. In terms of the real physical dimension, a value of A being 1 implies a microvoid on the order of nanometers. When varying the value of A from 1 to 1000 at $B = 50$ and 100, however, the transient curves of IFHF shown in Figure 8.14 *remain the same*. This shows the transient response is in fact *independent* of the microvoid size, at least for the present problem involving a microvoid free from the edge effect in the conducting medium. The size independence seems to be an extension of the steady-

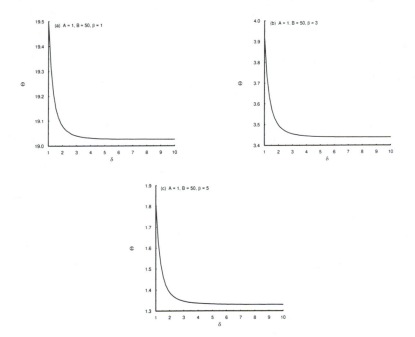

Figure 8.15 Transient distributions of temperature at (*a*) $\beta = 1$, (*b*) $\beta = 3$, and (*c*) $\beta = 5$ in the direction away from the microvoid surface.

state response shown by equation (8.133).

Intensification of heat fluxes around a microvoid in the short-time transient has been analyzed in this section. The small-scale effect of heat transport in space (microstructural interaction effect) interweaves with the small-scale effect in time (fast-transient effect of thermal inertia) in the short-time response, rendering extremely high heat fluxes in the vicinity of the microvoid. The highest value of IFHF is B, the ratio of τ_T to τ_q , which occurs at the initial time ($t = 0$) of the transient process. For femtosecond heat transport in metals, its value can be several tens times *higher* than that predicted by the classical diffusion or *CV* wave models. The macroscopic models (diffusion and *CV* wave) neglecting the microstructural interaction effect, more seriously, *underestimate* the high heat flux surrounding the microvoid.

In the course of approaching the steady-state value of IFHF, the linearized dual-phase-lag model accounting for the first-order effect of τ_T and τ_q displays a distinct pattern. Starting from the largest value (IFHF = B) at $t = 0$, the transient value of IFHF predicted by the dual-phase-lag model monotonically *decreases* to its steady-state value from *above*; see Figure 8.14. Both diffusion and *CV* wave models, on the other hand, predict that transient values monotonically *increase* to the steady-state value from *below*; see Figure 8.13. Such qualitatively reversed and quantitatively deviant results are a continuation of the large discrepancies of the macroscopic models in predicting the fast-transient temperature reported by Qiu and Tein (1992, 1993, 1994).

Under the same amount of laser heating provided to a metal film, accumulation of thermal energy (reflected by the IFHF) around the internal microvoid reduces the amount available to heat the electron gas near the film surfaces. Consequently, the resulting reflectivity change calculated from the temperature change of the electron gas at the film surfaces would be less than that estimated on the basis of a perfect (defect free) film. Since the value of IFHF is very large especially at short times (several picoseconds), the amount of reduction from the result assuming a perfect film is expected to be significant. This supports the suspicion (Aspnes et al., 1980; Qiu and Tien, 1994) that the exaggeration of the estimated spectrally-dependent reflectivity change during early times in femtosecond laser heating may be due to the presence of microvoids in the gold film sample. A quantitative analysis that actually implements microvoids in a gold film subjected to short-pulse laser heating is still needed to study the exact amount of reduction.

Formation of high heat fluxes around microvoids is dominated by the ratio of two phase lags, $B = \tau_T / \tau_q$. In terms of the microscopic properties, this implies dominance of the ratio of heat capacities (C_l/C_e) in phonon-electron interaction and of the ratio of relaxation times (τ_N/τ_R) in phonon scattering. The size (A) of microvoids appears in the expression for the transient IFHF, but it has no effect on the transient response of IFHF for microvoids on the nanometer scale.

NINE

THERMOMECHANICAL COUPLING

When a significant temperature occurs in a solid, deformation occurs due to thermal expansion. Depending on the way in which the solid is constrained, such a thermally induced deformation may either elongate, warp, bend the structure, or develop a highly elevated stress within the solid if deformation is prevented. While previous chapters investigated the lagging behavior of heat transport in rigid conductors, this chapter is devoted to the effect of lagging behavior on the straining of *deformable* bodies. Since both the microstructural interaction effect and the fast-transient effect of thermal inertia are special behaviors in time, a one-dimensional solid in space will be sufficient to illustrate the salient features in the short-time response. The deformation and stress waves, in practice, are the major causes for thermal damage in laser processing of materials. For characterizing the fundamental behavior of lagging waves, we shall restrict the discussion to the elastic (Hookean) body in studying the short-time propagation of stresses and strains.

9.1 THERMAL EXPANSION

As illustrated in Figure 9.1, the heat flux vector ($\bar{q}$) flowing into an elastic medium through the system boundary results in the change of internal energy within the system ($\dot{\varepsilon}$). The Reynolds's transport theorem used for describing the entropy flux vector in Figure 3.2 in Section 3.1 can be extended in a straightforward manner, with (1) the entropy flux vector ($\bar{J}$) replaced by the heat flux vector ($\bar{q}$) and (2) the time-rate of change of specific entropy per unit volume ($\rho\dot{s}$) by the time-rate of change of specific enthalpy per unit volume ($\rho\dot{\varepsilon}$):

$$\int_V \rho\dot{h}\, dV = \int_S \bar{q}\bullet\bar{n}\, dS \,. \tag{9.1}$$

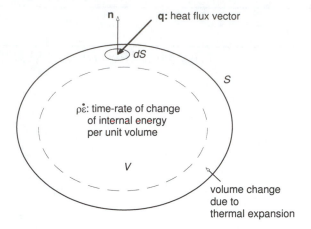

Figure 9.1 Time-rate of change of internal energy within the system resulting from the heat flux vector flowing through the system boundary.

The time-rate of change of enthalpy is used to include both effects of time-rate of changes of the internal energy and the volume-expanding work under constant pressure. For a better focus on the effect of thermomechanical coupling, the energy production rate within the body has been neglected in equation (9.1). The surface integral, likewise, is related to the volume integral by the divergence theorem,

$$\int_S \vec{q} \bullet \vec{n} \, dS = \int_V \nabla \bullet \vec{q} \, dV , \qquad (9.2)$$

resulting in

$$- \nabla \bullet \vec{q} = \rho \dot{h} \qquad (9.3)$$

from equation (9.1). The negative sign reflects the opposite direction of the heat flux vector to the unit normal of the differential surface, referring to Figure 9.1.

For a rigid conductor, the specific enthalpy is a function of temperature. For a deformable body, referring to Section 3.1, the specific enthalpy is a function of both temperature and volume *increase* of the body. Mathematically,

$$h \equiv h(T,e), \quad \text{with} \quad e = e_{11} + e_{22} + e_{33} \cong \frac{\Delta V}{V} . \qquad (9.4)$$

The Cauchy strain tensor is denoted by e_{ij}, as noted in Chapter 3. The sum of its three normal components defines the volume change of deformable body in small deformation, i.e., $e_{11}, e_{22}, e_{33} \ll 1$. The time-rate of change of the specific enthalpy is thus

$$\rho\dot{h} = \rho\left(\frac{\partial h}{\partial T}\right)_e \dot{T} + \rho\left(\frac{\partial h}{\partial e}\right)_T \dot{e} \qquad (9.5)$$

The first term, $\rho(\partial h/\partial T)_e$, is recognized as the volumetric specific heat, C_p under constant pressure, in classical thermodynamics. The second term, $\rho(\partial h/\partial e)_T \equiv C_\kappa$, can be defined as another type of *specific heat* in a parallel manner. It measures the energy required for the *isothermal* change in volume per unit volume of the continuum body. Since the volume change of a continuum body is a result of mechanical straining, the heat capacity C_κ serves as a bridge to the mechanical deformation. Like the volumetric specific heat, the specific heat C_κ is a function of temperature in general, $\rho(\partial h/\partial e)_T \equiv C_\kappa(T)$. Making the Taylor series expansion with respect to temperature gives:

$$\rho\left(\frac{\partial h}{\partial e}\right)_T \equiv C_\kappa(T) \cong C_\kappa(T_0) + \left(\frac{\partial C_\kappa}{\partial T}\right)_{T_0} (T - T_0) + \cdots. \qquad (9.6)$$

The heat capacity C_κ at the reference temperature, $C_\kappa(T_0)$, can be assumed zero without loss of generality, implying that the value of C_κ is measured with regard to its value at the reference temperature. For making contact with the coefficient of thermal expansion used in the *linear* theory of thermoelasticity (Fung, 1965; Boley and Weiner, 1960), moreover, the temperature change, $T - T_0$, is assumed to be of the same order of magnitude as the reference temperature, T_0, resulting in

$$\rho\left(\frac{\partial h}{\partial e}\right)_T \cong \left(\frac{\partial C_\kappa}{\partial T}\right)_{T_0} T_0 \qquad (9.7)$$

from equation (9.6). As will be shown below, this is a fair assumption because the thermomechanical coupling factor is indeed small in a wide range of temperatures from 10^2 to 10^3 K. Substituting equation (9.7) into (9.5), the time-rate of change of specific enthalpy per unit volume is obtained:

$$\rho\dot{h} = C_p\dot{T} + \left(\frac{\partial C_\kappa}{\partial T}\right)_0 T_0\dot{e}, \quad \text{with} \quad \left(\frac{\partial C_\kappa}{\partial T}\right)_0 \equiv \left(\frac{\partial C_\kappa}{\partial T}\right)_{T_0}. \qquad (9.8)$$

Substituting equation (9.8) into (9.8), finally,

$$-\nabla \bullet \vec{q} = C_p\dot{T} + \kappa_\sigma T_0\dot{e}, \quad \kappa_\sigma \equiv \left(\frac{\partial C_\kappa}{\partial T}\right)_0. \qquad (9.9)$$

The quantity $\kappa_\sigma \equiv (\partial C_\kappa/\partial T)_0$ retrieves the coefficient of thermal expansion in stress defined in the linear theory of thermoelasticity. In relation to the coefficient of thermal expansion in strain, κ_ε, which measures the change in strain per degree rise in temperature,

$$\kappa_\sigma = 3K\kappa_\varepsilon, \quad 3K = 3\lambda + 2\mu = \frac{E}{1-2\nu} \tag{9.10}$$

with K denoting the elastic bulk modulus, λ and μ the Lamé constants, E Young's modulus, and ν Poisson's ratio. The value for κ_σ has been measured and widely used as a thermomechanical property of materials. Typical values are shown in Table 9.1 for several metals and ceramics at room temperature. Note also that the coefficient of thermal expansion in strain (κ_ε) can be viewed as a definition for thermal strain. Because it measures the strain increase per degree rise in temperature, thermal strains at a temperature T (measured from the absolute zero degree Kelvin) in a continuum element are

$$e_{11}^{(T)} = \kappa_\varepsilon T, \quad e_{22}^{(T)} = \kappa_\varepsilon T, \quad e_{33}^{(T)} = \kappa_\varepsilon T, \tag{9.11}$$

with superscripts (T) denoting the strain components from the thermal effect alone. Thermal straining is an effect of dilatation contributing to the volume change of the continuum element. It does not affect the shear components of stress (σ_{ij}, for $i \neq j$) or strain (e_{ij}, for $i \neq j$) according to the Duhamel-Neumann generalization of Hooke's law.

The applicable regime of equation (9.9), in a mathematical sense, is limited to a temperature rise of the same order of magnitude as the reference temperature. For excessively high temperatures, the reference temperature T_0 in the last term of equation (9.9) is replaced by T, the instantaneous temperature in response. Equation (9.9) in this case involves *nonlinear* coupling between thermal (T) and mechanical ($e = e_{11} + e_{22} + e_{33}$) fields. From a physical point of view, however, importance of the mechanical effect ($\kappa T\dot{e}$) is weighed relative to the thermal effect ($C_p\dot{T}$) in equation (9.9). A change of temperature from T_0 to T does *not* sensitively vary the relative magnitude under "regular" conditions involving moderate mechanical and thermal strain rates.

9.1.1 Mechanically Driven Cooling Phenomenon

The mechanically driven cooling phenomenon is one of the salient features revealed by equation (9.9). In the absence of heat flow in the continuum body, including the experimental coupon used in the uniaxial tensile test maintained at a time-varying, uniform temperature, equation (9.9) results in

$$\dot{T} = -\left(\frac{\kappa_\sigma T_0}{C_p}\right)\dot{e}, \quad \text{implying} \quad \frac{T}{T_0} = 1 - \left(\frac{\kappa_\sigma \dot{e}}{C_p}\right)t. \tag{9.12}$$

Under a constant strain rate ($\dot{e}$ = constant > 0) in tension, a popular condition enforced in the uniaxial tensile test characterizing the mechanical properties of materials, equation (9.12) indicates a linearly *decreasing* temperature in time, revealing a cooling phenomenon in the initial stage of tensile loading. This is the

Table 9.1 Typical values of the coefficients of thermal expansion in strain (κ_ε) and stress (κ_σ), where Pa = N/m^2 = J/m^3, Gpa = 10^9 Pa, MJ = 10^6 J.

	E, Gpa	ν	K, Gpa	κ_ε, $\times 10^{-6}$, 1/K	κ_σ, MJ/m^3 K
Carbon steel (med.)	207	0.3	172.5	11.3	5.85
Aluminum	69	0.33	67.6	23.6	4.79
Copper	110	0.35	122.2	16.5	6.05
Brass	110	0.35	122.2	20	7.33
Bronze	110	0.35	122.2	19.2	6.67
Magnesium	45	0.29	35.7	27	2.89
Nickel	207	0.31	544.7	13.3	21.70
Silver	76	0.37	292.3	19	16.70
Titanium	107	0.34	111.5	9	3.01
Alumina (Al$_2$O$_3$)	393	0.27	284.8	9.8	7.52
Magnesia (MgO)	207	0.36	246.4	13.5	9.98
Fused silica (SiO$_2$)	75	0.16	36.8	0.5	0.06
Soda-lime glass	69	0.23	42.6	9.0	1.15

well-known Kelvin's cooling phenomenon in the elastic region, as illustrated by the representative point E in Figures 9.2(a) and 9.2(b). The occurrence of plastic straining, and hence the plastic energy dissipation rate, heats the specimen at the onset of yielding, point Y in Figure 9.2, bringing the temperature to the reference level and beyond, point P in the post-yielding region. The temperature trough in the cooling-heating curve, point Y in Figure 9.2, provides a precise measurement for the yield stress, abandoning the traditional 2%-offset approximation. Sih and Tzou (1986, 1987), Sih et al. (1987), and Tzou (1987) made a series of quantitative

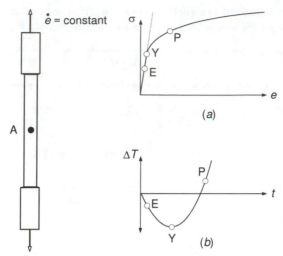

(a)

(b)

Figure 9.2 (*a*) Stress and strain curve at a representative point *A* in a uniaxial tensile coupon. (*b*) Temperature cooling in the elastic region (*E*), onset of heating at the yielding point (*Y*), and persistent heating in the post-yielding region (*P*).

studies of the heating preceded by cooling phenomenon in aluminum-6061 and 4340-steel, with extensive applications to thermomechanical coupling in stable crack growth and determination of the onset of material failure by yielding.

9.1.2 Thermomechanical Coupling Factor

An illuminating form of equation (9.9) can be written as

$$-\nabla \bullet \vec{q} = C_p \dot{T}\left[1 + \frac{\kappa_\sigma T_0 \dot{e}}{C_p \dot{T}}\right] = C_v \dot{T}\left[1 + \left(\frac{3KT_0 \kappa_\varepsilon^2}{C_p}\right)\left(\frac{\dot{e}}{\kappa_\varepsilon \dot{T}}\right)\right], \tag{9.13}$$

or, in a more organized form,

$$-\nabla \bullet \vec{q} = C_p \dot{T}\left[1 + \eta\left(\frac{\dot{e}}{\kappa_\varepsilon \dot{T}}\right)\right], \quad \text{with} \quad \eta = \frac{3KT_0 \kappa_\varepsilon^2}{C_p}. \tag{9.14}$$

The thermomechanical coupling factor, η in equation (9.14), which is dimensionless, measures the relative effect of two heat capacities (note that $3K\kappa_\varepsilon = \kappa_\sigma$). The ratio in parentheses, $\dot{e}/(\kappa_\varepsilon \dot{T})$, also dimensionless, measures the *mechanical* strain rate ($\dot{e}$) relative to the *thermal* strain rate ($\kappa_\varepsilon \dot{T}$, referring to equation (9.11)). As the thermomechanical coupling factor (material-dependent) or the strain rate ratio (loading-dependent) is small, the product term in equation (9.14) is negligible, comparing to 1. Consequently, equation (9.14) reduces to the energy equation for a *rigid* conductor. The thermomechanical coupling factor is indeed small for engineering materials under regular conditions. Typical values for several metals and ceramics are shown in Table 9.2 at room temperature, $T_0 = 300$ K. Note that even in high temperature operations, the temperature increase $(T - T_0)$ remains of the same order of magnitude as the ambient temperature (T_0). Small values of thermomechanical coupling factors shown in Table 9.2, therefore, cover a wide range of temperatures.

The ratio between mechanical strain rate and thermal strain rate, $\dot{e} / (\kappa_\varepsilon \dot{T})$, depends on the thermomechanical response of the system and is more difficult to predict. In high strain rate situations, such as high-speed penetration (Tzou and Li, 1994b) and dynamic crack propagation (Tzou, 1990a, b, 1992a), the mechanical strain rate in the vicinity of the penetrator head or the rapidly propagating crack tip

Table 9.2 Typical values of thermomechanical coupling factors for metals and ceramics, where $\eta = 3KT_0\kappa_\varepsilon^2/C_p$.

	K, GPa	κ_ε, $\times 10^{-6}$, 1/K	ρ, kg/m^3	C_p, MJ/m^3 K	η, $\times 10^{-3}$
Carbon steel (med.)	172.5	11.3	7850	3.61	5.49
Aluminum	67.6	23.6	2710	2.44	13.90
Copper	122.2	16.5	8940	3.45	9.68
Brass	122.2	20	8530	3.20	13.76
Nickel	544.7	13.3	8900	3.94	7.33
Silver	292.3	19	10490	2.47	12.84
Alumina (Al$_2$O$_3$)	284.8	9.8	3970	3.08	6.45
Magnesia (MgO)	246.4	13.5	3580	3.37	12.01
Fused silica (SiO$_2$)	36.8	0.5	2200	1.63	0.005
Soda-lime glass	42.6	9.0	2500	2.10	1.48

may reach as high as 10^5 s^{-1} (Freund, 1990). The local time-rate of increase of temperature may reach 10^8 K/s (Tzou, 1992c, f, g), rendering a local thermal strain rate of the order of 10^3 s^{-1}. The ratio between mechanical and thermal strain rates, $\dot{e}/(\kappa_\varepsilon \dot{T})$, therefore, is of the order of 10^2. Along with the thermomechanical coupling factor multiplied in the front, consequently,

$$\eta\left(\frac{\dot{e}}{\kappa_\varepsilon \dot{T}}\right) \cong O(10^{-1}) \quad \text{to} \quad O(10^0). \tag{9.15}$$

Comparing to the value of 1, with reference to equation (9.14), the thermomechanical coupling effect may become noticeable in this case.

 As a general trend, for more brittle materials, which promote the mechanical strain rate in the load-time history, thermomechanical coupling becomes more important in the transient process. For low-conducting media, which tend to develop high temperatures in the short-time transient, on the other hand, the thermomechanical coupling effect becomes less important. This is the behavior from the point of view of thermomechanical properties alone. For energy and momentum transport by wave propagation, including both stress waves and thermal waves in the short-time response, the mechanical strain rate and temperature rate in the vicinity of the wavefront are both large. Whether the thermomechanical coupling effect is negligible or not is a result of the analysis. An example is given in Section 9.3 for developing a better understanding of the thermomechanical coupling effect.

9.1.3 Apparent Thermal Conductivity

Fourier's law of heat conduction is sufficient to illustrate the *apparent* thermal conductivity attributed to the rate effect of mechanical deformation. Introducing

$$\vec{q} = -k\nabla T$$

into equation (9.14) gives

$$k\nabla^2 T = C_v \dot{T}\left[1 + \eta\left(\frac{\dot{e}}{\kappa_\varepsilon \dot{T}}\right)\right], \quad \text{or} \quad k^*\nabla^2 T = C_v \dot{T} \tag{9.16}$$

with

$$k^* = k - \frac{C_p \eta \dot{e}}{\kappa_\varepsilon\left(\nabla^2 T\right)} \tag{9.17}$$

being the apparent thermal conductivity accounting for the effect of thermomechanical coupling. An increase in the mechanical strain rate or the thermomechanical coupling factor, obviously, decreases the effective value of the thermal conductivity, rendering a higher temperature in the deformable Fourier solid.

9.2 THERMOELASTIC DEFORMATION

While the energy equation (9.14) involves the effect of deformation rate, the equation of equilibrium describing momentum transfer depends on temperature as well. This is the Duhamel-Neumann generalization of Hooke's law for nonisothermal deformation. The concept can be easily illustrated by elastic deformation in a Hookean body. Assuming that the *increase* of strain due to thermal expansion is described by equation (9.11), the stress and strain relation in a one-dimensional solid is

$$e_{11} = \frac{\sigma_{11}}{E} + \kappa_\varepsilon T \qquad (9.18a)$$

where e_{11} is the one-dimensional strain and σ_{11} is the one-dimensional stress. The subscript "1" refers to the x_1 direction along which deformation occurs. The reversed expression of stress in terms of strain is

$$\sigma_{11} = E e_{11} - E\kappa_\varepsilon T, \qquad (9.18b)$$

implying a thermal *relaxation* in stress due to thermal expansion. For multi-axial deformation where the lateral strains, e_{22} and e_{33}, are transmitted to the strain in the x_1 direction (e_{11}) by the Poisson effect,

$$e_{11} = \left(\frac{\sigma_{11}}{E} + \kappa_\varepsilon T\right) - \nu\left(\frac{\sigma_{22}}{E} + \kappa_\varepsilon T\right) - \nu\left(\frac{\sigma_{33}}{E} + \kappa_\varepsilon T\right)$$

$$= \frac{1}{E}\left[(1+\nu)\sigma_{11} - \nu(\sigma_{11} + \sigma_{22} + \sigma_{33})\right] + (1-2\nu)\kappa_\varepsilon T. \qquad (9.19a)$$

The minus signs in front of ν reflect contraction of strains in the lateral direction (x_1) when loaded in the longitudinal directions (x_2 and x_3). Also, an isotropic deformation has been assumed in equation (9.19a), implying the sufficiency of invoking two elastic moduli, E and ν, in describing the elastic response in all directions. In the x_2 and x_3 directions, similarly,

$$e_{22} = \left(\frac{\sigma_{22}}{E} + \kappa_\varepsilon T\right) - \nu\left(\frac{\sigma_{11}}{E} + \kappa_\varepsilon T\right) - \nu\left(\frac{\sigma_{33}}{E} + \kappa_\varepsilon T\right)$$

$$= \frac{1}{E}\left[(1+\nu)\sigma_{22} - \nu(\sigma_{11} + \sigma_{22} + \sigma_{33})\right] + (1-2\nu)\kappa_\varepsilon T \qquad (9.19b)$$

$$e_{33} = \left(\frac{\sigma_{33}}{E} + \kappa_\varepsilon T\right) - \nu\left(\frac{\sigma_{11}}{E} + \kappa_\varepsilon T\right) - \nu\left(\frac{\sigma_{22}}{E} + \kappa_\varepsilon T\right)$$

$$= \frac{1}{E}\left[(1+\nu)\sigma_{33} - \nu(\sigma_{11} + \sigma_{22} + \sigma_{33})\right] + (1-2\nu)\kappa_\varepsilon T. \qquad (9.19c)$$

For the case of $\nu = 1/2$, the deformation involves no volume change, and the effect of thermal expansion vanishes in equations (9.19a) to (9.19c). It is thus evident that thermal expansion is an effect of dilatation, supporting the use of the same form of Hooke's law in describing the shear response:

$$e_{ij} = \frac{\sigma_{ij}}{2\mu} = \frac{(1+\nu)}{E}\sigma_{ij} \quad \text{for} \quad i \neq j, \quad i,j = 1,2,3, \qquad (9.19d)$$

with μ denoting the shear modulus. Since $[(1+\nu)/E]$ is a common factor in both normal and shear strain components, a combination of the second expressions in equations (9.19a) to (9.19d) facilitates a universal expression for the stress and strain relation:

$$e_{ij} = \left\{-\frac{3\nu}{E}\sigma_m + (1-2\nu)\kappa_\varepsilon T\right\}\delta_{ij} + \left(\frac{1+\nu}{E}\right)\sigma_{ij}, \quad \text{for} \quad i,j = 1,2,3 \qquad (9.20)$$

where δ_{ij} denotes the Kronecker delta. Converting the shear modulus (μ) to Young's modulus (E) and Poisson's ratio (ν) in equation (9.19d) is necessary because only two independent moduli (E and ν used here) are allowed in describing the isotropic elastic deformation.

The reversed expression for stresses in terms of strains can be obtained by dividing both sides of equation (9.20) by $(1+\nu)/E$:

$$\sigma_{ij} = \left\{\left(\frac{3\nu}{1+\nu}\right)\sigma_m - \kappa_\varepsilon E\left(\frac{1-2\nu}{1+\nu}\right)T\right\}\delta_{ij} + \left(\frac{E}{1+\nu}\right)e_{ij}. \qquad (9.21)$$

The conversion becomes complete if the mean stress σ_m can be expressed in terms of strains. Setting $i = j$ in equation (9.21) and noting that a repeated index implies summation, $\delta_{ii} = 3$, $\sigma_{ii} = 3\sigma_m$, and $e_{ii} = 3e_m$,

$$\sigma_m = \left(\frac{E}{1-2\nu}\right)e_m - E\kappa_\varepsilon T, \quad e_m = \frac{e_{11} + e_{22} + e_{33}}{3}. \qquad (9.22)$$

Substituting equation (9.22) into (9.21) and simplifying the result,

$$\sigma_{ij} = \left\{ \left[\frac{3E\nu}{(1+\nu)(1-2\nu)} \right] e_m - \frac{\kappa_\varepsilon E}{1-2\nu} T \right\} \delta_{ij} + \left(\frac{E}{1+\nu} \right) e_{ij} \quad \text{for} \quad i, j = 1, 2, 3.$$

(9.23)

In the absence of thermomechanical coupling, $\kappa_\varepsilon = 0$, equations (9.20) and (9.23) reduce to Hooke's law in isothermal elasticity. In the presence of the thermal effect, equations (9.20) and (9.23) are the Duhamel-Neumann generalization of Hooke's law. Simpler expressions can be obtained by further relating Young's modulus and Poisson's ratio to the lamé constants, λ and μ. Equations (9.20) and (9.23), however, suffice for our purpose.

The thermoelastic formulation we have made so far, in summary, includes (1) the mechanical strain rate effect on energy transport, equation (9.14), (2) the effect of thermal expansion in strains, equation (9.20), and (3) the effect of thermal relaxation in stresses, equation (9.23). The mechanical effect in the thermal field and the thermal effect in the mechanical field are implemented on different physical bases. In equation (9.14), the mechanical strain rate effect enters the thermal field through *field* coupling. In equations (9.20) and (9.23), on the other hand, the thermal expansion and relaxation effect enters the deformation field through *constitutive* coupling. The elastic medium undergoing a nonisothermal deformation, in addition, is assumed to be a *simple* substance, reflected by equation (9.4), where the specific enthalpy is assumed to depend on two state variables, T and e, only.

The formulation becomes complete if (1) the *constitutive behavior* describing the way in which the heat flux vector varies with the temperature gradient is specified, for equation (9.14) describing the field coupling, and (2) the *equation of equilibrium* (the field equation guaranteeing satisfaction of conservation of momentum in transferring load) is specified, for equation (9.20) or (9.23) describing constitutive coupling. This is the convention used in thermoelasticity.

A more general coupled response was derived in Chapter 3, equation (3.60). The assumption of a simple substance is removed, thereby allowing for the possible precedence switch between the heat flux vector and the temperature gradient in energy transport and the precedence switch between the stress and the strain in momentum transport. It also describes a more consistent approach where the thermal effect on deformation and the deformation effect on the thermal response are incorporated through the constitutive equations. Combining these coupled equations with the conservation equations for energy (the energy equation) and momentum (the equation of equilibrium), a system of field equations result, with a much more complicated appearance than those resulting from the classical approach. This more general and consistent approach becomes more appropriate should the phase lags describing the relaxation behavior in the thermal and deformation fields be determined for engineering materials.

9.3 MECHANICALLY DRIVEN COOLING WAVES

While continuously exploring the unique features in the lagging response, it is also important to make frequent contacts with the existing models to show the effects of phase lags in transporting heat. For this purpose the classical framework, namely, equation (9.14) describing field coupling and equations (9.20) and (9.23) describing constitutive coupling shall be used in this section to retrieve the well-known thermoelastic solutions obtained by Boley and Tolins (1962) for thermal diffusion and Lord and Shulman (1967) for thermal waves.

The lagging response in heat transport is a special behavior in *time*. Consideration of a one-dimensional solid in *space* is thus sufficient. As an example illustrating the effect of lagging behavior on thermal stress propagation, let us consider a one-dimensional thin rod shown in Figure 9.3. The rod is so thin that there exists no neighboring medium in the lateral direction (the directions perpendicular to the x axis), implying the absence of Poisson's effect, $\nu = 0$. Equation (9.23) becomes

$$\sigma = Ee - E\kappa_\varepsilon T, \quad \text{with} \quad e = \frac{\partial u}{\partial x}. \tag{9.24}$$

Substituting equation (9.24) into the one-dimensional equation of equilibrium,

$$\frac{\partial \sigma}{\partial x} = \rho \frac{\partial^2 u}{\partial t^2}, \tag{9.25}$$

gives

$$\frac{\partial^2 u}{\partial x^2} - \kappa_\varepsilon \frac{\partial T}{\partial x} = \frac{1}{C_E^2} \frac{\partial^2 u}{\partial t^2}, \quad \text{with} \quad C_E = \sqrt{\frac{E}{\rho}}. \tag{9.26}$$

The quantity C_E is referred to as the dilatational wave speed. The temperature dependence in equation (9.26) necessitates the consideration of energy equation (9.14). In a one-dimensional situation, equation (9.14) reduces to

$$-\frac{\partial q}{\partial x} = C_p \dot{T} + \left(\frac{C_p \eta}{\kappa_\varepsilon}\right) \frac{\partial^2 u}{\partial x \partial t}. \tag{9.27}$$

For describing a generalized lagging behavior containing both phase lags in the heat flux vector (τ_q) and the temperature gradient (τ_T), the heat flux vector in equation (9.27) is

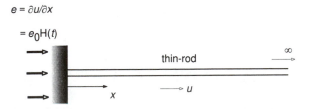

Figure 9.3 An infinitely long rod subjected to a strain impact at $x = 0$.

$$q(x,t) + \tau_q \frac{\partial q}{\partial t}(x,t) = -k\left\{\frac{\partial T}{\partial x}(x,t) + \tau_T \frac{\partial^2 T}{\partial t \partial x}(x,t)\right\}, \qquad (9.28)$$

which is the one-dimensional form of equation (2.7). The heat flux vector shall be eliminated from equation (9.27) and (9.28) for obtaining the temperature representation of the energy equation in a *deformable* conductor, in correspondence with equation (2.10) for a *rigid* conductor. For this purpose, differentiating equation (9.28) with respect to x, yielding

$$\frac{\partial q}{\partial x} + \tau_q \frac{\partial^2 q}{\partial t \partial x} = -k\frac{\partial^2 T}{\partial x^2} - k\tau_T \frac{\partial^3 T}{\partial t \partial x^2}. \qquad (9.29)$$

The first term, $\partial q/\partial x$, is already given by equation (9.27) in terms of derivatives of temperature. The second term immediately follows by taking the derivative of equation (9.27) with respect to time (t), resulting in

$$-\frac{\partial^2 q}{\partial x \partial t} = C_p \frac{\partial^2 T}{\partial t^2} + \left(\frac{C_p \eta}{\kappa_\varepsilon}\right)\frac{\partial^3 u}{\partial x \partial t^2}. \qquad (9.30)$$

Substituting equations (9.27) and (9.30) into equation (9.29), finally gives

$$\frac{\partial^2 T}{\partial x^2} + \tau_T \frac{\partial^3 T}{\partial x^2 \partial t} = \frac{1}{\alpha}\frac{\partial T}{\partial t} + \frac{\tau_q}{\alpha}\frac{\partial^2 T}{\partial t^2} + \left(\frac{C_p \eta}{k\kappa_\varepsilon}\right)\frac{\partial^2 u}{\partial x \partial t} + \left(\frac{C_p \eta \tau_q}{k\kappa_\varepsilon}\right)\frac{\partial^3 u}{\partial x \partial t^2}. \quad (9.31a)$$

Equation (9.31a) is the energy equation describing the lagging behavior of heat transport in a deformable conductor. The displacement field u involved is described by the equation of equilibrium, equation (9.26), which warrants conservation of momentum in transferring load. Although the lagging behavior in the mechanical deformation is not accounted for, evidenced by the use of Hooke's law (equation (9.24)) in describing the *instantaneous* response between stress and strain, a variety of situations result from equation (9.31a):

  1. A diffusion behavior in heat conduction, $\tau_T = \tau_q = 0$:

$$\frac{\partial^2 T}{\partial x^2} = \frac{1}{\alpha}\frac{\partial T}{\partial t} + \left(\frac{C_p \eta}{k\kappa_\varepsilon}\right)\frac{\partial^2 u}{\partial x \partial t}. \qquad (9.31b)$$

This is the energy equation considered by Boley and Tolins (1962) in studying the thermomechanical coupling effect in a relatively long time response.

  2. A wave behavior in heat propagation, $\tau_T = 0$ and $\tau_q = \alpha/C^2$, with C denoting the thermal wave speed:

$$\frac{\partial^2 T}{\partial x^2} = \frac{1}{\alpha}\frac{\partial T}{\partial t} + \frac{\tau_q}{\alpha}\frac{\partial^2 T}{\partial t^2} + \left(\frac{C_p \eta}{k\kappa_\varepsilon}\right)\frac{\partial^2 u}{\partial x \partial t} + \left(\frac{C_p \eta \tau_q}{k\kappa_\varepsilon}\right)\frac{\partial^3 u}{\partial x \partial t^2}, \quad \tau_q = \frac{\alpha}{C^2}. \quad (9.31c)$$

This is the basic equation derived by Lord and Shulman (1967) in establishing the dynamic theory of thermoelasticity accounting for the wave phenomenon in heat propagation. In view of the dual-phase-lag model, however, it describes only the fast-transient effect of thermal inertia (τ_q). The additional inertia effect in thermomechanical coupling, the last term in equation (9.31c), is noteworthy. This term was assumed negligible in Lord and Shulman's approach for obtaining an analytical solution characterizing the short-time behavior.

In the presence of both phase lags, with τ_T reflecting the time delay in heat transport due to the occurrence of microstructural interactions, equation (9.31a) further extends the dynamic theory of thermoelasticity into a new era. Owing to the use of Hooke's law in describing the mechanical response, however, equation (9.31a) still bears the assumption of an instantaneous response in deformation. The mechanical field (stress and strain) is assumed to stabilize faster than the thermal field in this version.

The energy equation (9.31a) and the equation of equilibrium (9.26) describe thermomechanical coupling in a deformable conductor. Mathematically, they present two coupled partial differential equations to be solved for temperature (T) and displacement (u) simultaneously.

An example is given to illustrate the effect of lagging response in heat transport. The thin rod shown in Figure 9.3 is disturbed from a stationary state, both thermally and mechanically, implying that

$$T = T_0, \quad \frac{\partial T}{\partial t} = 0, \quad u = 0, \quad \frac{\partial u}{\partial t} = 0 \quad \text{at} \quad t = 0, \tag{9.32}$$

with u denoting the one-dimensional displacement in the x direction and e the one-dimensional Cauchy strain for small deformation. At $t = 0^+$, a suddenly increased strain e_0 is imposed at the end of the rod,

$$e = \frac{\partial u}{\partial x} = e_0 H(t) \quad \text{at} \quad x = 0, \tag{9.33}$$

causing a *temperature change* in the rod because of thermomechanical interaction. The suddenly imposed strain is described by the unit-step function, $H(t)$. At a distance far from the boundary, both thermal and mechanical disturbances are assumed vanishing,

$$T \rightarrow T_0, \quad u \rightarrow 0 \quad \text{as} \quad x \rightarrow \infty. \tag{9.34}$$

The formulation is now completed. The momentum and energy transport are described by equations (9.26) and (9.31a), while the initial and boundary conditions are equations (9.32) to (9.34). Owing to complicated thermomechanical properties involved in the governing system, a dimensionless analysis is necessary to identify the dominant parameters. Introducing

$$\delta = \frac{x}{\left(\alpha / C_E\right)}, \quad \beta = \frac{t}{\left(\alpha / C_E^2\right)}, \quad \theta = \frac{T - T_0}{T_0}, \quad U = \frac{u}{\left(\alpha \kappa_\varepsilon T_0 / C_E\right)},$$

$$z_T = \frac{\tau_T}{\left(\alpha / C_E^2\right)}, \quad z_q = \frac{\tau_q}{\left(\alpha / C_E^2\right)}, \tag{9.35}$$

equations (9.26) and (9.31) to (9.34) become

$$\frac{\partial^2 U}{\partial \delta^2} - \frac{\partial^2 U}{\partial \beta^2} = \frac{\partial \theta}{\partial \delta} \quad \text{(momentum)}, \tag{9.36}$$

$$\frac{\partial^2 \theta}{\partial \delta^2} + z_T \frac{\partial^3 \theta}{\partial \delta^2 \partial \beta} - \frac{\partial \theta}{\partial \beta} - z_q \frac{\partial^2 \theta}{\partial \beta^2} = \eta \frac{\partial^2 U}{\partial \delta \partial \beta} + \eta z_q \frac{\partial^3 U}{\partial \delta \partial \beta^2} \quad \text{(energy)}, \tag{9.37}$$

$$\theta = 0, \quad \frac{\partial \theta}{\partial \beta} = 0, \quad U = 0, \quad \frac{\partial U}{\partial \beta} = 0 \quad \text{at} \quad \beta = 0, \tag{9.38}$$

$$\frac{\partial U}{\partial \delta} = e_0 H(\beta) \quad \text{at} \quad \delta = 0, \qquad \theta, U \to 0 \quad \text{as} \quad \delta \to \infty. \tag{9.39}$$

The thermomechanical response is thus characterized by three parameters, z_T (dimensionless phase lag of the temperature gradient), z_q (dimensionless phase lag of the heat flux vector), and η (thermomechanical coupling factor). In this version of the dimensionless phase lags, note that the quantity α/C_E^2 used in equation (9.35), with C_E being the dilatational wave speed, does not have a physical meaning. It does, however, have a dimension of time, and its value is of the order of picoseconds for metals at room temperature ($\alpha \sim 10^{-6}$ m^2/s and $C_E \sim 10^3$ m/s).

Equations (9.36) and (9.37) can be made decoupled in the Laplace transform domain. Taking the Laplace transforms of temperature and displacement,

$$\overline{\theta}(\delta; p) = \int_0^\infty \theta(\delta, \beta) e^{-p\beta} d\beta, \qquad \overline{U}(\delta; p) = \int_0^\infty U(\delta, \beta) e^{-p\beta} d\beta, \tag{9.40}$$

and making use of the initial conditions in equation (9.38), equations (9.36) and (9.37) become

$$\frac{d^2 \overline{\theta}}{d\delta^2} - \left[\frac{p\left(1 + p z_q\right)}{1 + p z_T}\right] \overline{\theta} = \eta \left[\frac{p\left(1 + p z_q\right)}{1 + p z_T}\right] \frac{d\overline{U}}{d\delta} \tag{9.41a}$$

$$\frac{d^2 \overline{U}}{d\delta^2} - p^2 \overline{U} = \frac{d\overline{\theta}}{d\delta}. \tag{9.41b}$$

In order to obtain uncoupled equations for temperature and displacement, take the second-order derivative of equation (9.41a) and the first-order derivative of equation (9.41b) with respect to δ, resulting in

$$\frac{d^4\overline{\theta}}{d\delta^4} - \left[\frac{p(1+pz_q)}{1+pz_T}\right]\frac{d^2\overline{\theta}}{d\delta^2} = \eta\left[\frac{p(1+pz_q)}{1+pz_T}\right]\frac{d^3\overline{U}}{d\delta^3} \tag{9.42a}$$

$$\frac{d^3\overline{U}}{d\delta^3} - p^2\frac{d\overline{U}}{d\delta} = \frac{d^2\overline{\theta}}{d\delta^2}. \tag{9.42b}$$

The expressions of $(d\overline{U}/d\delta)$ and $(d^3\overline{U}/d\delta^3)$ in terms of temperature and its derivatives are obtained from equations (9.41a) and (9.42a). Substituting the results into (9.42b), we have

$$\frac{d^4\overline{\theta}}{d\delta^4} - \left\{\frac{p\left[(1+\eta)(1+pz_q)+p(1+pz_T)\right]}{1+pz_T}\right\}\frac{d^2\overline{\theta}}{d\delta^2} + \left\{\frac{p^3(1+pz_q)}{1+pz_T}\right\}\overline{\theta} = 0 \tag{9.43a}$$

$$\frac{d^4\overline{U}}{d\delta^4} - \left\{\frac{p\left[(1+\eta)(1+pz_q)+p(1+pz_T)\right]}{1+pz_T}\right\}\frac{d^2\overline{U}}{d\delta^2} + \left\{\frac{p^3(1+pz_q)}{1+pz_T}\right\}\overline{U} = 0 \tag{9.43b}$$

The equations governing the temperature and the displacement have an identical form. Regardless of the complicated coefficients, these equations can be solved in a straightforward manner. The solutions for temperature and displacement satisfying the pulsed strain and regularity conditions, equation (9.39), are

$$\overline{\theta}(\delta;p) = \frac{\left(p^2 - F_2^2\right)\left(p^2 - F_1^2\right)}{p^3\left(F_2^2 - F_1^2\right)}\left[e^{-F_1\delta} - e^{-F_2\delta}\right], \tag{9.44a}$$

$$\overline{U}'(\delta;p) = \frac{\left(p^2 - F_2^2\right)\left(p^2 - F_1^2\right)}{p^3\left(F_2^2 - F_1^2\right)}\left[-\left(\frac{F_1^2}{p^2 - F_1^2}\right)e^{-F_1\delta} + \left(\frac{F_2^2}{p^2 - F_2^2}\right)e^{-F_2\delta}\right], \tag{9.44b}$$

where

$$F_1 = \sqrt{\frac{b - \sqrt{b^2 - 4c}}{2}}, \quad F_2 = \sqrt{\frac{b + \sqrt{b^2 - 4c}}{2}},$$

$$b = \frac{p\left[(1+\eta)(1+pz_q) + p(1+pz_T)\right]}{1+pz_T}, \quad c = \frac{p^3(1+pz_q)}{1+pz_T}. \quad \text{(9.44c Cont.)}$$

Strain is obtained in equation (9.44b) because it is more representative than displacement in characterizing the deformation field. The stress, for example, has exactly the same expression except for a normalized factor being $(E\kappa_\varepsilon T_0)$. Though more complicated than ever before, the transformed temperature and strain, equations (9.44a) and (9.44b) with the coefficients defined in equations (9.44c), can be inverted by the use of the Riemann-sum approximation. The function subroutine, FUNC(P), in the Appendix is replaced by either equation (9.44a) for obtaining temperature or equation (9.44b) for obtaining strain.

9.3.1 Heat Transport by Diffusion

Let us first examine the cooling wave driven by diffusion in transporting heat. This is the result obtained by Boley and Tolins (1962), which is a special case of $z_T = z_q = 0$ in equations (9.44a) to (9.44c). At a representative instant of time, $\beta = 1$, Figure 9.4 shows the temperature and strain distributions at various value of η, the thermomechanical coupling factor. The values of η are selected on the order of 10^{-2}, the threshold values shown in Table 9.2 for representative metals and ceramics. The strain pulse applied at the boundary induces a cooling wave of temperature ($\theta < 0$ implies $T < T_0$). This is Kelvin's cooling phenomenon illustrated in equation (9.12) and Figure 9.2 for thermoelastic bodies. The temperature, however, will not exceed the ambient value because no heating mechanism, such as the energy dissipation due to plasticity, is modeled in thermo*elasticity*. Due to heat propagation through a deformable body in which momentum transfer is a wave phenomenon, referring to equations (9.26) or (9.36), a *discontinuity* exists in the temperature field at exactly the same location as the mechanical wavefront (at $x = C_E t$ or $\delta = \beta$ according to equation (9.35)), although diffusion is assumed for heat transport. Traditionally, a

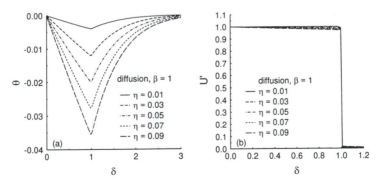

Figure 9.4 (*a*) Temperature and (*b*) strain distributions in the strain-pulsed solid showing the effect of the thermomechanical coupling factor (η). The thermal "wavefront" follows the strain wavefront located at $\delta = \beta = 1$. The case of diffusion occurs at $\tau_T = \tau_q = 0$.

thermal wavefront is defined as an infinitesimal boundary separating the heat-affected zone from the thermally undisturbed zone. If we broaden the definition of the wavefront to include the location where a discontinuity exists in the temperature *gradient*, the discontinuity existing at $\delta = \beta = 1$ in Figure 9.4(a) can be viewed as a thermal wavefront. The temperature distribution ahead of the wavefront ($\delta > 1$) is due to diffusion, which is assumed to propagate at an infinite speed, extending the heat-affected zone to infinity. The temperature distribution behind the wavefront, through thermomechanical coupling, is a combined result of diffusion and wave motion of the elastic medium. This domain is expanding at the same rate as the dilatational wave speed defined in equation (9.26). We explicitly demonstrate, in Section 9.3.2 below and in Chapter 10 with a more general treatment, that the energy transport employing diffusion *does have* a thermal wavefront that is coincident with the mechanical wavefront in a deformable body.

The thermomechanical coupling factor has a great influence on the amplitude of temperature cooling. The amount of cooling sensitively increases with the value of η, as shown in Figure 9.4(a). Coupling with the thermal field assuming diffusion, however, has an almost negligible effect on the strain wave, as shown in Figure 9.4(b). This is the same trend as that observed by Boley and Tolins (1962). Increasing the value of η by 1 order of magnitude decreases the strain level behind the mechanical wavefront by no more than 1%.

Figure 9.5 shows the evolution of temperature and strain waves as the transient time lengthens. The amount of cooling increases with time, as shown in Figure 9.5(a), but the peak value gradually tapers off when approaching the steady state. The location of the discontinuity propagates with the mechanical wavefront, $\delta = \beta$ at various times. As the transient time lengthens, the mechanically disturbed zone extends *beyond* the wavefront, as shown Figure by 9.5(b). This is due to the thermal expansion caused by diffusion, which penetrates into the solid at an infinite speed. At a longer time, say $\beta = 10$ in Figure 9.5, the temperature gradient ahead of the thermal wavefront (at $\delta = \beta = 1$) becomes more exaggerated. It thus induces a more pronounced thermal strain *ahead* of the mechanical wavefront, developing an extended disturbed zone that is approximately 10 to 20% larger than that in an isothermal elastic medium. Note also that the thermomechanical coupling effect partially destroys the sharp mechanical wavefront. The strain gradient is still

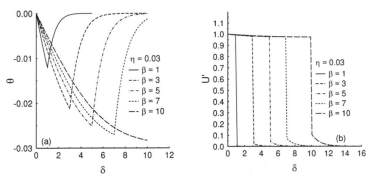

Figure 9.5 (*a*) Temperature and (*b*) strain waves at $\beta = 1$, 3, 5, 7, and 10 for $\eta = 0.03$, $\tau_T = \tau_q = 0$ (diffusion).

discontinuous at $\delta = \beta$, but the infinitely large gradient smooths to a finite gradient in transition to the extended zone.

9.3.2 Heat Transport by Thermal Waves

Thermomechanical coupling employing thermal wave behavior in heat transport is given by equations (9.44a) to (9.44c) with $z_T = 0$. The dimensionless phase lag of the heat flux vector, z_q, which dictates the thermal wave speed (referring to equation (9.31c)), becomes a parameter in addition to the thermomechanical coupling factor η.

Lord and Shulman's solution. Lord and Shulman (1967) obtained the analytical inversions of equations (9.44a) and (9.44b) by neglecting the inertia term in the energy equation (9.31c), $(C_p \eta \tau_q / k \kappa_\varepsilon)(\partial^3 u / \partial x \partial t^2) = 0$. In the present formulation, this is equivalent to dropping all the terms containing the product of (ηz_q) from equations (9.41a), (9.42a), (9.43a), (9.43b), and (9.44c) while retaining the individual terms containing η or z_q alone. Accordingly, the temperature and strain solutions remain the same as those shown by equations (9.44a) and (9.44b), but the coefficient b defined in equation (9.44c) changes to

$$b = \frac{p\left[(1 + \eta + p z_q) + p(1 + p z_T)\right]}{1 + p z_T},$$

(without mechanical inertia in energy transport)   (9.45)

In addition, analytical solutions are only possible at a particular value of z_q, $z_q = 1/(1+\eta)$ or $\eta = (1/z_q) - 1$. Under these two assumptions, the solutions for dimensionless temperature and strain can be written as (Lord and Shulman, 1967):

$$\theta = -(1 - z_q)\left\{ F_3 H\left(\beta - \sqrt{z_q}\,\delta\right) + \left[F_4 - F_5\right]H(\beta - \delta)\right\},$$ (9.46a)

$$U' = z_q \left\{ F_3 H\left(\beta - \sqrt{z_q}\,\delta\right) + \left[F_4 + \left(\frac{1 - z_q}{z_q}\right)F_5\right]H(\beta - \delta)\right\},$$ (9.46b)

where

$$F_3 = 1 - e^{-\frac{\beta - \sqrt{z_q}\,\delta}{z_q(1 - z_q)}},$$ (9.46c)

$$F_4 = e^{-\frac{\beta}{z_q(1-z_q)}}\left\{ e^{\frac{\delta(1+z_q)}{2z_q(1-z_q)}} + \frac{\delta}{2z_q}\int_\delta^\beta \frac{e^{\frac{y(1+z_q)}{2z_q(1-z_q)}}}{\sqrt{y^2 - \delta^2}} \times I_1\left[\frac{\sqrt{y^2 - \delta^2}}{2z_q}\right] dy \right\},$$ (9.46d)

$$F_5 = e^{-\frac{\delta}{2z_q}} + \frac{\delta}{2z_q} \int_\delta^\beta \frac{e^{-\frac{y}{2z_q}}}{\sqrt{y^2 - \delta^2}} \times I_1 \left[\frac{\sqrt{y^2 - \delta^2}}{2z_q} \right] dy, \quad \text{with} \quad z_q = \frac{1}{1+\eta}.$$

(9.46e)

Though having an analytical form, evaluation of equations (9.46a) to (9.46e) is nontrivial due to involvement of two integrals containing the modified Bessel function of the first kind of order 1. Numerical integration is needed, provided that the *singularity* at $y = \delta$ in equations (9.46d) and (9.46e) is removed by a further transformation from y to w, $w = (y^2 + \delta^2)^{1/2}$. Equations (9.46a) and (9.46b) do reveal two wavefronts in thermomechanical coupling. One is at $\delta = \beta$, the other is at $\delta = \beta/\sqrt{z_q}$. These are not the general results, however, because equations (9.46a) to (9.46e) are only valid at the particular value of $z_q = 1/(1+\eta)$, referring to equation (9.46e).

By using the transformed temperature and strain in equations (9.44a) and (9.44b) with the coefficients b defined in equation (9.45) and $\beta = 1/(1+\eta)$, the results employing the Riemann-sum approximation for the Laplace inversion are compared to the analytical solutions shown by equations (9.46a) and (9.46b) in Figure 9.6. The value of η is taken to be 0.03, the same as that chosen by Lord and Shulman (1967). The excellent agreement fully supports the accuracy of the Riemann-sum approximation even when two wavefronts exist in the complicated functions.

The Riemann-sum approximation, to reiterate, results from the decomposition of the Fourier integral representing the Laplace inversion, referring to equations (2.45) and (2.46) in Section 2.5. In fact, the definite integrals shown in equations (9.46d) and (9.46e) are of the same type as those shown in Section 2.5 resulting from the Bromwich contour integrations. Should the integral be decomposed into a certain finite sum in the numerical evaluation, which is the basis for any algorithm in numerical integrations, the analytical approach performs

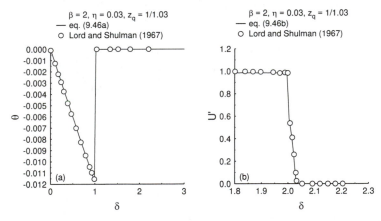

Figure 9.6 Comparison of the Riemann-sum approximation for the Laplace inversion and the analytical solution obtained by Lord and Shulman (1967). (*a*) Temperature distribution at $\beta = 1$ and (*b*) strain distribution at $\beta = 2$ for $\eta = 0.03$ and $z = 1/(1+\eta) = 1/1.03$.

essentially the same task. The inversion procedure employing the Riemann-sum approximation developed in Chapter 2 introduces this step at an early stage, avoiding the complexity and difficulty to a great extent. It is in this sense that the Riemann- sum approximation is analytical in nature, and should be distinguished from the other numerical methods of Laplace inversion.

Coupling effect with inertia. With the assistance of the Riemann-sum approximation, the inertia term in the energy equation can be reinstated and the special condition of $z_q = 1/(1+\eta)$ can be released for a general treatment. To focus attention on the fast-transient effect of thermal inertia (τ_q or z_q), the phase lag of the temperature gradient is set to zero for the time being, $\tau_T = z_T = 0$. The temperature and strain solutions are represented by equations (9.44a) and (9.44b), with the coefficient b defined in equation (9.44c). They allow an independent variation between η and z_q in the fast-transient response.

Under exactly the same conditions that Figure 9.6(a) describes, Figure 9.7 displays the temperature distribution incorporating the inertia term, $(C_p\eta\tau_q/k\kappa_\varepsilon)$ $(\partial^3 u/\partial x\partial t^2)$ in equation (9.31c), which Lord and Shulman (1967) neglected for obtaining an analytical solution. From a physical point of view, as seen in Figure 9.7, the effect of mechanical inertia in energy transport not only results in a peak temperature of cooling of about 7 times larger than that neglecting the inertia effect, but, most important, it induces *two* thermal wavefronts in heat propagation. Both mechanical and thermal waves are present in this problem. The location of the thermal wavefront in a *rigid* solid is at

$$x = Ct = \left(\sqrt{\frac{\alpha}{\tau_q}}\right) t \quad \text{or} \quad \delta = \left(\frac{1}{\sqrt{z_q}}\right)\beta, \tag{9.47a}$$

while the location of the stress or strain wavefront in an *isothermal* solid is at

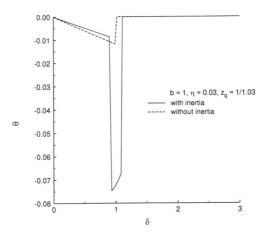

Figure 9.7 Large deviation in the temperature profile caused by the mechanical inertia effect in heat transport showing *dual* wavefronts in the temperature wave.

$$x = C_E t = \left(\sqrt{\frac{E}{\rho}}\right) t \quad \text{or} \quad \delta = \beta, \tag{9.47b}$$

according to the dimensionless scheme defined in equation (9.35). For $z_q = 1/1.03 \cong 0.971$ adopted in Figure 9.7, their locations should be at 0.985 (equation (9.47a)) for the thermal wavefront) and 1.0 (equation (9.47b) for the mechanical wavefront). A closer inspection of Figure 9.7, however, reveals that the two wavefronts are *not* located at either of these two locations. One wavefront is close to 0.9, the other is close to 1.1.

Wavefront analysis. Generation from one to two wavefronts associated with activation of the mechanical inertia effect in heat propagation necessitates an analytical study. The partial expansion technique developed in Section 2.5 provides a straightforward and powerful tool for this purpose. The transformed solution at a large value of p resembles the short-time solution of temperature,

$$\lim_{\beta \to 0} \theta(\delta, \beta) \sim \lim_{p \to \infty} \overline{\theta}(\delta; p) = \lim_{p \to \infty} \left\{ \frac{\left(p^2 - F_2^2\right)\left(p^2 - F_1^2\right)}{p^3\left(F_2^2 - F_1^2\right)} \left[e^{-F_1\delta} - e^{-F_2\delta}\right] \right\}. \tag{9.48}$$

As the value of p approaches infinity, according to the coefficients defined in equation (9.44c), the asymptotic behavior of the factors involved in equation (9.48) is summarized as follows

$$\lim_{p \to \infty}\left(p^2 - F_2^2\right) \sim d_3 p^2, \quad \lim_{p \to \infty}\left(p^2 - F_1^2\right) \sim d_4 p^2, \quad \lim_{p \to \infty} p^3\left(F_2^2 - F_1^2\right) \sim \sqrt{d_1}\, p^5,$$

$$\lim_{p \to \infty} e^{-F_1\delta} \sim e^{-\left(\sqrt{1-d_4}\right)p\delta}, \quad \lim_{p \to \infty} e^{-F_2\delta} \sim e^{-\left(\sqrt{1-d_3}\right)p\delta}, \tag{9.49a}$$

where

$$d_1 = d_2^2 - 4z_q, \quad d_2 = 1 + (1+\eta)z_q, \quad d_3 = 1 - \frac{d_2 + \sqrt{d_1}}{2}, \quad d_4 = 1 - \frac{d_2 - \sqrt{d_1}}{2} \tag{9.49b}$$

are all dimensionless and depend only on the thermomechanical coupling factor (η) and the phase lag of the heat flux vector (z_q). Substituting equation (9.49a) into (9.48) results in

$$\lim_{\beta \to 0} \theta(\delta, \beta) \sim \lim_{p \to \infty} \left\{ \frac{\left(p^2 - F_2^2\right)\left(p^2 - F_1^2\right)}{p^3\left(F_2^2 - F_1^2\right)} e^{-F_1\delta} - \frac{\left(p^2 - F_2^2\right)\left(p^2 - F_1^2\right)}{p^3\left(F_2^2 - F_1^2\right)} e^{-F_2\delta} \right\}$$

$$= \lim_{p \to \infty} \frac{d_3 d_4}{\sqrt{d_1}} \left[\frac{e^{-\left(\sqrt{1-d_4}\right)p\delta}}{p} - \frac{e^{-\left(\sqrt{1-d_3}\right)p\delta}}{p} \right]. \qquad \text{(9.50 Cont.)}$$

The short-time temperature distribution is thus

$$\lim_{\beta \to 0} \theta(\delta, \beta) \sim \frac{d_3 d_4}{\sqrt{d_1}} \left\{ L^{-1} \left[\frac{e^{-\left(\sqrt{1-d_4}\right)p\delta}}{p} \right] - L^{-1} \left[\frac{e^{-\left(\sqrt{1-d_3}\right)p\delta}}{p} \right] \right\}$$

$$= \frac{d_3 d_4}{\sqrt{d_1}} \left\{ H\left(\beta - \sqrt{\frac{d_2 - \sqrt{d_1}}{2}} \, \delta \right) - H\left(\beta - \sqrt{\frac{d_2 + \sqrt{d_1}}{2}} \, \delta \right) \right\}, \qquad \text{(9.51)}$$

where $H(\bullet)$ denotes the unit-step function and the Laplace inversion result has been used in equation (2.40).

Equation (9.51) indicates a temperature distribution *bounded* between

$$\beta = \sqrt{\frac{d_2 - \sqrt{d_1}}{2}} \, \delta \quad \text{and} \quad \beta = \sqrt{\frac{d_2 + \sqrt{d_1}}{2}} \, \delta. \qquad \text{(9.52)}$$

In terms of the physical space (x) and time (t), equation (9.52) can be expressed as

$$x = C^{(1)} t \quad \text{and} \quad x = C^{(2)} t, \quad \text{with}$$

$$\frac{C^{(1),(2)}}{C_E} = \sqrt{\frac{2}{1 + (1 + \eta)z_q \mp \sqrt{1 + 2(\eta - 1)z_q + \left[(1 + \eta)z_q\right]^2}}}. \qquad \text{(9.53)}$$

The coefficients d_1 and d_2 have been replaced by η and z_q according to equation (9.49b). The expression with a minus sign in the denominator refers to the thermal wave speed $C^{(1)}$, while that with a plus sign refers to the thermal wave speed $C^{(2)}$. Equation (9.53) provides the closed-form solutions for the two wave speeds. They are identical to the results obtained by Popov (1967) through a cumbersome wavefront expansion method and by Achenbach (1968) in terms of the jump conditions across the stress and temperature discontinuities. Obviously, the partial expansion technique used here provides a much simpler and intuitive approach.

In the absence of thermomechanical coupling, $\eta = 0$, equation (9.53) gives

$$\frac{C^{(1),(2)}}{C_E} = \sqrt{\frac{2}{1 + z_q \mp |z_q - 1|}} = \begin{cases} 1, \dfrac{1}{\sqrt{z_q}} & \text{for } z_q \geq 1 \\[2ex] \dfrac{1}{\sqrt{z_q}}, 1 & \text{for } z_q \leq 1 \end{cases} \qquad (\eta = 0). \qquad (9.54a)$$

Thermal and mechanical waves propagate at their own speeds in this case, yielding the same results as equations (9.47a) and (9.47b). In the case of diffusion ($z_q = 0$) with the thermomechanical coupling effect ($\eta \neq 0$), on the other hand, equation (9.53) becomes

$$\frac{C^{(1),(2)}}{C_E} = 0,1 \quad \text{or} \quad C^{(1)} = 0 \quad \text{and} \quad C^{(2)} = C_E \qquad (z_q = 0). \qquad (9.54b)$$

Only one wavefront remains in this case, which is identical to the mechanical wavefront due to the wave motion in the elastic medium.

At a typical value of $z_q = 1.0$, Figure 9.8 shows the effect of η on the thermal wave speeds. The wave speed $C^{(1)}$ increases with the value of η, while the wave speed $C^{(2)}$ decreases as the value of η increases. When the thermomechanical coupling becomes stronger, in other words, one thermal wave becomes faster while the other becomes slower.

Figure 9.9 shows the effect of phase lag of the heat flux vector, z_q, on the wave speeds. The value of η is taken as 0.05, an averaged value shown in Table 9.2. The difference between the two wave speeds is the smallest in the neighborhood of z_q = 1. As the value of z_q approaches zero, on revival of the diffusion behavior, the wave speed $C^{(1)}$ approaches infinity ($C^{(1)} \rightarrow \infty$), while the wave speed $C^{(2)}$ approaches the dilatational wave speed C_E ($C^{(2)}/ C_E \rightarrow 1$). This indicates that one

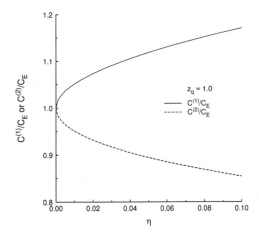

Figure 9.8 Effect of thermomechanical coupling factor on the two wave speeds in heat propagation.

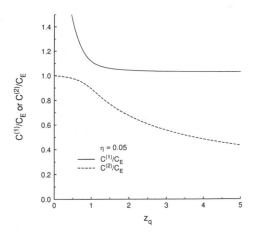

Figure 9.9 Effect of phase lag of the heat flux vector (z_q) on the two wave speeds in heat propagation.

branch of thermal disturbance propagates at an infinite speed ($C^{(1)}$) like diffusion, while the other propagates together with the stress or strain wave ($C^{(2)} = C_E$) due to wave motion in the elastic medium. As the value of z_q approaches infinity, as another extreme, the value of $C^{(1)}$ approaches C_E (thermal disturbance carrying by the elastic wave) while the value of $C^{(2)}$ approaches zero (completely relaxed thermal disturbance). Owing to the long relaxation behavior in the second branch described by $C^{(2)}$, the thermal disturbance becomes stationary in this limiting case.

Regardless of the complicated evolution of wave patterns, at least one wavefront exists in thermomechanical coupling due to the wave motion of the elastic medium. The second wavefront is present should an additional wave behavior activate in heat conduction, such as the CV-wave for $z_q \neq 0$. When this happens, the stress wave speed in an isothermal medium and the thermal wave speed in a rigid conductor vary from their original values. Equation (9.53) describes the way in which such thermomechanical wave speeds vary with the coupling factor and the phase lag.

The transformed strain shown by equation (9.44b) has a similar structure to the transformed temperature in equation (9.44a), except that two additional coefficients appear in front of the exponential functions. As the value of p approaches infinity, however, these two coefficients become independent of p,

$$\lim_{p \to \infty} \left[\frac{F_1^2}{p^2 - F_2^2} \right] = \frac{d_2 - \sqrt{d_1}}{2d_3}, \quad \lim_{p \to \infty} \left[\frac{F_2^2}{p^2 - F_2^2} \right] = \frac{d_2 + \sqrt{d_1}}{2d_3}, \quad (9.55)$$

implying that the strain wave has exactly the same wave speeds as the temperature wave shown by equations (9.52) and (9.53).

Temperature and strain waves. Since the effect of mechanical inertia in the energy equation has been shown to be significant, referring to the comparison shown in Figure 9.7, it shall be incorporated into equations (9.44a) to (9.44c) for evaluating the transient temperature and strain waves. At a constant value of $\eta = 0.05$ and a representative time $\beta = 1$, Figure 9.10 shows the effect of z_q (the phase lag of the heat flux vector) on the temperature cooling waves and strain waves. The response curves for diffusion with $z_q = 0$ is included for comparison, which preserves a discontinuity in the temperature gradient at the strain wavefront, $\delta = \beta = 1$ (referring to Section 9.3.1). As the value of z_q deviates from zero, Figure 9.10(a), the single wavefront splits into two wavefronts. The thermal wave speeds are dictated by equation (9.53), and the thermal wavefronts are located at

$$\delta = \left(\frac{C^{(1)}}{C_E} \right)\beta \quad \text{and} \quad \delta = \left(\frac{C^{(2)}}{C_E} \right)\beta. \tag{9.56}$$

At a certain instant of time (β), the physical domain between the two wavefronts decreases with the difference between the two wave speeds, $C^{(1)} - C^{(2)}$. The two wave speeds, as shown in Figure 9.9, become the closest and, consequently, the distance between the two wavefronts becomes the smallest as the value of z_q approaches 1. The physical domain between the two wavefronts increases as the value of z_q exceeds 1 and thereafter. This is shown by the plateaus in the temperature curves in Figure 9.10(a). Since the difference between the two wave speeds stabilizes at large values of z_q, the resulting increase of the physical domain between the two wavefronts is not as sensitive as that at smaller values of z_q.

The time-rate of changes of temperature, $\partial\theta/\partial\beta$, are large at the two wavefronts. This can be seen by the delta functions resulting from the time derivatives of equation (9.51). When the two wavefronts approach each other, i.e., the value of z_q approaching 1 (from zero), the physical domain between the two wavefronts shortens, and the large cooling rates of temperature carried by the wavefronts serve as additional sources for temperature cooling. As shown by the response curves of $z = 0.5$ and 1 in Figure 9.10(a), consequently, a most exaggerated cooling temperature results at $z_q = 1$. This provides another example illustrating the rate effect discussed in Section 2.7. As the value of z_q exceeds 1, the two

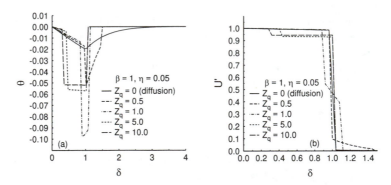

Figure 9.10 Effect of phase lag of the heat flux vector (z_q) on (a) the temperature cooling wave and (b) the strain wave at $\beta = 1$ and $\eta = 0.05$.

wavefronts separate from each other owing to the larger difference in the wave speeds. The rate effect in temperature cooling decreases, rendering a smaller amount of temperature cooling, as shown by the curves of $z_q = 5$ and 10.

The strain wave shown in Figure 9.10(b) basically possesses the same behavior. Note the formation of two wavefronts as the value of z_q increases from zero (single wavefront at $\delta = \beta = 1$) to 0.5. Relative to the location of the single wavefront for the case of $z_q = 0$, the first wavefront shown by the first expression in equation (9.56) stretches ahead owing to the effect of thermomechanical coupling. The other, shown by the second expression in equation (9.56), on the other hand, withdraws owing to the fast-transient effect of thermal inertia. The case of $z_q = 1$ still provides the most distinguishable response owing to the closest wave speeds (and hence the smallest physical domain between the two wavefronts).

Figure 9.11 shows the effect of the thermomechanical coupling factor on the temperature and strain waves. The most sensitive case of $z_q = 1$ is selected in the illustration. As shown by Figure 9.11(a), the physical domain between the two wavefronts, the temperature level in the entire thermomechanically disturbed zone, and the amount of temperature cooling, all increase with the value of η because of the stronger coupling response with the mechanical field. This is also a clear trend in the pattern of strain waves shown in Figure 9.11(b). Unlike the temperature response, however, the strain level in the thermomechanically disturbed zone behind the slower wavefront is not sensitive to the value of η.

9.3.3 Lagging Behavior in Heat Transport

The wave behavior resulting from the phase lag of the heat flux vector, as shown by Figures 9.8 to 9.11, yields completely different wave patterns from those predicted by the use of diffusion in describing the process of heat transport. In general, the formation of two wavefronts and the rate effect on temperature cooling are special features that require more detailed study for possible applications. Temperature and strain responses in the physical domain between the two wavefronts are of primary interest in thermal processing of materials. Equation (9.53) provides a convenient tool to determine these important criteria.

In the presence of an additional time delay due to the microstructural

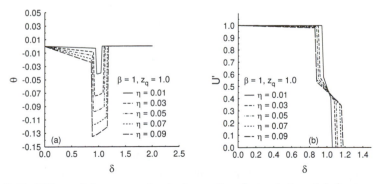

Figure 9.11 Effect of thermomechanical coupling factor (η) on (a) the temperature cooling wave and (b) the strain wave at $\beta = 1$ and $z_q = 1.0$.

interaction effect, z_T, the wave patterns actually become simpler because of the diminution of the thermal wavefront. Diminution of a sharp wavefront by the effect of z_T (τ_T) was discussed in Figure 2.7 in Section 2.5. For the case of the temperature gradient preceding the heat flux vector in heat transport, $z_T < z_q$, Figure 9.12 shows how two wavefronts collapse onto each other as the value of z_T deviates from zero. Note that in the presented case of $z_q = 1$, the curve with $z_T = 0$ reflects the CV-wave behavior and the curve with $z_T = 1$ ($\tau_q = \tau_T$) describes diffusion in heat transport. Since the microstructural interaction effect ($z_T \neq 0$) destroys the sharp wavefront in heat transport, only one wavefront exists, which is solely due to the wave motion of the elastic medium. The location of the wavefront, $\delta = \beta = 1$, as a result, is the same as that in the case of diffusion, Figures 9.4 and 9.5. The phase lag of the temperature gradient imposes strong damping to the temperature wave, shown by the smaller value of cooling in temperature at a larger value of z_T. The temperature level, however, is still more pronounced than that predicted by diffusion ($z_T = 1 = z_q$). Strain waves shown in Figure 9.12(b) display the same behavior. The two wavefronts collapse onto each other as the value of z_T deviates from zero. In the thermomechanically disturbed zone behind the wavefront, the strain level increases with the value of z_T. In the physical domain ahead of the wavefront, on the other hand, the strain level decreases as the value of z_T increases. The amount of change, however, is less than 10 % in both regions. The major effect of z_T, again, is on the temperature response rather than strain waves.

The amount of cooling continuously decreases as the value of z_T exceeds 1, the value assumed for z_q, $z_T > z_q$. The heat flux vector precedes the temperature gradient in this case. For metals where the value of τ_T (z_T) is about 1 to 2 orders of magnitude greater than that of τ_q (z_q), referring to Table 5.1 in Chapter 5, Figure 9.13 displays the temperature and strain waves at the same instant of time, $\beta = 1$. The large value of z_T not only diminishes cooling, Figure 9.13(a), but also reduces the gradient of temperature before and after the discontinuity. Consequently, as shown by Figure 9.13(b), strain waves (which relate to the temperature *gradient*) become indistinguishable at large values of z_T.

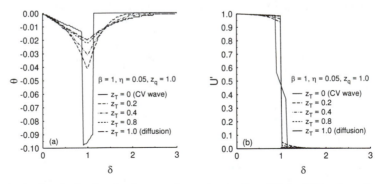

Figure 9.12 Revival of a single wavefront due to the effect of the phase lag of the temperature gradient (z_T). (*a*) Temperature cooling waves and (*b*) strain waves at $\beta = 1$, $\eta = 0.05$, and $z_q = 1.0$. Response curves for the case of gradient precedence are $0 \leq z_T \leq 1$, $z_T < z_q$.

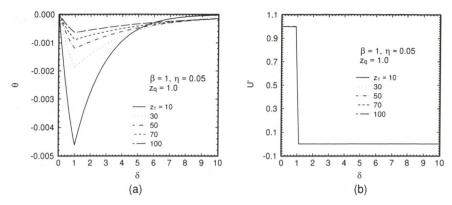

Figure 9.13 (*a*) Temperature cooling waves and (*b*) strain waves at $\beta = 1$, $\eta = 0.05$, and $z_q = 1.0$. The case of flux precedence with $z_T / z_q \gg 1$.

9.4 THERMAL STRESSES IN RAPID HEATING

The temperature cooling in a slender, one-dimensional solid discussed in Section 9.3 results from the thermoelastic coupling induced by a mechanical (strain) pulse. As a result, the temperature change is usually less than 1 K, which is negligible in most cases with engineering significance.

Thermal stresses induced by intensified heating are conjugate to the cooling phenomenon (of temperature) driven by mechanical waves. In thermal processing of materials, however, excessive thermal stresses developed in the workpiece owing to thermal localization are the direct cause for thermal failure and are far more important than the cooling phenomenon of temperature in practical applications.

This section studies the problem of thermal stresses induced by rapid heating applied at the boundary of a *half-space*, as shown in Figure 9.14. In contrast to the previous case of a slender rod, thermoelastic deformation in a half-space involves the Poisson effect from the neighboring media. It results from equation (9.23) that the only nonzero stress in the *x* direction is

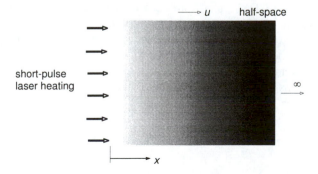

Figure 9.14 A semi-infinite half-space subjected to short-pulse heating.

$$\sigma = \frac{E(1-v)}{(1+v)(1-2v)}\frac{\partial u}{\partial x} - \frac{E\kappa_\varepsilon}{(1-2v)}T \tag{9.57}$$

where the one-dimensional Cauchy strain, $e_{11} = \partial u/\partial x$, has been substituted and the temperature T is measured from the reference temperature (T_0). In the absence of the Poisson effect, $v = 0$, equation (9.57) reduces to equation (9.18b), the previous case of a slender rod. Substituting equation (9.57) into equation (9.25), the equation of equilibrium governing the nonisothermal deformation becomes

$$\frac{\partial^2 u}{\partial x^2} - C_1\kappa_\varepsilon\frac{\partial T}{\partial x} = \left(\frac{1}{C_2}\right)\frac{1}{C_E^2}\frac{\partial^2 u}{\partial t^2}, \quad \text{with} \quad C_1 = \frac{1+v}{1-v}, \quad C_2 = \frac{1-v}{(1+v)(1-2v)},$$

$$\tag{9.58}$$

and C_E denoting the dilatational wave speed, equation (9.26).

The rapid heating applied at the boundary of the half-space is modeled by the energy absorption rate in the solid, as shown by equation (5.23). The energy equation describing the lagging response in the presence of thermomechanical coupling is

$$\frac{\partial^2 T}{\partial x^2} + \tau_T\frac{\partial^3 T}{\partial x^2\partial t} + \frac{1}{k}\left(S + \tau_q\frac{\partial S}{\partial t}\right)$$

$$= \frac{1}{\alpha}\frac{\partial T}{\partial t} + \frac{\tau_q}{\alpha}\frac{\partial^2 T}{\partial t^2} + \left(\frac{C_p\eta}{k\kappa_\varepsilon}\right)\frac{\partial^2 u}{\partial x\partial t} + \left(\frac{C_p\eta\tau_q}{k\kappa_\varepsilon}\right)\frac{\partial^3 u}{\partial x\partial t^2} \tag{9.59}$$

where

$$S(x,t) = S_0 e^{-gx - \frac{a|t-2t_p|}{t_p}}, \quad S_0 = 0.94J\left(\frac{1-R}{t_p}\right)g, \tag{5.23'}$$

with g being the reciprocal of the penetration depth of the laser energy into the solid. Recall that equation (5.23) results from equation (5.7b), which simulates the laser light intensity in Qiu et al.'s experiment, as shown in Figure 5.3.

Equations (9.58) and (9.59) are the momentum and energy equations describing thermomechanical coupling in the short-time transient. The two unknowns, displacement and temperature, have to be solved under certain initial and boundary conditions. The initial conditions remain the same as those specified in equation (9.32),

$$T = T_0, \quad \frac{\partial T}{\partial t} = 0, \quad u = 0, \quad \frac{\partial u}{\partial t} = 0 \quad \text{at} \quad t = 0. \tag{9.32'}$$

They describe thermomechanical disturbances from a stationary state. The remote boundary condition as x approaches infinity is the same as equation (9.34),

$$T \rightarrow T_0, \quad u \rightarrow 0 \quad \text{as} \quad x \rightarrow \infty. \tag{9.34'}$$

They are simply the regularity conditions for problems involving an infinite extent. The boundary conditions at $x = 0$ are twofold. Thermodynamically, neglecting the heat loss from the front surface in short times, a parallel treatment to equation (5.25) results in

$$\frac{\partial T}{\partial x} = 0 \quad \text{at} \quad x = 0. \tag{9.60}$$

Mechanically, deformation at the front surface occurs without any constraint, implying a stress-free condition at all times in the transient process,

$$\sigma = 0 \quad \text{or} \quad \frac{\partial u}{\partial x} = C_1 \kappa_\varepsilon T \quad \text{at} \quad x = 0 \tag{9.61}$$

from equation (9.57). Comparing the previous problem in Section 9.3, the present problem also contains thermomechanical coupling in the boundary condition.

In the absence of the Poisson effect, $\nu = 0$, $C_1 = C_2 = 1$ in equation (9.58). Equation (9.58) thus reduces to equation (9.26) for thermomechanical coupling in a one-dimensional slender rod. In the absence of heating, on the other hand, equation (9.59) reduces to equation (9.31a). The heat source term and the coefficients C_1 and C_2 do not alter the fundamental structure (the highest order differentials) in the momentum and energy equations. All the salient features discussed in Section 9.3, including the evolution of thermomechanical wavefronts, thus remain.

Employing the same scheme for the dimensionless variables, equation (9.35), along with the dimensionless volumetric heating rate defined as

$$Q(x,t) = \frac{S(x,t)}{C_E^2 k T_0 / \alpha^2}, \tag{9.62}$$

equations (9.58) and (9.59) become

$$\frac{\partial^2 U}{\partial \delta^2} - \frac{1}{C_2} \frac{\partial^2 U}{\partial \beta^2} = C_1 \frac{\partial \theta}{\partial \delta} \quad \text{(momentum)}, \tag{9.63}$$

$$\frac{\partial^2 \theta}{\partial \delta^2} + z_T \frac{\partial^3 \theta}{\partial \delta^2 \partial \beta} + \left(Q + z_q \frac{\partial Q}{\partial \beta} \right) - \frac{\partial \theta}{\partial \beta} - z_q \frac{\partial^2 \theta}{\partial \beta^2}$$

$$= \eta \frac{\partial^2 U}{\partial \delta \partial \beta} + \eta z_q \frac{\partial^3 U}{\partial \delta \partial \beta^2} \quad \text{(energy)}. \tag{9.64}$$

The volumetric heating in equation (9.64), in terms of the dimensionless space and time variables, can be expressed as

$$Q = \frac{S_0}{C_E^2 k T_0 \,/\, \alpha^2} e^{-g_0 \delta - \frac{a|\beta - 2\beta_0|}{\beta_0}}, \quad \text{with} \quad g_0 = \frac{g}{C_E \,/\, \alpha}, \quad \beta_0 = \frac{t_p}{\alpha \,/\, C_E^2}. \tag{9.65}$$

The initial and boundary conditions, equations (9.32), (9.34), (9.60) and (9.61), take the form

$$\theta = 0, \quad \frac{\partial \theta}{\partial \beta} = 0, \quad U = 0, \quad \frac{\partial U}{\partial \beta} = 0 \quad \text{at} \quad \beta = 0, \tag{9.66}$$

$$\frac{\partial U}{\partial \delta} = C_1 \theta, \quad \frac{\partial \theta}{\partial \delta} = 0 \quad \text{at} \quad \delta = 0, \tag{9.67}$$

$$\theta \to 0, \quad U \to 0 \quad \text{as} \quad \delta \to \infty. \tag{9.68}$$

Similar to equations (9.41a) and (9.41b), the transformed temperature and displacement are obtained from equations (9.63), (9.64), and (9.66):

$$\left[\frac{1 + p z_T}{\eta p \left(1 + p z_q\right)} \right] \frac{d^2 \bar{\theta}}{d\delta^2} - \left(\frac{1}{\eta}\right) \bar{\theta} + \left[\frac{\bar{Q}}{\eta p} - \frac{Q_0 z_q}{\eta p \left(1 + p z_q\right)} \right] = \frac{d\bar{U}}{d\delta}, \tag{9.69a}$$

$$\frac{d^2 \bar{U}}{d\delta^2} - \frac{p^2}{C_2} \bar{U} = C_1 \frac{d\bar{\theta}}{d\delta}. \tag{9.69b}$$

where the Laplace transform of the volumetric heating, equation (9.65), is

$$\bar{Q}(\delta; p) = \bar{Q}_b e^{-g_0 \delta}, \quad \bar{Q}_b = \frac{S_0 \beta_0}{C_E^2 k T_0 \,/\, \alpha^2} \left[\frac{e^{-2a} - e^{-2p\beta_0}}{p\beta_0 - a} + \frac{e^{-2p\beta_0}}{p\beta_0 + a} \right] \tag{9.70a}$$

and Q_0 is the value of Q (equation (9.65)) at $\beta = 0$:

$$Q_0 = Q(\delta, 0) = \frac{S_0}{C_E^2 k T_0 \,/\, \alpha^2} e^{-g_0 \delta - 2a}. \tag{9.70b}$$

Again, equations (9.69a) and (9.69b) can be made decoupled by eliminating the displacement field from the energy equation and the temperature field from the momentum equation. The results are

$$\frac{d^4\overline{\theta}}{d\delta^4} - b\frac{d^2\overline{\theta}}{d\delta^2} + c\overline{\theta} = f_1, \quad \frac{d^4\overline{U}}{d\delta^4} - b\frac{d^2\overline{U}}{d\delta^2} + c\overline{U} = f_2 \quad \text{with} \tag{9.71a}$$

$$b = \frac{p\left[C_2(1 + C_1\eta)(1 + pz_q) + p(1 + pz_T)\right]}{C_2(1 + pz_T)}, \quad c = \frac{p^3(1 + pz_q)}{C_2(1 + pz_T)}, \tag{9.71b}$$

$$f_1 = -\frac{1 + pz_q}{1 + pz_T}\left[\frac{d^2\overline{Q}}{d\delta^2} - \left(\frac{z_q}{1 + pz_q}\right)\frac{d^2Q_0}{d\delta^2}\right] + \frac{p^2(1 + pz_q)}{C_2(1 + pz_T)}\left[\overline{Q} - \frac{z_q Q_0}{1 + pz_q}\right],$$

$$f_2 = \frac{C_1(1 + pz_q)}{1 + pz_T}\left[\left(\frac{z_q}{1 + pz_q}\right)\frac{dQ_0}{d\delta} - \frac{d\overline{Q}}{d\delta}\right]. \tag{9.71c}$$

in correspondence with equations (9.43a) and (9.43b). From equation (9.70a), the derivatives of the energy absorption rate in the Laplace transform domain are

$$\frac{d\overline{Q}}{d\delta} = -\left(\overline{Q}_b g_0\right)e^{-g_0\delta}, \quad \frac{d^2\overline{Q}}{d\delta^2} = \left(\overline{Q}_b g_0^2\right)e^{-g_0\delta}. \tag{9.71d}$$

The solutions of equations (9.71a) satisfying the transformed initial and boundary conditions (9.66) to (9.68) are straightforward,

$$\overline{\theta} = A_1 e^{-F_1\delta} + A_2 e^{-F_2\delta} + B_1 e^{-g_0\delta}, \quad \overline{U} = D_1 A_1 e^{-F_1\delta} + D_2 A_2 e^{-F_2\delta} + B_2 e^{-g_0\delta}, \tag{9.72a}$$

where, with b and c defined in equation (9.71b),

$$F_1 = \sqrt{\frac{b - \sqrt{b^2 - 4c}}{2}}, \quad F_2 = \sqrt{\frac{b + \sqrt{b^2 - 4c}}{2}}, \tag{9.72b}$$

$$B_1 = \frac{C_2 g_0^2 - p^2}{C_2\left(g_0^4 - bg_0^2 + c\right)}\left[\left(\frac{S_0 e^{-2a}}{C_E^2 kT_0 / \alpha^2}\right)\left(\frac{z_q}{1 + pz_T}\right) - \overline{Q}_b\left(\frac{1 + pz_q}{1 + pz_T}\right)\right];$$

$$B_2 = -\frac{C_1 g_0}{\left(g_0^4 - bg_0^2 + c\right)}\left[\left(\frac{S_0 e^{-2a}}{C_E^2 kT_0 / \alpha^2}\right)\left(\frac{z_q}{1 + pz_T}\right) - \overline{Q}_b\left(\frac{1 + pz_q}{1 + pz_T}\right)\right], \tag{9.72c}$$

$$A_1 = \frac{C_1 B_1(g_0 - F_2) + g_0 F_2(D_2 B_1 - B_2)}{C_1(F_2 - F_1) + F_1 F_2(D_1 - D_2)}, \quad A_2 = -\frac{A_1 F_1 + B_1 g_0}{F_2}, \tag{9.72d}$$

$$D_1 = \frac{C_1 C_2 F_1}{p^2 - C_2 F_1^2}, \quad D_2 = \frac{C_1 C_2 F_2}{p^2 - C_2 F_2^2}. \tag{9.72e}$$

The temperature and displacement fields directly give the thermal stress in the solid. From equation (9.57),

$$\sigma = EC_2 \frac{\partial u}{\partial x} - EC_3 \kappa_\varepsilon T, \quad \text{with} \quad C_3 = \frac{1}{1 - 2\nu}, \tag{9.73a}$$

or

$$\Sigma = C_2 \frac{\partial U}{\partial \delta} - C_3 \theta, \quad \text{where} \quad \Sigma = \frac{\sigma}{E \kappa_\varepsilon T_0} \tag{9.73b}$$

according to equation (9.35). From equation (9.72a), more explicitly,

$$\overline{\Sigma} = -\left[\left(D_1 A_1 C_2 F_1 + A_1 C_3 \right) e^{-F_1 \delta} + \left(D_2 A_2 C_2 F_2 + A_2 C_3 \right) e^{-F_2 \delta} + \left(C_2 g_0 B_2 + C_3 B_1 \right) e^{-g_0 \delta} \right]$$

$$\tag{9.74}$$

The transformed temperature and stress, equations (9.72a) and (9.74), are complicated functions of p (the Laplace transform parameters) and the thermomechanical properties of the material. Regardless of their complicated appearance, Laplace inversion for these functions can be performed likewise. The function subroutine, FUNC(P), in the Appendix is replaced by equation (9.72a) for computing the temperature and equation (9.74) for computing the stress in the physical domain of time.

9.4.1 Diffusion

Thermomechanical coupling with a diffusion behavior in heat transport is first analyzed. Equations (9.72a) to (9.74) are used with $z_T = z_q = 0$ ($\tau_T = \tau_q = 0$, no phase lag in heat transport) to obtain the temperature and stress distributions. At a representative instant of time, $\beta = 4$, Figure 9.15 displays the effect of the thermomechanical coupling factor (η) on the temperature and stress distributions in the semi-infinite half-space. As shown by Figure 9.15(a), the temperature in the near field (with $\delta \leq 8$) decreases as the thermomechanical coupling factor increases. In the far field with $\delta > 8$, however, the temperature is not sensitive to the change of the thermomechanical coupling factor. A distinct *wavefront* exists at

$$x = \sqrt{C_2 C_E} t \quad \text{or} \quad \delta = \sqrt{C_2} \beta \tag{9.75}$$

owing to the wave motion in the elastic medium. It coincides with the mechanical wavefront shown in equation (9.58), rendering a value of $\sqrt{C_2} \cong 1.16$ for $\nu = 0.3$. As the thermomechanical coupling factor increases, the discontinuity of temperature gradient existing at the wavefront increases accordingly.

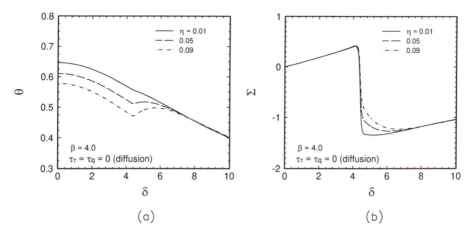

Figure 9.15 (*a*) Temperature and (*b*) stress distributions in a half-space induced by rapid heating at $a = 1.88$, $t = 100$ fs, $J = 5000$ J/m^2, $g = 5.76 \times 10^7$ 1/m, $v = 0.3$, $\alpha = 1.0 \times 10^{-6}$ m^2/s, $k = 100$ W/m K, $\beta = 4.0$. The case of diffusion occurs at $\tau_T = \tau_q = 0$.

The stress distribution for a right-running wave is shown in Figure 9.15(b). Note that a positive value of Σ refers to tension while a negative value refers to compression. At $\beta = 4$, the wavefront is located at $\delta \cong 4.64$. The thermally induced stress rapidly switches from tension ($\Sigma > 0$) to compression ($\Sigma < 0$) across the wavefront. The physical domain behind the wavefront ($0 \le \delta \le 4$) is in tension, while that in front of the wavefront ($\delta > 4$) is in compression. Since regular metals have much higher strength in compression than in tension, the tensile region right behind the wavefront deserves special attention in failure prevention. In the physical domain in front of the wavefront, $5 \le \delta \le 7$ approximately, the magnitude of compressive stress decreases as the thermomechanical coupling factor increases. This is a trend consistent with the effect on the temperature response shown in Figure 9.15(a). The stress level is not sensitive to the thermomechanical coupling factor in the tensile region behind the wavefront. This implies that the material failure due to thermal stresses (in tension) does not strongly depend on the thermomechanical coupling factor.

At the same value of $\eta = 0.05$, Figure 9.16 shows the evolution of temperature and stress waves at various times, $\beta = 2.0$, 4.0, and 6.0. Clearly, the wavefront advances with time according to $\delta \cong 1.16\beta$. A dramatic switch from tension to compression across the wavefront remains at all times. The temperature level decreases as the transient time lengthens, which is an intrinsic behavior in diffusion. The thermally induced stress, on the other hand, gradually develops into a stationary wave as time lengthens, with the difference between the maximum tension (behind the wavefront) and compression (in front of the wavefront) approaching a constant.

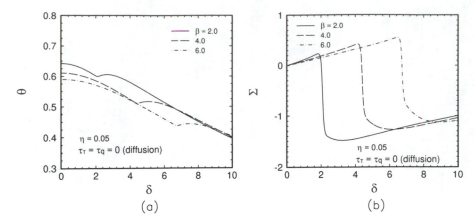

Figure 9.16 Evolution of (*a*) temperature and (*b*) stress waves in the time history at $\eta = 0.05$. The other parameters remain the same as those in Figure 9.15. The case of diffusion occurs at $\tau_T = \tau_q = 0$.

9.4.2 *CV* Waves

In the case of $\tau_T = 0$ and $\tau_q \neq 0$, equations (9.72a) to (9.74) reduce to the solutions for temperature, displacement, and stress employing the *CV*-wave behavior in heat transport. Two thermomechanical waves exist in the short-time response due to the co-existence of wave behavior in heat and momentum transport, as shown in Figure 9.17. Like the situations shown in Figures 9.8 to 9.10, thermomechanical coupling gives rise to two waves containing the thermal wave (of speed $(\alpha/\tau_q)^{1/2}$ in a *rigid* conductor) and the mechanical stress wave (of speed $(\sqrt{C_2})C_E$ in an isothermal body) in between. As the thermomechanical coupling factor increases, as shown in Figure

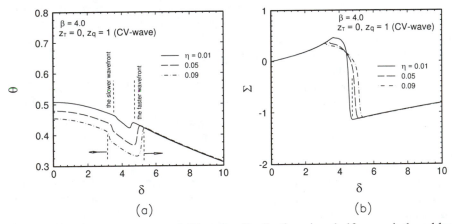

Figure 9.17 (*a*) Temperature and (*b*) stress distributions in a half-space induced by rapid heating for the case of a *CV* wave with $z_T = 0$ and $z_q = 1$. All system parameters are identical to those shown in Figure 9.15.

9.17(a), one of the thermomechanical waves moves faster, while the other moves slower. The physical domain between the two wavefronts, consequently, increases with the thermomechanical coupling factor.

Figure 9.18 shows the effect of z_q (τ_q) on the temperature and stress waves at the same instant of time, $\beta = 4$. Similar to the effect of the thermomechanical coupling factor, the physical domain between the two wavefronts increases with the phase lag of the heat flux vector, z_q or τ_q. Both wavefronts slow down as the value of τ_q increases, resulting in the same trend as that shown in Figure 9.9. The first (faster) wavefront follows the mechanical stress wavefront in an isothermal body and does not vary much with the value of τ_q. The slight drawback at a larger value of τ_q results from the thermomechanical coupling. The second (slower) wavefront follows the thermal wave speed in a rigid conductor, $(\alpha/\tau_q)^{1/2}$, which decreases as the value of τ_q increases. Compared with the effect of the thermomechanical coupling factor, Figure 9.17, increasing the value of τ_q produces a more pronounced amplitude of stress wave as shown in Figure 9.18(b). The physical domain in which tension abruptly switches to compression dramatically decreases as the value of τ_q increases.

Figure 9.19 shows the time history of (a) temperature and (b) stress waves at $\beta = 2.0$, 4.0, and 6.0. Both Figures 9.18(a) (for the temperature wave) and 9.18(b) (for the stress wave) clearly show that the discontinuities at the slower wavefront gradually diminish as the transient time (β) lengthens. Because the mechanical stress wave is induced by the thermal wave in this case, the discontinuity across the stress wavefront is not as sharp as that across the strain wavefront shown previously in Figures 9.10(b) and 9.11(b). Compression preceded by tension is still a clear pattern in stress wave propagation. The maximum tensile stress does not occur at the surface subjected to rapid heating. Rather, it occurs at a distance underneath the free surface, advocating a *subsurface* type of thermal failure in material processing.

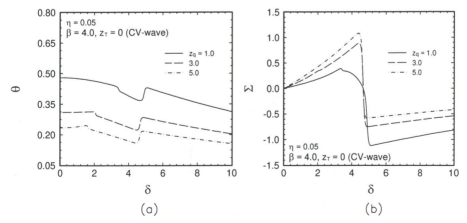

Figure 9.18 Effect of phase lag of the heat flux vector, z_q (τ_q), on (*a*) temperature and (*b*) stress waves for the case of a *CV* wave with $z_T = 0$ and $\beta = 4$. All system parameters are identical to those shown in Figure 9.15.

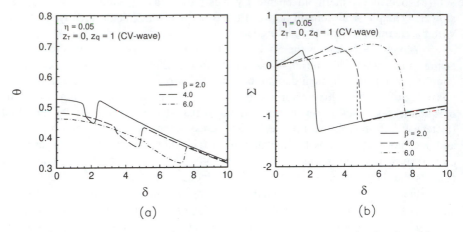

Figure 9.19 Evolution of (*a*) temperature and (*b*) stress waves in the time history for the case of a *CV*-wave with $z_T = 0$, $z_q = 1$, and $\eta = 0.05$. All system parameters remain the same as those in Figure 9.15.

9.4.3 Lagging Behavior

In the presence of the phase lag of the temperature gradient, $\tau_T \neq 0$, the lagging behavior due to the microstructural interaction effect interweaves with the delayed response owing to the fast-transient effect of thermal inertia.

Figure 9.20 shows the effect of the thermomechanical coupling factor in the presence of both phase lags. For $\tau_q = 1$, the ratio of z_T/z_q is fixed at 10 ($\tau_T = 10$) to represent the threshold value for metals. The phase lag of the temperature gradient ($z_T \neq 0$) destroys the wavefront resulting from the phase lag of the heat flux vector,

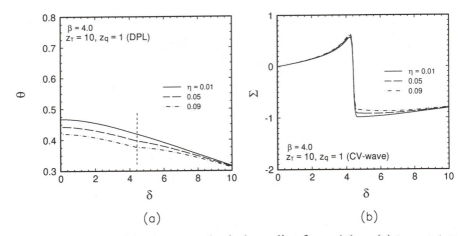

Figure 9.20 Effect of the thermomechanical coupling factor (η) on (*a*) temperature and (*b*) stress waves using the dual-phase-lag (DPL) model with $z_T = 10$, $z_q = 1$, and $\beta = 4$. All system parameters are identical to those shown in Figure 9.15.

rendering a single wavefront in the propagation of temperature and stress waves, Figures 9.20(a) and 9.20(b). The wavefront is dominated by the mechanical dilatational wave, with a modification factor of $\sqrt{C_2}$ shown by equation (9.75). This is similar to the phenomenon of wavefront coalescence shown in Figures 9.12 (for z_T < z_q) and 9.13 ($z_T > z_q$).

Figure 9.21 shows the effect of the phase lag of the temperature gradient (z_T) on the temperature and stress waves. For $z_q = 1$, the case of $\tau_T = 0.1$ is close to the situation employing the CV-wave model ($\tau_T = 0$) in heat transport. There exists only one wavefront, but abrupt changes of temperature and stress *gradients* exist in the vicinity of the original thermal wavefront (in a rigid conductor) and mechanical wavefront (in an isothermal body). The physical domain in which stress switches from tension to compression, as shown by Figure 9.21(b), becomes wider as the value of z_T decreases. The case of diffusion is retrieved as $z_T = z_q$ ($\tau_T = \tau_q$). Two discontinuities in the temperature and stress distributions collapse onto one (Figure 9.21(a)), rendering a narrow band in which tension suddenly switches to compression (Figure 9.21(b)). As the value of z_T becomes large, exemplified by the curves with $z_T = 10$ in Figure 9.21(a), the location of the wavefront does not change, but the amount of discontinuity across the wavefront decreases. A similar trend is observed in the compressive stress ahead of the wavefront. At a sufficiently large value of z_T, however, the tensile stress behind the wavefront is not sensitive to the value of z_T.

Figure 9.22 shows the time histories of the temperature and stress waves. The wavefront advances as the transient time lengthens, as shown in Figure 9.22(a), with the discontinuity across the wavefront decaying with time. At $\beta = 6$, the discontinuity across the wavefront at $\delta \cong 6.6$ is so weak that the distribution of temperature is almost continuous. Comparing to Figure 9.16(b) for the case of diffusion ($\tau_T = \tau_q = 0$) and Figure 9.19(b) for the case of a CV-wave ($z_T = 0$), the phase lag of the temperature gradient, z_T or τ_T, results in a higher tensile stress behind the wavefront, as shown in Figure 9.22(b). When the delayed time due to the microstructural interaction becomes pronounced, evidently, the specimen risks

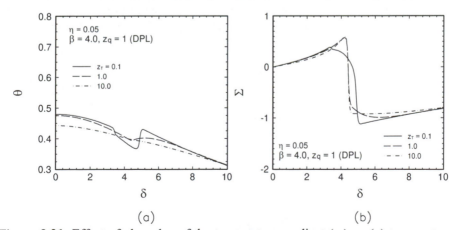

Figure 9.21 Effect of phase lag of the temperature gradient (z_T) on (*a*) temperature and (*b*) stress waves using the dual-phase-lag (DPL) model with $z_q = 1$, $\eta = 0.05$, and $\beta = 4$. All system parameters are identical to those shown in Figure 9.15.

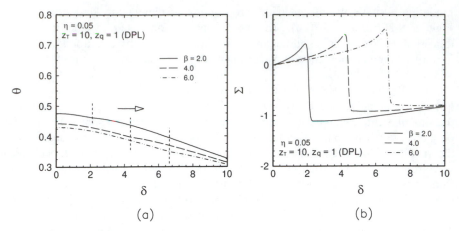

Figure 9.22 Evolution of (*a*) temperature and (*b*) stress waves in the time history using the dual-phase-lag (DPL) model with $z_T = 10$, $z_q = 1$, $\eta = 0.05$. All system parameters are identical to those shown in Figure 9.15.

mechanical failure by excessive tension.

Thermal and mechanical failure must be distinguished in thermal processing of materials. Thermal failure is caused by the excessively high temperature developed in the specimen, resulting in phase change should the temperature level reach the threshold for melting. According to Figures 9.16(a), 9.19(a), and 9.22(a), the temperature level in the heated specimen decreases with time, implying that thermal damage of the specimen is a major concern at an earlier stage in the transient process. Mechanical failure, on the other hand, is caused by the excessive tensile stress established in the specimen. This type of failure includes both yielding (in more ductile media such as metals) and fracture (in more brittle media such as glass). According to Figures 9.16(b), 9.19(b), and 9.22(b), the tensile stress behind the wavefront *increases* with time, implying that mechanical failure by yielding or fracture is a major concern at a later stage in the transient process.

HIGH-ORDER EFFECT AND NONLOCAL BEHAVIOR

Capricious behavior in heat conduction results as the fast-transient effect of thermal inertia interweaves with the microstructural interaction effect in the short-time transient. A diffusion-like behavior is present if the lagging behavior due to the microstructural interaction overcomes that due to the fast-transient effect of thermal inertia. On the other hand, a wave-type response occurs should the fast-transient effect dominate over the microstructural interaction effect. A monotonically decaying response in time is no longer a monopoly of diffusion behavior, neither is a sharp wavefront an exclusive possession of *CV* waves.

This chapter studies some high-order effects that render other types of energy equations in heat transport. Special emphases are on the distinct structures of the *T* wave accounting for the second-order effect of τ_q, the associated resonance phenomenon under high-frequency excitations, and the *single* energy equation describing heat transport in *deformable* conductors. When these refined structures enter the short-time response, the energy equation evolves, depicting a combined behavior of high-order waves and diffusion.

10.1 INTRINSIC STRUCTURES OF *T* WAVES

The alternating behavior of waves and diffusion from the various order effects of τ_T and τ_q, as discussed in Section 2.10, is a salient feature in the generalized lagging response. While the linearized dual-phase-lag model retaining the first-order effects of τ_T and τ_q describes several representative macroscopic and microscopic behaviors in the same framework, the nonlinear response involving the particular expansion of the *first*-order effect in τ_T and the *second*-order effect in τ_q deserves special attention. Equation (2.7) in this case becomes

$$\bar{q}(\bar{x},t) + \tau_q \frac{\partial \bar{q}}{\partial t}(\bar{x},t) + \frac{\tau_q^2}{2} \frac{\partial^2 \bar{q}}{\partial t^2}(\bar{x},t) \cong -k\left\{\nabla T(\bar{x},t) + \tau_T \frac{\partial}{\partial t}\left[\nabla T(\bar{x},t)\right]\right\}, \quad (10.1)$$

with τ_T and τ_q regarded as constant delay times characterizing the fast-transient process in small scale. In terms of the ratio of the response time (t) to the phase lag of the heat flux vector (τ_q), the left side of equation (10.1) can be written as

$$\bar{q} + \frac{\partial \bar{q}}{\partial(t/\tau_q)} + \frac{1}{2}\frac{\partial^2 \bar{q}}{\partial(t/\tau_q)^2} = \cdots. \tag{10.2}$$

Clearly, the importance of considering the second-order effect of τ_q depends on transient times in physical responses. The τ_q^2 effect becomes important if $(t/\tau_q)^2 \ll (t/\tau_q)$, a condition that generally holds in the short-time transient as t approaches zero.

Along with the energy equations (2.5),

$$-\nabla \bullet \bar{q}(x,t) + S(x,t) = C_p \frac{\partial T}{\partial t}(x,t), \tag{10.3}$$

similarly, equation (10.1) furnishes two equations for determining two unknowns, the heat flux vector and the temperature. Eliminating the heat flux vector from equations (10.1) and (10.3) results in

$$\nabla^2 T + \tau_T \frac{\partial}{\partial t}\left[\nabla^2 T\right] + \frac{1}{k}\left(S + \tau_q \frac{\partial S}{\partial t} + \frac{\tau_q^2}{2}\frac{\partial^2 S}{\partial t^2}\right)$$

$$= \frac{1}{\alpha}\frac{\partial T}{\partial t} + \frac{\tau_q}{\alpha}\frac{\partial^2 T}{\partial t^2} + \frac{\tau_q^2}{2\alpha}\frac{\partial^3 T}{\partial t^3}. \tag{10.4}$$

As $(t/\tau_q)^2 \gg 1$ but $(t/\tau_q) \sim 1$, i.e., if the transient time is much longer than the phase lag (delay times) of the heat flux vector in a square sense, equation (10.4) reduces to equation (2.10) employing the linear relationship between τ_T and τ_q. The τ_q^2 effect introduces a third-order term of transient temperature as well as a second-order term describing the apparent heating. It is thus expected to influence the fundamental behavior of lagging both qualitatively and quantitatively.

Characteristics of the temperature are governed by the highest order differentials in equation (10.4),

$$\tau_T \frac{\partial}{\partial t}\left[\nabla^2 T\right] + \cdots = \frac{\tau_q^2}{2\alpha}\frac{\partial^3 T}{\partial t^3} + \cdots, \tag{10.5}$$

or

$$\frac{\partial}{\partial t}\left[\nabla^2 T - \frac{1}{C_T^2}\frac{\partial^2 T}{\partial t^2}\right] + \text{ low order terms} = 0, \quad \text{with} \quad C_T = \frac{\sqrt{2\alpha\tau_T}}{\tau_q}. \tag{10.6}$$

Equation (10.6) represents another type of thermal wave, called T wave in contrast to the CV wave, with C_T denoting the wave speed. Compared to the speed of the CV wave, $C_v = (\alpha/\tau_q)^{1/2}$,

$$\frac{C_T}{C_v} = \sqrt{2\left(\frac{\tau_T}{\tau_q}\right)}. \qquad (10.7)$$

Whether the T wave is faster than the CV wave or not depends on the ratio of τ_T to τ_q. For the medium with $\tau_T > \tau_q$ in transporting heat (the flux precedence), the speed of the T wave is faster than that of the CV wave. Femtosecond laser heating on metal films discussed in Chapter 5 falls into this category. A ratio of (τ_T/τ_q) of the order of 10^2 renders a T wave that propagates faster than the CV wave by an order of magnitude. For the medium with $\tau_T < \tau_q$ in transporting heat (the gradient precedence), on the other hand, the speed of the T wave is slower than that of the CV wave. Heat propagation in superfluid liquid helium at extremely low temperatures, Chapter 4, is an example in this category. The T wave in this case is slower than the CV wave by 1 to 2 orders of magnitude (for $\tau_T/\tau_q \cong 10^{-3}$).

10.1.1 Ballistic Heat Transport in the Electron Gas

The T wave represented by equation (10.4) provides a perfect correlation to the *hyperbolic* two-step model accounting for the ballistic heat transport through the electron gas. This mode of heat transport was derived by Qiu and Tien (1993) based on the macroscopic averages of the electric and heat currents carried by electrons in the momentum space. In the absence of an electric current during laser heating, they derived three coupled equations describing the one-dimensional energy exchange between phonons and electrons:

$$C_e \frac{\partial T_e}{\partial t} = -\frac{\partial q}{\partial x} - G(T_e - T_l) + S \qquad (10.8a)$$

$$C_l \frac{\partial T_l}{\partial t} = G(T_e - T_l), \qquad (10.8b)$$

$$\tau_F \frac{\partial q}{\partial t} + K \frac{\partial T_e}{\partial x} + q = 0 \qquad (10.8c)$$

In equation (10.8c), the value of τ_F has been assumed small and the second-order and higher terms are neglected (Qiu and Tien, 1993). Like the parabolic two-step model, the externally supplied photons (the source term S) first increase the temperature of the electron gas as represented by equation (10.8a). Through the phonon-electron interactions, the second step, the hot electron gas then heats up the metal lattice as represented by equation (10.8b). Equation (10.8c) describes the way in which heat propagates through the electron gas, the constitutive equation. Distinct from the Cattaneo-Vernotte equation for *macroscopic* thermal waves, however, equations (10.8a) and (10.8c) describe *microscale* heat transport through the electron gas. The thermal conductivity (K) in the phonon-electron system, consequently, may vary with the microscopic quantities such as the electron temperature. The quantity τ_F is the relaxation time evaluated at the Fermi surface:

$$\tau_F = (2)^{4/3} \Lambda^{-1} \left(\frac{T_D}{T_l} \right) E_0 E^{3/2} \tag{10.9}$$

where E_0 is the Fermi energy of electrons at 0 K, T_D is the Debye temperature, and Λ is a constant defined as

$$\Lambda = \frac{3\pi^2 P^2 (m/2)^{1/2}}{M\kappa T_D} \left(\frac{3}{4\pi\Delta} \right)^{1/3} \tag{10.10}$$

with P standing for the transient matrix element, m the effective mass of electrons, M the atomic mass, κ the Boltzmann constant, and Δ the averaged volume of the unit cell (Qiu and Tien, 1993). The energy exchange between phonons and electrons is characterized by the coupling factor G, as shown by Qiu and Tien (1992),

$$G = \frac{\pi^4 (n_e v_s \kappa)^2}{K}. \tag{10.11}$$

It depends on the number density of free electrons per unit volume (n_e), the Boltzmann constant (κ), and the speed of sound v_s:

$$v_s = \frac{\kappa}{2\pi h} \left(6\pi^2 n_a \right)^{-\frac{1}{3}} T_D. \tag{10.12}$$

The phonon-electron coupling factor, through the speed of sound, further depends on the Planck constant (h), the atomic number density per unit volume (n_a), and the Debye temperature (T_D). Qiu and Tien (1992) showed that the *s*-band approximation provides an accurate estimate for the number density of free electrons in pure metals. The volumetric heat capacities of the electron gas and the metal lattice, C_e and C_l in equations (10.8a) and (10.8b), respectively, are functions of the electron temperature (T_e) and the lattice temperature (T_l). Qiu and Tien (1992, 1993) numerically solved equation (10.8a) by specifying the heat source term, $S(x, t)$, as the energy absorption rate in a gold film with the laser wavelength in the visible light range. The film thickness is 0.1 μm and the laser pulse duration is 100 femtoseconds. They predicted that temperature change of the electron gas established in picoseconds agrees very well with the experimental data. The classical diffusion and the thermal wave models, owing to the absence of modeling the microstructural effect in the short-time transient, predicted a *reversed* trend for the surface reflectivity at the rear surface of the thin film. The analysis supports well the validity of the hyperbolic two-step model when used for describing the heat transfer mechanisms during short-pulse laser heating of metals.

For exploring the *wave* structure of temperatures behind equations (10.8a) to (10.8c), let us first focus our attention on the metal-lattice temperature (T_l) by eliminating the temperature of the electron gas (T_e) from equations (10.8a) to (10.8c). Since the temperature-dependent properties, such as the volumetric heat capacity of the electron gas, only affect the quantitative behavior of the temperature waves while the fundamental behavior remains the same, all the thermal properties are assumed constant in the following treatment. Differentiating equation (10.8a) with respect to t and equation (10.8c) with respect to x, and combining the results with equation (10.8b) to eliminate the term $\partial^2 q / \partial x \partial t$, the following equation results:

$$K \frac{\partial^2 T_e}{\partial x^2} + (S + \tau_F \frac{\partial S}{\partial t}) = C_e \frac{\partial T_e}{\partial t} + C_e \tau_F \frac{\partial^2 T_e}{\partial t^2}$$

$$+ G(T_e - T_l) + \tau_F G \frac{\partial}{\partial t}(T_e - T_l). \tag{10.13}$$

The quantities T_e and $(T_e - T_l)$ in this equation can be related to the lattice temperature by equation (10.8b):

$$T_e = T_l + \frac{C_l}{G} \frac{\partial T_l}{\partial t}, \quad \text{consequently} \quad T_e - T_l = \frac{C_l}{G} \frac{\partial T_l}{\partial t}. \tag{10.14}$$

Substituting equation (10.14) into (10.13) gives

$$\frac{\partial^2 T}{\partial x^2} + \left(\frac{C_l}{G} \right) \frac{\partial^3 T}{\partial x^2 \partial t} + \frac{1}{K} \left(S + \tau_F \frac{\partial S}{\partial t} \right)$$

$$= \tau_F \left(\frac{C_e C_l}{KG} \right) \frac{\partial^3 T}{\partial t^3} + \left[\frac{\tau_F (C_e + C_l)}{K} + \frac{C_e C_l}{KG} \right] \frac{\partial^2 T}{\partial t^2} + \left[\frac{C_e + C_l}{K} \right] \frac{\partial T}{\partial t} \tag{10.15}$$

where $T \equiv T_l$, with the subscript "l" omitted for the sake of convenience. In the case of $\tau_F = 0$, no ballistic behavior of heat transport in the electron gas, equation (10.15) reduces to that in the parabolic two-step equation. The mixed-derivative term involving the second-order derivative in space and the first-order derivative in time, ($\partial^3 T / \partial x^2 \partial t$), is a common feature existing in both parabolic and hyperbolic two-step models. In the presence of τ_F, most important, (1) the time derivative in the energy equation is raised to the *third order* and (2) an *apparent* heat source term containing the time derivative of the real heat source applied to the body, ($\partial S / \partial t$), exists. While the third-order time-derivative intrinsically alters the fundamental structure of the temperature solution, the apparent heating in equation (10.15) resembles that in the classical thermal wave model (Frankel et al., 1985; Tzou, 1989a, b, 1992a).

Along with the relaxation time of the electron gas (τ_F), the phonon-electron coupling factor G is the most important factor characterizing equation (10.15). In the case that τ_F approaches zero and G approaches infinity, implying that either the number density of free electrons (n_e) approaches infinity (according to equation (10.11)) or the speed of sound approaches infinity (the atomic number density per unit volume n_a approaches zero according to equation (10.12)), equation (10.15) reduces to the classical diffusion equation. Fourier's law embedded in diffusion thus inherits all these assumptions.

10.1.2 Lagging Behavior

The complicated coefficients in equation (10.15) describing the hyperbolic two-step model can be interpreted as the delay times in the fast-transient process. Since the heat source term, $S(x, t)$, does not affect the fundamental behavior of temperature, it shall be dropped from equations (10.4) (the *macroscopic* dual-phase-lag model) and (10.15) (the *microscopic* hyperbolic two-step model) for the purpose of characterization.

In a one-dimensional situation, equation (10.4), based on the phase-lag concept, reduces to

$$\frac{\partial^2 T}{\partial x^2} + \tau_T \frac{\partial^3 T}{\partial x^2 \partial t} = \frac{\tau_q^2}{2\alpha} \frac{\partial^3 T}{\partial t^3} + \frac{\tau_q}{\alpha} \frac{\partial^2 T}{\partial t^2} + \frac{1}{\alpha} \frac{\partial T}{\partial t}. \tag{10.16}$$

In the absence of heating, on the other hand, equation (10.15) describing the microscopic, ballistic heat transport in the electron gas takes the form of

$$\frac{\partial^2 T}{\partial x^2} + \left(\frac{C_l}{G} \right) \frac{\partial^3 T}{\partial x^2 \partial t} = \tau_F \left(\frac{C_e C_l}{KG} \right) \frac{\partial^3 T}{\partial t^3}$$

$$+ \left[\frac{\tau_F (C_e + C_l)}{K} + \frac{C_e C_l}{KG} \right] \frac{\partial^2 T}{\partial t^2} + \left[\frac{C_e + C_l}{K} \right] \frac{\partial T}{\partial t}. \tag{10.17}$$

Equations (10.16) and (10.17) have an identical form, facilitating the following correspondence between the two approaches:

$$\alpha = \frac{K}{C_e + C_l}, \quad \tau_T = \frac{C_l}{G} \tag{10.18a}$$

$$\frac{\tau_q}{\alpha} = \frac{\tau_F (C_e + C_l)}{K} + \frac{C_e C_l}{KG}, \quad \frac{\tau_q^2}{2\alpha} = \left(\frac{C_e C_l}{KG} \right) \tau_F. \tag{10.18b}$$

Equation (10.18a) expresses the *macroscopic* properties, α and τ_T, in the dual-phase-lag model in terms of the *microscopic* properties, G, C_e, and C_l, in the hyperbolic two-step model. The two conditions shown in equation (10.18b) seem to overdetermine the remaining property τ_q, but they are essentially *the same* within the context of the Taylor series expansion of the same order. To demonstrate this important result for a unique correlation, combining equations (18a) (for α) and (18b) gives

$$\frac{\tau_q^2}{2\alpha} = \frac{\tau_q}{2}\left(\frac{\tau_q}{\alpha}\right) = \frac{1}{2}\frac{\tau_F^2(C_e + C_l)}{K} + \left(\frac{C_e C_l}{KG}\right)\tau_F + \frac{(C_e C_l)^2}{2KG^2(C_e + C_l)}. \tag{10.18c}$$

The first term on the right side of (10.18c) is negligibly small because it is of the order of $\tau_F{}^2$; refer to the consistent treatment in equation (10.8c). The third term on the right side of equation (10.18c), on the other hand, can be arranged into the following form:

$$\frac{(C_e C_l)^2}{2KG^2(C_e + C_l)} = \frac{C_e^2}{2K(C_e + C_l)}\tau_T^2. \tag{10.18d}$$

It is proportional to $\tau_T{}^2$, which is again negligible, consistent with the Taylor series expansion made in equation (10.1) with respect to τ_T. The remaining expression in equation (10.18c) is thus identical to the second expression in equation (10.18b). By using equations (10.18b) and the α expression in (10.18a), the relation between the phase lag of the heat flux vector, τ_q, and the microscopic properties in the hyperbolic two-step model becomes

$$\tau_q = \tau_F + \frac{1}{G}\left(\frac{1}{C_e} + \frac{1}{C_l}\right)^{-1}. \tag{10.19}$$

In the presence of ballistic heat transport in the electron gas, equations (10.18) and (10.19) are the expressions for the phase lags and thermal diffusivity describing the lagging behavior in the fast-transient process. They are in correspondence with equation (5.20), which describes the phase lags and thermal diffusivity in correlation to the parabolic two-step model.

The dual-phase-lag model has successfully captured six representative macroscopic and microscopic models in its linear and nonlinear framework. They are summarized in Table 10.1 for an overview. Indeed, the dual-phase-lag model covers a wide range of physical responses in space and time, with τ_T, τ_q, and α reduced to various macroscopic and microscopic parameters in each correspondence. Note that the correlations to the parabolic and hyperbolic two-step models are close, except for an added relaxation time τ_F to τ_q. Since the value of τ_F is of the order of femtoseconds for metals (Qiu and Tien, 1993), its effect on the value of τ_q accounting for the ballistic behavior of heat transport is not significant, usually less than 6% for metals. Rather, its major impact lies in the introduction of a special wave behavior in heat propagation, the T wave shown by equation (10.4) or (10.16).

Based on the experimental data for heat capacities and the electron-phonon coupling factors obtained by Qiu and Tien (1992, 1993), see also Table 5.1, the values of τ_q accounting for the ballistic effect of heat transport, equation (10.19), and

Table 10.1 Correspondence of the dual-phase-lag (DPL) model to diffusion, *CV* wave, heat flux equation of Jeffreys type (HFE-JT), phonon-electron interactions (parabolic and hyperbolic), and phonon scattering field theory in terms of τ_T and τ_q ($\tau_R \equiv$ the relaxation time in the umklapp process, $\tau_N \equiv$ the relaxation time in the normal process, $\tau \equiv$ effective relaxation time in the Jeffreys model, $k_1 \equiv$ ratio of (τ_T/τ_q) in the dual-phase-lag model).

DPL	Diffusion	*CV* wave	HFE-JT	Phonon-electron interactions (parabolic)	Phonon-electron interactions (hyperbolic)	Phonon scattering field
τ_q	0	α/C^2	τ	$\dfrac{1}{G}\left(\dfrac{1}{C_e}+\dfrac{1}{C_l}\right)^{-1}$	$\tau_F+\dfrac{1}{G}\left(\dfrac{1}{C_e}+\dfrac{1}{C_l}\right)^{-1}$	τ_R
τ_T	0	0	$k_1\tau$	$\dfrac{C_l}{G}$	$\dfrac{C_l}{G}$	$\dfrac{9}{5}\tau_N$
α	α	α	α	$\dfrac{K}{C_e+C_l}$	$\dfrac{K}{C_e+C_l}$	$\dfrac{c^2\tau_R}{3}$

the resulting speed of the *T* wave, equation (10.6), are calculated for copper (Cu), silver (Ag), gold (Au), and lead (Pb) and shown in Table 10.2. The effect of ballistic heat transport, in terms of the relaxation time τ_F for the electron gas, affects the phase lag of the heat flux vector, τ_q, and consequently, the speed of the *T* wave. The phase lag of the temperature gradient, τ_T, is not affected by this special behavior. For the four representative metals shown in Table 10.2, the value of τ_T is approximately 2 orders of magnitude *larger* than that of τ_q, implying preservation of the flux precedence in heat transport. The apparent heating in equation (10.4) (the dual-phase-lag model) contains an additional term, $\partial^2 S/\partial t^2$, in comparison with that in equation (10.15) (the hyperbolic two-step model). This term, however, is led by τ_q^2, which is of the order of 10^{-24} seconds, according to Table 10.2. For transient times of the order of picoseconds, the contribution from this additional term is negligibly small for the femtosecond laser heating developed to date.

10.1.3 Propagation of *T* Waves

The *T* wave represented by equation (10.16) displays a new type of wave equation in heat conduction.   It does carry a sharp wavefront in heat propagation, but no confusion with the classical *CV* wave should arise because of the completely different physical mechanisms involved. The *T* wave is caused by the delayed response due to the microstructural interaction and the fast-transient effects of

Table 10.2 Thermal diffusivity (α), phase lags (τ_T and τ_q), and speed of T wave (C_T) ($C_e = 2.1 \times 10^4$ J/m^3 K at room temperature, ps $\equiv$ picosecond, ns $\equiv$ nanosecond).

	K, W/m K	C_l, J/m^3 K $\times 10^6$	G, W/m^{3}K $\times 10^{16}$	α, m^2/s $\times 10^{-4}$	τ_F, ps	τ_T, ps	τ_q, ps	C_T, m/s $\times 10^5$
Cu	386	3.4	4.8	1.1283	0.03	70.833	0.4648	2.7201
Ag	419	2.5	2.8	1.6620	0.04	810.286	0.7838	2.1979
Au	315	2.5	2.8	1.2495	0.04	810.286	0.7838	1.9058
Pb	35	1.5	12.4	0.2301	0.005	12.097	0.1720	1.3718

thermal inertia. It describes the microscale effect of heat transport in space in terms of the resulting delayed response in time. The classical CV wave model, on the other hand, completely neglects heat transport by the microstructural agencies. It focuses attention on the fast-transient effect alone on the basis of a macroscopically averaged behavior.

Like the linearized dual-phase-lag model, the salient feature of the T wave lies in its responses in time. A one-dimensional problem in space is thus sufficient to illustrate the evolution of T waves in the time history. For this purpose, let us consider a semi-infinite solid initially maintained at a uniform temperature T_0. At $t = 0^+$, The boundary at $x = 0$ is suddenly raised to a constant temperature T_W, which produces a thermal disturbance propagating downstream, i.e., in the direction of $x > 0$. For characterizing the wave behavior, again, a dimensionless study is adopted for identifying the dominating groups. Introducing

$$\theta = \frac{T - T_0}{T_W - T_0}, \quad \beta = \frac{t}{\tau_q}, \quad \delta = \frac{x}{\sqrt{\alpha \tau_q}}, \tag{10.20}$$

equation (10.16) becomes

$$\frac{\partial^2 \theta}{\partial \delta^2} + B \frac{\partial^3 \theta}{\partial \delta^2 \partial \beta} = \frac{\partial \theta}{\partial \beta} + \frac{\partial^2 \theta}{\partial \beta^2} + \frac{1}{2} \frac{\partial^3 \theta}{\partial \beta^3}, \quad \text{with} \quad B = \frac{\tau_T}{\tau_q}. \tag{10.21}$$

Compared to the parabolic two-step model, exemplified by equation (4.16), equation (10.21) contains a third-order time derivative, which, along with the mixed-derivative term led by B, accounts for the wave behavior in heat propagation. The boundary conditions are

$$\theta = 1 \quad \text{at} \quad \delta = 0, \qquad \theta \to 0 \quad \text{as} \quad \delta \to \infty. \tag{10.22}$$

Three initial conditions are needed because of the presence of the third-order time-derivative in equation (10.21):

$$\theta = 0, \quad \frac{\partial \theta}{\partial \beta} = 0, \quad \frac{\partial^2 \theta}{\partial \beta^2} = 0 \quad \text{as} \quad \beta = 0. \tag{10.23}$$

Although the time-rate of change of temperature exaggerates the temperature response in a wave-type problem, as discussed in Section 2.7, zero rates are assumed here, i.e., $\partial\theta/\partial\beta = \partial^2\theta/\partial\beta^2 = 0$, in order not to overcomplicate the wave structure.

The Laplace transform solution satisfying equations (10.21) to (10.23) can be easily obtained:

$$\overline{\theta}(\delta; p) = \frac{e^{-\sqrt{\frac{p(2+2p+p^2)}{2(1+Bp)}}\,\delta}}{p}. \tag{10.24}$$

The partial expansion technique developed in Section 2.5 provides a direct and powerful approach to study the short-time response of T waves:

$$\lim_{\beta \to 0} \theta(\delta, \beta) \sim L^{-1}\left[\lim_{p \to \infty} \overline{\theta}(\delta; p)\right] \cong L^{-1}\left[\frac{e^{-\left(\frac{\delta}{\sqrt{2B}}\right)p}}{p}\right] = H\left(\beta - \frac{\delta}{\sqrt{2B}}\right), \tag{10.25}$$

with $H(\bullet)$ denoting the unit-step function. A *wavefront* clearly exists at

$$\beta = \frac{\delta}{\sqrt{2B}} \quad \text{or} \quad x = C_T t, \quad \text{with} \quad C_T = \frac{\sqrt{2\alpha\tau_T}}{\tau_q} \tag{10.26}$$

which is an identical result to that shown in equation (10.6). The wave structure and thus the speed of the T wave are thus confirmed on an analytical basis.

Equation (10.21) includes several macroscopic and microscopic models in special cases:

$$\frac{\partial^2 \theta}{\partial \delta^2} + B\frac{\partial^3 \theta}{\partial \delta^2 \partial \beta} = \frac{\partial \theta}{\partial \beta} + \frac{\partial^2 \theta}{\partial \beta^2}, \quad B = \frac{\tau_T}{\tau_q} \quad \text{(linearized dual-phase-lag model)}$$

$$\tag{10.27a}$$

$$\frac{\partial^2 \theta}{\partial \delta^2} = \frac{\partial \theta}{\partial \beta} + \frac{\partial^2 \theta}{\partial \beta^2} \quad \text{(macroscopic } CV \text{ wave model)} \tag{10.27b}$$

$$\frac{\partial^2 \theta}{\partial \delta^2} = \frac{\partial \theta}{\partial \beta} + \frac{\partial^2 \theta}{\partial \beta^2} \quad \text{(macroscopic diffusion model).} \tag{10.27c}$$

Their solutions in the Laplace transform domain are of the same form as equation (10.24), with the coefficients in p properly adjusted according to equations (10.27a) to (10.27c). As usual, equation (10.24) can be implemented into the FUNC(P) subroutine in the Appendix to obtain the inverse solution.

At $\beta = 1$ and a typical value of $B = 100$ for metals (referring to Chapter 5), Figure 10.1 compares the temperature distributions predicted by the T wave model accounting for the τ_q^2 effect (equation (10.21)), the linearized dual-phase-lag model accounting for the first-order effects of τ_T and τ_q (equation (10.27a)), the macroscopic CV wave model (equation (10.27b)), and the macroscopic diffusion model (equation (10.27c)). Heat transport accounting for the microstructural interaction effect $\tau_T \neq 0$ and, consequently, $B \neq 0$, significantly enlarge the heat-affected zone. Both T-wave and CV-wave models predict a sharp wavefront in heat propagation. They are located at

$$x = \begin{cases} C_T t, & \text{for} \quad T \text{ wave} \\ C_v t, & \text{for} \quad CV \text{ wave} \end{cases} \quad \text{or} \quad \delta = \begin{cases} \sqrt{2B}\beta, & \text{for} \quad T \text{ wave} \\ \beta, & \text{for} \quad CV \text{ wave} \end{cases} \tag{10.28}$$

in terms of the dimensionless space (δ) and time (β) shown in equation (10.20). For $\beta = 1$ and $B = 100$, the wavefronts are located at $\delta = 1$ (CV wave) and $\delta \cong 14.14$ (T wave). In comparison with the linearized dual-phase-lag model, the τ_q^2 effect introduces a sharp wavefront that not only shrinks the physical domain of the heat-affected zone but also increases the temperature level thereby. The time delay (τ_T) due to the microstructural interaction results in a *much larger* heat-affected zone and *much higher* temperature level in the heat-affected zone. These are the major reasons for the successful prediction of the subpicosecond surface reflectivity in

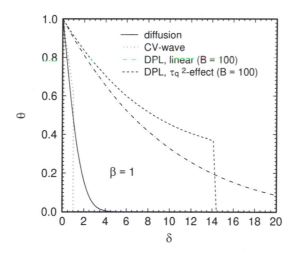

Figure 10.1   Temperature profiles resulting from the T wave model, equation (10.21), the linearized dual-phase-lag model, equation (10.27a), the macroscopic CV wave model, equation (10.27b), and the macroscopic diffusion model, equation (10.27c).

gold films, referring to Chapter 5. In a relative sense, interestingly, the classical *CV* wave model to diffusion is what the hyperbolic dual-phase-lag model (with the τ_q^2 effect) is to the parabolic (linearized) dual-phase-lag model.

Figure 10.2 shows the effect of *B*, the ratio of τ_T to τ_q, on the propagation of *T*-waves. The location of the *T*-wave front, to reiterate, is at $\delta = \sqrt{2B\beta}$. At the same instant of time β, the thermal penetration depth into the solid increases with the square root of *B*. This can also be viewed as the effect of time delays due to microstructural interactions because the ratio *B* is proportional to τ_T. The fast-transient effect of thermal inertia, on the other hand, is absorbed in the phase lag of the heat flux vector τ_q. Compared to the effect of τ_T, it provides a counterbalanced effect with regard to the evolution of the penetration depth.

Figure 10.3 displays the time history of *T* waves as the dimensionless time, β, advances from 1 to 4. The temperature level in the heat-affected zone increases with the value of *B*. The location of the *T*-wave front advances with time according to the relation $\delta = \sqrt{2B\beta}$. The wavefront almost diminishes at $\beta = 4$. For responses at longer times, evidenced by the distributions of $B = 10$ as $\beta = 3$ and 4, the temperature profile flattens when approaching the wavefront from the heat-affected zone. This is a unique behavior pertinent to the *T*-wave (the τ_q^2 effect) that is not found in the macroscopic *CV* wave model.

10.1.4 Effect of τ_T^2

There is no evidence showing the necessity of considering the τ_T^2 effect along with the τ_q^2 effect. From a mathematical point of view, however, the second-order effects of delay times include both terms of τ_T^2 and τ_q^2. To develop a complete picture of the second-order effect of τ_T and τ_q, therefore, it may be desirable to incorporate the τ_T^2 effect at this point and see how it does affect the structure of the *T* wave. Incorporating the second-order term of τ_T^2 in the Taylor series expansion for

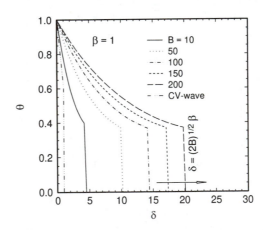

Figure 10.2 Penetration depth of the *T*-wave front increasing with the square root of *B*, where $B = \tau_T/\tau_q$ and $\beta = 1$.

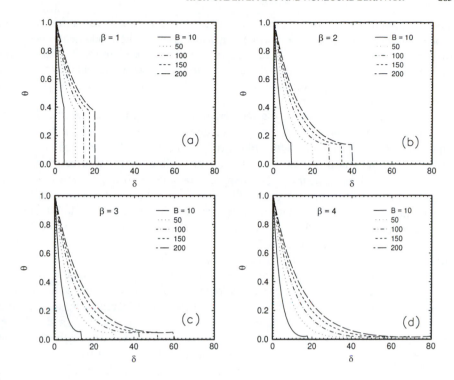

Figure 10.3 Evolution of T waves with time. (*a*) $\beta = 1$, (*b*) $\beta = 2$, (*c*) $\beta = 3$, and (*d*) $\beta = 4$.

equation (2.7) results in, for the same one-dimensional case,

$$q + \tau_q \frac{\partial q}{\partial t} + \frac{\tau_q^2}{2} \frac{\partial^2 q}{\partial t^2} \cong -k \left\{ \frac{\partial T}{\partial x} + \tau_T \frac{\partial^2 T}{\partial t \partial x} + \frac{\tau_T^2}{2} \frac{\partial^3 T}{\partial t^2 \partial x} \right\}, \qquad (10.29)$$

in correspondence with equation (10.1). Combining with equation (10.3), the energy equation, the new term involving τ_T^2 induces an additional fourth-order derivative in temperature, rendering

$$\frac{\partial^2 T}{\partial x^2} + \tau_T \frac{\partial^3 T}{\partial x^2 \partial t} + \frac{\tau_T^2}{2} \frac{\partial^4 T}{\partial x^2 \partial t^2} = \frac{1}{\alpha} \frac{\partial T}{\partial t} + \frac{\tau_q}{\alpha} \frac{\partial^2 T}{\partial t^2} + \frac{\tau_q^2}{2\alpha} \frac{\partial^3 T}{\partial t^3}. \qquad (10.30)$$

in correspondence with equation (10.4). Under the same dimensionless scheme, equation (10.20), equation (10.30) becomes

$$\frac{\partial^2 \theta}{\partial \delta^2} + B \frac{\partial^3 \theta}{\partial \delta^2 \partial \beta} + \frac{B^2}{2} \frac{\partial^4 \theta}{\partial \delta^2 \partial \beta^2} = \frac{\partial \theta}{\partial \beta} + \frac{\partial^2 \theta}{\partial \beta^2} + \frac{1}{2} \frac{\partial^3 \theta}{\partial \beta^3}, \quad \text{with} \quad B = \frac{\tau_T}{\tau_q}.$$

$$(10.31)$$

The additional effect of τ_T^2 introduces a fourth-order derivative of temperature, twice in time and twice in space, which intrinsically alters the fundamental behavior of the T wave. Owing to the presence of the fourth-order term, the third-order differentials in the case of the T wave no longer dictate the characteristics of equation (10.31). Instead, the fourth-order differential describes a *parabolic* behavior in heat propagation.

Figure 10.4 compares the temperature profiles predicted by the T-wave, equation (10.21) containing the τ_q^2 effect alone, and the dual-phase-lag model, equation (10.31) containing both effects of τ_T^2 and τ_q^2. The presence of the τ_T^2 effect completely destroys the wavefront, extending the heat-affected zone to infinity like that in diffusion. The temperature level induced by the τ_T^2 effect, compared to those shown in Figure 10.1, however, is the highest among all cases. The monotonically decaying pattern of temperature shown in Figure 10.4, therefore, should not be confused with the classical diffusion model employing Fourier's law.

As the various orders of τ_T and τ_q are gradually taken into account, interchange between wave and diffusion-like behavior shown in Figures 10.1 and 10.4 clearly illustrates the alternating behavior in the generalized lagging response discussed in Section 2.10. As a general trend, the high-order waves and diffusion (corresponding to higher orders of τ_T and τ_q) gradually increase the temperature level in the heat-affected zone. When high-order waves are activated, evidenced by the result of the CV wave in Figure 10.1 and that of the T wave in Figure 10.4, the physical domain of the heat-affected zone gradually increases as well.

The space and time are normalized with respect to the relaxation time τ_q and the equivalent length $\sqrt{(\alpha\tau_q)}$. For metals, the value of α is of the order of 10^{-6} m²/s and the value of τ_q is of the order of 10^{-12} s. One unit of β ($\beta = 1$) and one unit of δ ($\delta = 1$) in Figures 10.1 to 10.4, therefore, correspond to a real time of 1 picosecond ($t = 1$ ps) and a real dimension of 1 nanometer ($x = 1$ nm).

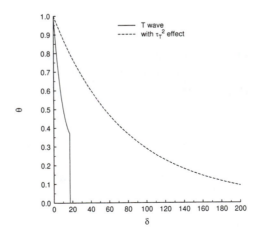

Figure 10.4 Diminution of the sharp wavefront in the T wave by the τ_T^2 effect for B = 150 and $\beta = 1$.

10.1.5 Effect of Microvoids on the Amplification of T-Waves

Since the T-wave behavior becomes more pronounced as the transient time (t) becomes much shorter than the phase lag of the heat flux vector (τ_q), it is desirable to study its effect on the early-time intensification of the heat flux vector (IFHF) around a microvoid. The value of IFHF, referring to equation (8.133) or (8.155), measures the local strength of the heat flux vector around the microvoid relative to the incoming heat flux vector. A value of IFHF larger than 1 implies the presence of flux localization in the neighborhood of the microvoid which is the major cause for hot-spot formation, causing thermal failure of the specimen in thermal processing. In the absence of T-wave behavior, as shown in Section 8.5, the phase lag of the temperature gradient (τ_T) dramatically increases the value of IFHF by 1 to 2 orders of magnitude compared to the values predicted by the classical diffusion and CV wave models, referring to Figures 8.13 and 8.14. Such an exaggerated response, to repeat, results from the time delay due to the microstructural interactions in transporting heat, which, in most cases, enhances heat transport in short times.

The fast-transient response around a microvoid is illustrated by Figure 8.12. In the presence of the τ_q^{2} effect, namely, the T-wave behavior, however, the energy equation (8.27b) becomes

$$\nabla^2 T + \tau_T \frac{\partial}{\partial t}\left[\nabla^2 T\right] = \frac{1}{\alpha}\frac{\partial T}{\partial t} + \frac{\tau_q}{\alpha}\frac{\partial^2 T}{\partial t^2} + \frac{\tau_q^{2}}{2\alpha}\frac{\partial^3 T}{\partial t^3} \tag{10.32}$$

which removes the heat source term from equation (10.4). The Laplacian operator in equation (10.32) is shown by equation (8.110),

$$\nabla^2 T = \frac{\partial^2 T}{\partial r^2} + \frac{2}{r}\frac{\partial T}{\partial r} + \frac{1}{r^2 \sin\theta}\frac{\partial}{\partial\theta}\left(\sin\theta \frac{\partial T}{\partial\theta}\right), \tag{8.110'}$$

for a spherical geometry with azimuthal symmetry in ϕ. The initial and boundary conditions remain the same, equations (8.111) to (8.113), but one more initial condition is needed owing to the presence of the third-order derivative with respect to time in equation (10.32). For a disturbance from a stationary state, using equation (8.111), we further impose

$$\frac{\partial^2 T}{\partial t^2} = 0 \quad \text{as} \quad t = 0. \tag{10.33}$$

Introducing the same dimensionless scheme shown in equation (8.114), the governing system from equations (8.115) to (8.118) becomes

$$\left(1 + B\frac{\partial}{\partial\beta}\right)\left[\frac{\partial^2\Theta}{\partial\delta^2} + \frac{2}{\delta}\frac{\partial\Theta}{\partial\delta} + \frac{1}{\delta^2 \sin\theta}\frac{\partial}{\partial\theta}\left(\sin\theta \frac{\partial\Theta}{\partial\theta}\right)\right]$$

$$= \frac{\partial\Theta}{\partial\beta} + \frac{\partial^2\Theta}{\partial\beta^2} + \frac{1}{2}\frac{\partial^3\Theta}{\partial\beta^3}, \quad \text{with} \quad B = \frac{\tau_T}{\tau_q}, \tag{10.34 Cont.}$$

$$\Theta = 0, \quad \frac{\partial\Theta}{\partial\beta} = 0, \quad \frac{\partial^2\Theta}{\partial\beta^2} = 0 \quad \text{as} \quad \beta = 0, \tag{10.35}$$

$$\eta_r = 0 \quad \text{at} \quad \delta = A, \tag{8.117$'$}$$

$$\eta_3 = -\eta_0 \quad \text{as} \quad \xi \to \infty, \quad \xi = \delta\cos\theta. \tag{8.118$'$}$$

Except for the additional initial condition in equation (10.35), the boundary conditions, equations (8.117) and (8.118), remain the same. Owing to the presence of the τ_q^2 effect, the constitutive equations (8.119) and (8.120) change to

$$\left[q_r, q_\theta, q_3\right] + \tau_q \frac{\partial}{\partial t}\left[q_r, q_\theta, q_3\right] + \frac{\tau_q^2}{2}\frac{\partial^2}{\partial t^2}\left[q_r, q_\theta, q_3\right]$$

$$= -k\left\{\left[\frac{\partial T}{\partial r}, \frac{1}{r}\frac{\partial T}{\partial\theta}, \frac{\partial T}{\partial x_3}\right] + \tau_T\frac{\partial}{\partial t}\left[\frac{\partial T}{\partial r}, \frac{1}{r}\frac{\partial T}{\partial\theta}, \frac{\partial T}{\partial x_3}\right]\right\}, \tag{10.36}$$

$$\left[\eta_r, \eta_\theta, \eta_3\right] + \frac{\partial}{\partial\beta}\left[\eta_r, \eta_\theta, \eta_3\right] + \frac{1}{2}\frac{\partial^2}{\partial\beta^2}\left[\eta_r, \eta_\theta, \eta_3\right]$$

$$= -\left[\frac{\partial\Theta}{\partial\delta}, \frac{1}{\delta}\frac{\partial\Theta}{\partial\theta}, \frac{\partial\Theta}{\partial\xi}\right] - B\left[\frac{\partial^2\Theta}{\partial\beta\partial\delta}, \frac{1}{\delta}\frac{\partial^2\Theta}{\partial\beta\partial\theta}, \frac{\partial^2\Theta}{\partial\beta\partial\xi}\right]. \tag{10.37}$$

With these modifications, including the linear decomposition into steady-state and transient components, the temperature distribution around the microvoid can be found in the same manner as that in Section 8.5. The result for the total temperature in the Laplace transform domain, in correspondence with the combination of equations (8.134), (8.148), and (8.151), is

$$\overline{\Theta}(\delta; p) = \eta_0\cos\theta\left\{\frac{\delta}{p} + \frac{p^2 + 2p + 2}{2p(1 + Bp)}\frac{(1 + D\delta)A^3}{\left[(DA)^2 + 2(DA) + 2\right]\delta^2}\right.$$

$$\times\left.\frac{\sinh(D\delta) - \cosh(D\delta)}{\sinh(DA) - \cosh(DA)}\right\}, \quad \text{with} \quad D = \sqrt{\frac{p(p^2 + 2p + 2)}{2(1 + Bp)}}, \tag{10.38}$$

while the ratio of the heat flux vector in the Laplace transform domain, in correspondence with equation (8.153), becomes

$$\frac{\overline{\eta}_\theta}{\eta_0} = \frac{2(1+Bp)}{p(p^2+2p+2)} \left\{ 1 + \frac{p^2+2p+2}{2(1+Bp)} \frac{1+D\delta}{(DA)^2+2(DA)+2} \right.$$

$$\left. \times \left(\frac{A}{\delta}\right)^2 \times \frac{\sinh(D\delta)-\cosh(D\delta)}{\sinh(DA)-\cosh(DA)} \right\} \sin\theta . \qquad (10.39)$$

Likewise, the maximum heat flux occurs at the microvoid surface at $\delta = A$ and $\theta = \pm\pi/2$, rendering from equation (10.39),

$$\left(\frac{\overline{\eta}_\theta(p)}{\eta_0}\right)_{max} = \left(\frac{\overline{q}_\theta(p)}{q_0}\right)_{max}$$

$$= \frac{2(1+Bp)}{p(p^2+2p+2)} \left\{ 1 + \frac{p^2+2p+2}{2(1+Bp)} \times \frac{1+DA}{(DA)^2+2(DA)+2} \right\} \qquad (10.40)$$

in correspondence with equation (8.154). With this new expression for the ratio of the heat flux vector that accounts for the T-wave behavior in the short-time transient, the value of IFHF can be calculated from equation (8.155).

Limiting behavior. Before the general results of the Laplace inversion are presented, it is desirable to explore the limiting behavior of IFHF at long and short times. The short-time response, according to equation (8.156), is

$$\lim_{t\to 0^+}\left(\frac{q_\theta(t)}{q_0}\right) = \lim_{p\to\infty} p\left(\frac{\overline{q}_\theta(p)}{q_0}\right) = \frac{2B}{p}\left\{1+\frac{\left(\frac{A}{\sqrt{2B}}\right)^p}{\left(\frac{A^2}{2B}\right)^p p^2}\frac{p^2}{2Bp}\right\} \sim p^{-1} \to 0 . \quad (10.41)$$

Unlike the linearized dual-phase-lag model, equation (8.157), in which the short-time value of IFHF follows the value of B (τ_T/τ_q), the initial value of IFHF as time approaches zero is zero in the presence of T-wave behavior. This is the same result as that employing the CV wave model, referring to Figure 8.13, which reflects the general effect of finite speed of heat propagation.

The short-time behavior, according to equation (8.158), is

$$\lim_{t\to\infty}\left(\frac{q_\theta(t)}{q_0}\right) = \lim_{p\to 0} p\left(\frac{\overline{q}_\theta(p)}{q_0}\right) = \frac{2}{2}\left\{1+\frac{1}{2}\frac{2}{2}\right\} = \frac{3}{2} . \qquad (10.42)$$

This is an expected result at steady state (as $t \to \infty$) because the T-wave behavior only exists in the transient stage. As time approaches infinity, as a matter of fact, the classical theory of diffusion, the CV wave, the linearized dual-phase-lag model, and the T-wave model accounting for the τ_q^2 effect all approach the same steady-state value of 3/2.

Transient values of IFHF. The way in which the transient value of IFHF varies from zero (as $t \to 0$) to 3/2 ($t \to \infty$) can be analyzed by the use of equation (10.40) in replacement of the function subroutine FUNC(P) in the Appendix. The results are shown in Figure 10.5 for typical values of $B = 10$, 30, and 50. For all cases, the steady-state value arrives at $\beta \geq 6$. The T wave differs from the CV wave in many ways. In approaching the steady state, first of all, the IFHF predicted by the T wave increases to a peak value (about 65% of the value of B) at an earlier time (around $\beta = 1$). It then decays to the steady-state value of 3/2, with a slight oscillation existing for $3 \leq \beta \leq 5$. The CV wave model, on the other hand, predicts a monotonically increasing curve from zero to the steady-state value (3/2). No oscillation occurs, and the steady-state value is the maximum value in the transient response. This can be seen more clearly in Figure 8.13. The peak value of IFHF, roughly $0.65B$, is not as pronounced as that predicted by the linearized dual-phase-lag model, referring to Figure 8.14 and equation (8.157). The heat flux around the microvoid, however, is still highly localized, with an intensity of about 1 order of magnitude larger than that of the incoming heat flux vector.

The amount of post-peak oscillation increases with the value of B. As the value of B becomes sufficiently large, starting from $B \cong 60$ and thereafter, a negative value of $(q_\theta/q_0)_{max}$ appears between $\beta = 3$ and 5. A negative value of q_θ in this case implies reversal of heat flux in heat transport, which is a new phenomenon necessitating experimental support.

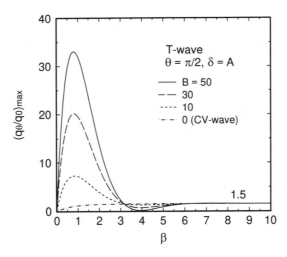

Figure 10.5 Transient response of IFHF resulting from the T-wave behavior (τ_q^2 effect).

10.2 THERMAL RESONANCE

Resonance is a common feature in any type of wave phenomenon. In addition to the displacement waves in mechanical vibrations, Tzou (1991b, c, 1992d, e) showed that the classical thermal wave (which is macroscopic and accounts for the fast-transient effect of thermal inertia only) can be excited to resonate should the oscillating frequency of the externally applied heat source couple with the modal frequency of the wave modes. In the presence of the microstructural interaction in the short-time transient, τ_q^2 effect in the dual-phase-lag model shown by equation (10.4), likewise, the resulting T wave may display a resonance phenomenon under the proper conditions. The τ_q^2 effect appears at two places in equation (10.4): the third-order time derivative of temperature constituting the fast-propagating T wave, and the second-order time derivative of the heat source resulting from the apparent heating due to the fast-transient effect of thermal inertia.

A one-dimensional analysis in space is sufficient because resonance is a special response in time. Imposing a body heat source oscillating at a frequency Ω,

$$S(x,t) = Qe^{i\Omega t}g(x) \tag{10.43}$$

with $g(x)$ describing the spatial distribution of the body heat source in a one-dimensional solid. Both ends of the solid at $x = 0$ and L are maintained at a zero temperature. Thermal resonance, by definition, occurs at a specific value of Ω under which the amplitude of the T wave depicted by equation (10.4) reaches a *maximum*. Since the temperature response reaches a maximum value under the same intensity of heating, thermal resonance provides an efficient way of conserving energy in thermal processing of materials.

To demonstrate the resonance phenomenon of T waves under frequency excitations, the temperature satisfying equation (10.4) is expressed in terms of the spatial eigenfunctions, $\phi_n(x)$, of the pure wave equation,

$$T(x,t) = \sum_{n=1}^{\infty} \Gamma_n(t)\phi_n(x), \tag{10.44}$$

where

$$\phi_n(x) = \sin\left(\frac{\omega_n x}{C_T}\right), \quad \omega_n = \frac{n\pi C_T}{L}, \quad C_T = \frac{\sqrt{2\alpha\tau_T}}{\tau_q}. \tag{10.45}$$

Note that the boundary conditions, $T(0) = T(L) = 0$, are satisfied by the ϕ_n function along with equation (10.44). The quantity ω_n for $n = 1, 2, 3, \ldots$, etc. is called the modal frequency of the nth wave mode.

The temperature wave shown in equation (10.44) is thus the linear combination of all the fundamental wave modes. Substituting equations (10.43) to (10.45) into equation (10.4), a third-order differential equation governing the time-dependent amplitude of the T-wave results:

$$\tau_q^2 \dddot{\Gamma}_n + 2\tau_q \ddot{\Gamma}_n + 2\left[1 + \alpha\tau_T\left(\frac{\omega_n}{C_T}\right)^2\right]\dot{\Gamma}_n + 2\alpha\left(\frac{\omega_n}{C_T}\right)^2\Gamma_n$$

$$= \frac{2Q\alpha}{k}e^{i\Omega t}\left[1 + i\Omega\tau_q - \frac{\tau_q^2\Omega^2}{2}\right]D_n \qquad (10.46)$$

where D_n is the Fourier coefficient of the spatial distribution $g(x)$,

$$D_n = \frac{\int_0^L g(x)\sin\left(\frac{\omega_n x}{C_T}\right)dx}{\int_0^L \sin^2\left(\frac{\omega_n x}{C_T}\right)dx}. \qquad (10.47)$$

It is independent of the driving frequency Ω. The time-varying part of the heat source term, $e^{i\Omega t}$, in equation (10.46) suggests a solution of the following form:

$$\Gamma_n(t) = H_n e^{i\Omega t}. \qquad (10.48)$$

Substituting equation (10.48) into (10.46), the amplitude of the T wave in response to the excitation of the oscillating heat source results:

$$H_n = \left(\frac{Q\alpha\tau_q D_n}{k}\right)H; \quad \text{with}$$

$$H = \frac{\left(1 - \frac{\Omega^{*2}}{2}\right) + i\Omega^*}{\left(\frac{\omega_n^{*2}}{2B} - \Omega^{*2}\right) + i\Omega^*\left(1 + \frac{\omega_n^{*2} - \Omega^{*2}}{2}\right)}, \quad i = \sqrt{-1}. \qquad (10.49)$$

where the frequencies Ω and ω_n have been normalized with respect to the relaxation time of the heat flux vector τ_q, $\Omega^* = \Omega\tau_q$ and $\omega_n^* = \omega_n\tau_q$. The quantity B denotes the ratio of τ_T/τ_q. The quantities in parentheses are independent of the applied frequency, implying that the maximum temperature response results as the norm of H,

$$H^2 = H\overline{H} = \frac{\Omega^{*4} + 4}{\Omega^{*6} - \left(2\omega_n^{*2}\right)\Omega^{*4} + \omega_n^{*2}\left[\omega_n^{*2} + 4\left(1 - \frac{1}{B}\right)\right]\Omega^{*2} + \frac{\omega_n^{*4}}{B^2}}. \qquad (10.50)$$

reaches a maximum.

Denoting the resonance frequency by $\Omega_{max}{}^*$, at which the amplitude of the T wave (H) reaches a maximum value, the stationary condition, $d(H^2)/d\Omega^* = 0$, gives a fourth-order algebraic equation for the determination of $\Omega_{max}{}^*$:

$$z^4 - \left[\omega_n^{*4} + 4\omega_n^{*2}\left(1 - \frac{1}{B}\right) - 8\right]z^2 - 2\omega_n^{*2}\left(8 + \frac{1}{B^2}\right)z$$

$$+ 4\left[\omega_n^{*4} + 4\left(1 - \frac{1}{B}\right)\omega_n^{*2} + 4\right] = 0, \quad z = \Omega_{max}^2 \qquad (10.51)$$

As shown in Table 10.2, the ratio of $B = \tau_T/\tau_q$ is of the order of 10^2 for metals. For the value of B in this range, the resonance frequency (z or Ω_{max}) resulting from equation (10.51) is almost *independent of B* because the $1/B$ and $1/B^2$ terms are negligibly small compared to the constants (1 and 8) in front of them. For metals exemplified in Table 10.2, therefore, the resonance frequency depends only on the modal frequencies ω_n.

In searching for the resonance frequency satisfying equation (10.51), note that (1) the root for z must be positive definite because the resonance frequency is real and (2) the *smallest* root of z is of primary interest because it is desirable to produce the thermal resonance at the *lowest* possible applied frequency. Bearing these clues in mind, it can be shown that positive roots for z (and hence for $\Omega_{max}{}^*$) exist only for $\omega_n{}^* \geq 1.3935$. For the values of $\omega_n{}^*$ smaller than this threshold, say $\omega_n{}^*$ = 1.2 shown in Figure 10.6, the amplitude H (from equation (10.50)) monotonically decreases as the driving frequency Ω^* increases, implying nonexistence of a stationary value.

The distribution is almost identical for any value of B greater than 50. A typical value of B = 150 for gold (referring to Table 10.2) is used in Figure 10.6. For the values of $\omega_n{}^*$ greater than the threshold of 1.3935, exemplifying by the curves of $\omega_n{}^*$ = 1.3937, 1.5, and 1.6, one minimum and one maximum exist in the positive domain of Ω^*. When the value of $\omega_n{}^*$ approaches the threshold value of 1.3935, the difference between the minimum and the maximum decreases, as shown by the values of Ω^* = 1.49834 (for the minimum value of H) and 1.51650 (for the maximum value of H) at $\omega_n{}^*$ = 1.3937. At the critical value of $\omega_n{}^*$ = 1.3935, the onset for the occurrence of thermal resonance, the two locations possessing the maximum and minimum values of H collapse onto each other at $\Omega_{max}{}^* \cong 1.5$, resulting in the degeneration of two stationary points into a single *inflection point*. Analytically, therefore, the onset of thermal resonance is described by

$$\frac{d(H^2)}{d\Omega^*} = 0 \quad \text{and} \quad \frac{d^2(H^2)}{d\Omega^{*2}} = 0 \qquad \text{(thermal resonance of } T \text{ waves)} \quad (10.52)$$

These two equations are to be solved for $\Omega_{max}{}^*$, the lowest frequency driving the thermal resonance to occur, and $\omega_n{}^*$, the lowest excitable wave mode to resonate.

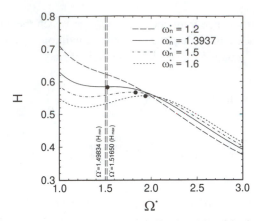

Figure 10.6 Variations of the wave amplitude (H) with the oscillating frequency (Ω^*) of the externally applied body heating. Thermal resonance occurs at Ω_{max}^*, where the values of H reach maxima. Here, $\omega_n^* = 1.2$, 1.3937, 1.5, and 1.6. An inflection point exists at the critical mode with $\omega_n^* = 1.3935$ at $\Omega^* \cong 1.5$ for $B = 150$.

The lowest resonance frequency (Ω_{max}) shown in Figure 10.6 is around ($1.5/\tau_q$), which is of the order of terahertz.

According to the *CV* wave model, as shown by Tzou (1992d, e), the resonance frequency (Ω_{max} or Ω_{max}^*) approaches the modal frequency (ω_n^*) in the high-frequency domain (as ω_n^* approaches a large value). This is no longer the case for the *T* wave including the τ_q^2 effect in the short-time transient. As shown in Figure 10.7, the *difference* between the resonance frequency and the modal frequency, namely $\Omega_{max}^* - \omega_n^*$, approaches a constant value of about 0.33. By the use of equation (10.51) at various values of ω_n^*, this unusual behavior is displayed in Figure 10.7. The constant difference arrives as $\omega_n^* \geq 1.8$. Also, the mixed-derivative term ($\tau_T \partial^3 T/\partial x^2 \partial t$) appearing on the left side of equation (10.16) provides a "negative" damping which balances the effect of diffusion on the right side of the equation. Consequently, the mixed-derivative term exaggerates the resonance amplitude even for the wave modes in the high-frequency domain. Compared to the *CV*-wave behavior, these are salient features resulting from the microstructural interaction effect (the τ_T effect) in the dual-phase-lag equation (10.4). The critical modal frequency, $\omega_n^* \geq 1.3935$, dictates the excitable mode of *T*-waves for thermal resonance. With the expression of ω_n shown in equation (10.45), this critical condition can be expressed in terms of the *critical* modal number n_c:

$$n_c \geq \frac{1.3935L}{\sqrt{2\alpha\tau_T}} \quad \text{or} \quad n_c \geq \frac{1.3935L}{\sqrt{2\left(\dfrac{K}{G}\right)\left(\dfrac{1}{1+R_c}\right)}}, \quad \text{with} \quad R_c = \frac{C_e}{C_l}, \tag{10.53}$$

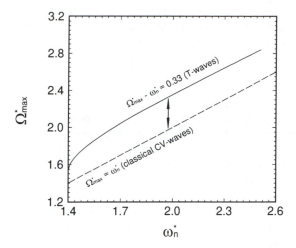

Figure 10.7 The constant difference, 0.33, between the resonance frequency $(\Omega_{max}{}^{*})$ and the modal frequency $(\omega_n{}^{*})$ in the high-frequency domain of T waves.

in terms of the microscopic parameters in the hyperbolic two-step model, equations (10.18a) and (10.19). The modal number dictating the excitable modes for thermal resonance, in other words, depends on the film thickness (L), the effective thermal diffusivity (α), and the phase lag of the temperature gradient (τ_T). In terms of the microscopic properties, alternatively, it depends on the ratio of the effective thermal conductivity to the phonon-electron coupling factor (K/G) and the ratio of heat capacities of the electron gas to the metal lattice (R_c). For the representative metals shown in Table 10.2, the value of n_c is 2 for Cu, 1 for Ag and Au, and 6 for Pb. The excitable modes occur early in the wave train shown by equation (10.44).

10.3 HEAT TRANSPORT IN DEFORMABLE CONDUCTORS

At least within the context of classical theories, equations of heat transport are established on the basis of rigid conductors. Rigidity of the conductor facilitates a sole consideration of conservation of energy across a material volume without agonizing over the thermal expansion/contraction associated with the occurrence of heat flow. The classical diffusion equation, the thermal wave equation, and the dual-phase-lag equation of heat transport we have considered so far are all in this category. Coupling with mechanical deformation, however, may become important should the thermally induced mechanical strain rate be sufficiently high. When this occurs, flexibility (or elasticity) of the conducting medium must be taken into account, rendering many salient features that cannot be described by the heat conduction theory assuming a rigid conductor. They include the presence of a thermal wavefront even in a *diffusive* type of heat transport due to wave motion in the elastic medium, and the evolution of wavefronts due to interaction between thermal and mechanical waves. Traditionally, as shown in Chapter 9, these phenomena are analyzed by way of the coupled formulation between thermal and mechanical fields. The two fields are coupled *via* the volumetric strain rate in the energy equation, equation (8.9), and the temperature gradient in the equation of equilibrium, equation (8.26). Mathematically, the coupled partial differential equations describing energy and momentum transport through the deformable and conducting medium must be solved simultaneously for temperature and displacement.

A simultaneous treatment of the energy and momentum equations, generally speaking, involves the solutions for four coupled partial differential equations, one for the temperature field and the other three for each component of the displacement vector. The mathematical details involved are nontrivial. Particular solutions characterizing the fundamental behavior of thermomechanical coupling may be found in special cases, exemplified by the thermal stresses around a rapidly moving heat source accommodating the fast-transient effect of thermal inertia (Tzou, 1989c, d, 1992b), but a general solution is not guaranteed, especially for problems involving finite boundaries. From a physical point view, moreover, the coupled equations between energy and momentum may cover some important mechanisms in thermomechanical coupling. The wave behavior in heat conduction due to wave motion in the elastic medium, and the evolution of wavefronts due to thermomechanical interactions simply cannot be foreseen prior to a profound analysis.

A natural question thus arises: Is it possible to establish a *single* energy equation for a *deformable* conductor that not only captures all the fundamental behavior in the thermal field but also accounts for the elasticity of the conductor? From the viewpoint of load transfer, alternatively, is it possible to establish a *single* equation of equilibrium that captures the effect of thermal expansion/contraction of the elastic medium while describing the stress and strain distributions? This philosophy is especially valuable for thermal engineers because the resulting single energy equation provides an effective means for characterizing the temperature field, which is a scalar, without looping around the displacement field, which is a three-component vector.

Development of a single energy equation accounting for the elasticity of conductors must be made on the basis of the existing thermoelasticity. While Chapter 9 provides an example for a one-dimensional problem, this chapter provides a rigorous formulation for three-dimensional problems.

10.3.1 Energy Equation

In the presence of thermomechanical coupling, as shown in Chapter 9, the energy equation (8.9) is

$$-\nabla \bullet \vec{q} + S = C_p \dot{T} + \kappa_\sigma T_0 \dot{e}, \quad \text{with} \quad \kappa_\sigma = \kappa_\varepsilon (3\lambda + 2\mu), \qquad (10.54)$$

where e denotes the dilatation measuring the change of volume per volume of a material volume, $e = u_{i,i} = u_{1,1} + u_{2,2} + u_{3,3} \equiv \partial u_1/\partial x_1 + \partial u_2/\partial x_2 + \partial u_3/\partial x_3$, and u_i are the displacement components in the x_i direction, for $i = 1, 2, 3$. The volumetric heat source S is included for the sake of generality. In relation to the mean strain e_m

defined in equation (8.22), $e_m = e/3$. The subscripts used here denote differentiation with respect to space. The quantity $u_{i,i}$, for example, is $\partial u_i/\partial x_i$.

The dilatation is determined by the equation of equilibrium,

$$\frac{\partial \sigma_{ij}}{\partial x_j} = \rho \frac{\partial^2 u_i}{\partial t^2}, \quad \text{for} \quad i, j = 1, 2, 3. \tag{10.55}$$

It represents three equations for $i = 1, 2, 3$, with the repeated index j summed up from 1 to 3 for every value of i. In the case of one-dimensional deformation, $\sigma_{11} \equiv \sigma$, $\sigma_{ij} = 0$ otherwise, and $x_1 \equiv x$, equation (10.55) reduces to equation (8.25).

For a three-dimensional deformation in general, equation (10.55) contains nine unknowns, six in symmetric stress components ($\sigma_{ij} = \sigma_{ji}$ for $i, j = 1, 2, 3$, implying the absence of a body moment in deformation) and three in displacement components (u_i for $i = 1, 2, 3$). For making the formulation well-posed, the mechanical constitutive equation relating stress to strain and the Cauchy strain tensor relating strain components to displacement gradients are needed. Equation (8.23), the mechanical constitutive equation in terms of Young's modulus E and Poisson ratio ν, is convenient to use for an explicit expression of the lateral effect (equation (8.19)) on longitudinal deformation (equation (8.18)). The representation in terms of the Lamé constants, λ and μ, with

$$E = \frac{\mu(3\lambda + 2\mu)}{\lambda + \mu}, \quad \nu = \frac{\lambda}{2(\lambda + \mu)}, \tag{10.56}$$

however, results in a more compact expression. This can be seen by a direct substitution of equation (10.56) into (8.23):

$$\sigma_{ij} = \left[\lambda e - \kappa_\varepsilon (3\lambda + 2\mu) T\right]\delta_{ij} + 2\mu e_{ij}, \quad \text{for} \quad i, j = 1, 2, 3 \tag{10.57}$$

where e_{ij} is the Cauchy strain tensor and e the dilatation,

$$e_{ij} = \frac{1}{2}\left(\frac{\partial u_i}{\partial x_j} + \frac{\partial u_j}{\partial x_i}\right), \quad e = e_{ii} = \frac{\partial u_1}{\partial x_1} + \frac{\partial u_2}{\partial x_2} + \frac{\partial u_3}{\partial x_3} = u_{i,i}. \tag{10.58}$$

For a further combination of equation (10.57) with (10.55), we first take the spatial derivative of equation (10.57) with respect to x_j, resulting in

$$\sigma_{ij,j} = \left[\lambda\left(u_{i,i}\right)_{,j} - \kappa_\varepsilon (3\lambda + 2\mu) T_{,j}\right]\delta_{ij} + \mu\left(u_{i,j} + u_{j,i}\right)_{,j}$$

$$= \mu u_{i,jj} + (\lambda + \mu) u_{j,ji} - \kappa_\varepsilon (3\lambda + 2\mu) T_{,i} \tag{10.59a}$$

where repeated indices imply summation and the results

$$T_{,j}\,\delta_{ij} = T_{,i}\,, \quad \left(u_{i\,,ij}\right)\delta_{ij} = u_{j\,,ji}\,, \quad \text{and} \quad u_{j\,,ji} = u_{j\,,ij} \tag{10.59b}$$

have been used. Substituting equation (10.59a) into equation (10.55),

$$\mu u_{i\,,jj} + (\lambda + \mu)u_{j\,,ji} - \kappa_\varepsilon (3\lambda + 2\mu)T_{,i} = \rho\ddot{u}_i \quad \text{for} \quad i,j = 1,2,3 \tag{10.60}$$

which now provides three coupled differential equations to be solved for *three* unknowns, u_i for i = 1, 2, and 3. In terms of a vector notation, note that $u_{i,jj} \equiv \nabla^2 u_i$ in equation (10.60) because the index j is repeated, implying a summation from 1 to 3.

Equation (10.54) is the general form of the energy equation incorporating the effect of elasticity in a deformable conductor. At this stage, it still allows for any relation between the heat flux vector and the temperature gradient. The deformation rate can be further eliminated from the energy equation (10.54) with the assistance of the momentum equation (10.60). From equation (10.54), the deformation rate is related to the heat flux vector and the time-rate of change of temperature by

$$\dot{e} = -\frac{\nabla \bullet \vec{q} + C_p \dot{T} - S}{(3\lambda + 2\mu)\kappa_\varepsilon T_0}, \quad \text{resulting in} \quad \ddot{e} = -\frac{\nabla \bullet \dot{\vec{q}} + C_p \ddot{T} - \dot{S}}{(3\lambda + 2\mu)\kappa_\varepsilon T_0}, \tag{10.61}$$

where the superscripts dots denote differentiation with respect to time. The equation governing the dilatation and its space and time derivatives can be obtained from equation (10.60). Taking the spatial derivative of equation (10.60) with respect to x_i, $\partial(10.60)/\partial x_i$ gives

$$\mu\left(u_{i\,,i}\right)_{,jj} + (\lambda + \mu)\left(u_{j\,,j}\right)_{,ii} - \kappa_\varepsilon (3\lambda + 2\mu)T_{,ii} = \rho\frac{\partial^2\left(u_{i\,,i}\right)}{\partial t^2}. \tag{10.62}$$

Since $u_{i,i} \equiv e$, referring to equation (10.58), equation (10.62) is simplified to

$$(\lambda + 2\mu)\nabla^2 e - \kappa_\varepsilon (3\lambda + 2\mu)\nabla^2 T = \rho\ddot{e}. \tag{10.63}$$

Taking an additional time derivative of equation (10.63), resulting in

$$(\lambda + 2\mu)\nabla^2 \dot{e} - \kappa_\varepsilon (3\lambda + 2\mu)\nabla^2 \dot{T} = \rho\dddot{e}, \tag{10.64}$$

a direct substitution of equation (10.61) into (10.64) gives

$$\nabla^2 R - \frac{1}{C_L^2}\frac{\partial^2 R}{\partial t^2} + C_p\eta\nabla^2 \dot{T} = 0, \quad \text{with} \tag{10.65a}$$

$$R = \nabla \bullet \vec{q} + C_p \dot{T} - S, \quad \eta = \frac{\kappa_\varepsilon^2 T_0 (3\lambda + 2\mu)^2}{C_p (\lambda + 2\mu)}, \quad C_L = \sqrt{\frac{\lambda + 2\mu}{\rho}}. \qquad (10.65b)$$

The quantity η is the thermomechanical coupling factor, referring to equation (8.14) for its degenerated form with $v = 0$. C_L is the longitudinal elastic wave speed. It reduces to the dilatational wave speed in the case of $v = 0$, from equation (10.56). Recognizing that

$$\nabla \bullet \vec{q} + C_p \dot{T} - S = 0 \qquad (10.66)$$

is in fact the energy equation for a *rigid* conductor without deformation, equation (10.65a) describes transport of the *rigidity propagator R* in a deformable conductor. The rigidity propagator R transports at the same speed as the longitudinal wave speed in an elastic body.

Fourier behavior. Equation (10.65a) is the energy transport equation in a deformable conductor. It contains two unknowns, heat flux vector and temperature, necessitating consideration of the constitutive equation between the heat flux vector and the temperature gradient to furnish the formulation. Fourier's law in heat conduction provides the simplest example:

$$\vec{q} = -k\nabla T. \qquad (10.67a)$$

The resulting rigidity propagator R, according to equation (10.65b), is simply the diffusion operator,

$$R = -k\nabla^2 T + C_p \dot{T} - S. \qquad (10.67b)$$

Direct differentiations of equation (10.67b) give

$$\nabla^2 R = -k\nabla^4 T + C_p \nabla^2 \dot{T} - \nabla^2 S, \quad \ddot{R} = -k\nabla^2 \ddot{T} + C_p \dddot{T} - \ddot{S}. \qquad (10.67c)$$

Substituting equation (10.67c) into (10.65a), a *single* energy equation governing heat transport in a deformable conductor is obtained:

$$\nabla^4 T - \left(\frac{1 + \eta}{\alpha}\right)\nabla^2 \dot{T} = \frac{1}{C_L^2}\left[\nabla^2 \ddot{T} - \left(\frac{1}{\alpha}\right)\dddot{T}\right] - \frac{1}{k}\left[\nabla^2 S - \frac{1}{C_L^2}\ddot{S}\right]. \qquad (10.68)$$

Having the appearance of an energy equation with temperature as the dependent variable, equation (10.68) absorbs the effect of mechanical deformation in the coefficients C_L (the longitudinal wave speed) and η (thermomechanical coupling factor). In the absence of mechanical deformation due to the thermal effect, $\kappa_\varepsilon \to 0$ implying $\eta \to 0$, equation (10.68) reduces to

$$\nabla^2 \left[\nabla^2 T - \frac{1}{\alpha} \dot{T} + \frac{1}{k} S \right] = 0, \quad \text{implying that} \quad \nabla^2 T - \frac{1}{\alpha} \dot{T} + \frac{1}{k} S = 0 \qquad (10.69)$$

as a particular solution. Clearly, equation (10.69) is the classical diffusion equation describing heat transport in a rigid conductor.

Equation (10.68) has a new appearance in heat transport. Its fundamental properties are characterized by the highest order differentials,

$$\nabla^2 \left[\nabla^2 T - \frac{1}{C_L^2} \ddot{T} \right] + \text{low order terms} = 0 \quad \text{or} \quad \nabla^2 T = \frac{1}{C_L^2} \frac{\partial^2 T}{\partial t^2}. \qquad (10.70)$$

Equation (10.70) clearly indicates a temperature *wave*, even though assuming diffusion in heat transport, with a thermal wave speed of C_L. Since diffusion does not possess a wave behavior, the thermal wavefront depicted by equation (10.70) is induced by the elastic wave in the deformable conductor. This is also evidenced by the coincidence of the thermal wave speed to the longitudinal wave speed of elastic waves.

Characteristics of equation (10.68) can be revealed by considering the short-time transient in a one-dimensional problem. For this purpose, let us consider the wave propagation in a half-space as shown in Figure 9.14. Equation (10.68) reduces to

$$\frac{\partial^4 T}{\partial x^4} - \left(\frac{1+\eta}{\alpha} \right) \frac{\partial^3 T}{\partial x^2 \partial t} = \frac{1}{C_L^2} \frac{\partial^4 T}{\partial x^2 \partial t^2} - \frac{1}{\alpha C_L^2} \frac{\partial^3 T}{\partial t^3} \qquad (10.71)$$

in this case, with the heat source terms neglected, since they do not alter the fundamental characteristics of the solution. The fourth-order derivative in space and third-order derivative in time require four boundary conditions and three initial conditions. The half-space is assumed to be disturbed from a stationary state,

$$T = T_0, \quad \frac{\partial T}{\partial t} = 0, \quad \text{and} \quad \frac{\partial^2 T}{\partial t^2} = 0 \quad \text{as} \quad t = 0. \qquad (10.72)$$

At the boundary of the half-space, $x = 0$, the temperature is suddenly raised to T_w from a zero slope,

$$T = T_w, \quad \frac{\partial T}{\partial x} = 0 \quad \text{at} \quad x = 0. \qquad (10.73a)$$

Allowance for the co-existence of a temperature- and a gradient-specified condition at the same boundary is a special feature in this problem. It results from the high-order derivative of temperature with respect to space in equation (10.71). The condition of zero slope in equation (10.73a) implies no heat loss from the boundary after the temperature is raised to T_w (an insulated boundary condition).

Consequently, the temperature in the vicinity of $x = 0$ may exceed the boundary temperature (T_w) owing to energy accumulation near the surface. The regularity conditions as x approaches infinity, likewise, read

$$T \rightarrow T_0, \quad \frac{\partial T}{\partial x} \rightarrow 0 \quad \text{as} \quad x \rightarrow \infty. \tag{10.73b}$$

A dimensionless analysis is performed to characterize the dominant groups. Introducing

$$\theta = \frac{T - T_0}{T_w - T_0}, \quad \delta = \frac{x}{(\alpha / C_L)}, \quad \beta = \frac{t}{(\alpha / C_L^2)}, \tag{10.74}$$

which are identical to those introduced earlier in equations (2.24) or (2.67), equations (10.71) to (10.73b) become

$$\frac{\partial^4 \theta}{\partial \delta^4} - (1 + \eta) \frac{\partial^3 \theta}{\partial \delta^2 \partial \beta} = \frac{\partial^4 \theta}{\partial \delta^2 \partial \beta^2} - \frac{\partial^3 \theta}{\partial \beta^3} \tag{10.75a}$$

$$\theta = 0, \quad \frac{\partial \theta}{\partial \beta} = 0, \quad \frac{\partial^2 \theta}{\partial \beta^2} = 0 \quad \text{as} \quad \beta = 0, \tag{10.75b}$$

$$\theta = 1, \quad \frac{\partial \theta}{\partial \delta} = 0 \quad \text{at} \quad \delta = 0, \tag{10.75c}$$

$$\theta \rightarrow 0, \quad \frac{\partial \theta}{\partial \delta} \rightarrow 0 \quad \text{as} \quad \delta \rightarrow \infty. \tag{10.75d}$$

Taking the Laplace transform of equations (10.75a), (10.75c), and (10.75d), and making use of the initial conditions in equation (10.75b), the transformed temperature is governed by

$$\frac{d^4 \overline{\theta}}{d \delta^4} - b \frac{d^2 \overline{\theta}}{d \delta^2} + c = 0, \quad \text{with} \quad b = p[p + (1 + \eta)], \quad c = p^3, \tag{10.76a}$$

$$\overline{\theta} = \frac{1}{p}, \quad \frac{d \overline{\theta}}{d \delta} = 0 \quad \text{at} \quad \delta = 0 \tag{10.76b}$$

$$\overline{\theta} \rightarrow 0, \quad \frac{d \overline{\theta}}{d \delta} \rightarrow 0 \quad \text{as} \quad \delta \rightarrow \infty. \tag{10.76c}$$

Equation (10.76a) can be integrated in a direct fashion, resulting in

$$\bar{\theta}(\delta; p) = C_1 e^{-F_1\delta} + C_2 e^{-F_2\delta}, \quad \text{with}$$

$$C_1 = -\frac{F_2}{p(F_1 - F_2)}, \quad C_2 = \frac{F_1}{p(F_1 - F_2)}, \quad F_{1,2} = \sqrt{\frac{b \mp \sqrt{b^2 - 4c}}{2}}. \tag{10.77}$$

The Laplace inversion for temperature can then be performed by replacing the function subroutine, FUNC(P), in the Appendix with the transformed temperature shown in equation (10.77).

Figure 10.8 shows the temperature wave represented by equation (10.77) at $\beta = 1$ and $\eta \in [0.01, 0.09]$, typical values of the thermomechanical coupling factor for metals. Under the boundary and initial conditions specified in equations (10.75b) to (10.75d), the temperature profile is not sensitive to the value of the thermomechanical coupling factor. The discontinuity, namely, the thermal wavefront defined in a more general sense, exists at $\delta = \beta = 1$, which is caused by the mechanical wavefront located at $x = C_L t$ according to equation (10.74). Although diffusion is assumed for heat transport, to reiterate, a distinct wavefront is still present owing to the motion of elastic waves in a deformable conductor.

The exaggerated response of temperature, $\theta > 1$ ($T > T_w$) for $\eta \in [0, 1.4]$, in Figure 10.8 results from the insulated boundary condition as the temperature is suddenly raised to T_w at the boundary, referring to equation (10.75c). The temperature overshooting disappears should the insulated boundary condition be replaced by the higher-order derivatives, such as

$$\frac{\partial^2 \theta}{\partial \delta^2} = 0 \quad \text{at} \quad \delta = 0 \quad \text{or} \quad \frac{\partial^3 \theta}{\partial \delta^3} = 0 \quad \text{at} \quad \delta = 0. \tag{10.78}$$

This condition replaces the second condition in equation (10.76b), rendering

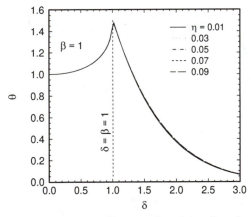

Figure 10.8 Temperature profiles predicted by the single energy equation in a deformable conductor, equation (10.75a) at $\beta = 1$, showing results for the boundary condition $\partial\theta/\partial\delta = 0$ at $\delta = 0$.

$$C_1 = -\frac{F_2^2}{p\left(F_1^2 - F_2^2\right)}, \quad C_2 = \frac{F_1^2}{p\left(F_1^2 - F_2^2\right)} \quad \text{(for } \frac{\partial^2 \theta}{\partial \delta^2} = 0 \quad \text{at} \quad \delta = 0 \text{)},$$

$$(10.79a)$$

$$C_1 = -\frac{F_2^3}{p\left(F_1^3 - F_2^3\right)}, \quad C_2 = \frac{F_1^3}{p\left(F_1^3 - F_2^3\right)} \quad \text{(for } \frac{\partial^3 \theta}{\partial \delta^3} = 0 \quad \text{at} \quad \delta = 0 \text{)}.$$

$$(10.79b)$$

The transformed solution for temperature and the F functions remain the same as those given in equation (10.77). A physical interpretation in terms of heat transport alone for the high-order, zero derivatives in equation (10.78) is not clear. A nonzero, first-order derivative, however, does allow heat transfer across the boundary at $x = 0$. Figure 10.9 shows the temperature profiles resulting from equations (10.79a) (a zero second-order derivative, Figure 10.9(a)) and (10.79b) (a zero third-order derivative, Figure 10.9(b)). The field temperature becomes everywhere lower than the boundary temperature in both cases, supporting the fact that the overshooting phenomenon shown in Figure 10.8 is indeed caused by the insulated boundary condition (the second expression in equation (10.75c)). The discontinuity of the temperature gradient across the wavefront decreases as the order of the zero derivative of the temperature specified at the boundary $\delta = 0$ increases. In transition from the boundary condition of $\partial^2 \theta / \partial \delta^2 = 0$ to $\partial^3 \theta / \partial \delta^3 = 0$, as shown by Figures 10.9(a) and 10.9(b), the discontinuity at $\delta = \beta = 1$ quickly diminishes, and the temperature distribution becomes much smoother in Figure 10.9(b) than in Figure 10.9(a).

Lagging behavior. Equation (10.68), the single energy equation for a deformable conductor, results from the combination of equation (10.65a) describing transport of the rigidity propagator and equation (10.67a) describing Fourier's law in heat conduction. In the presence of a general lagging behavior in heat transport,

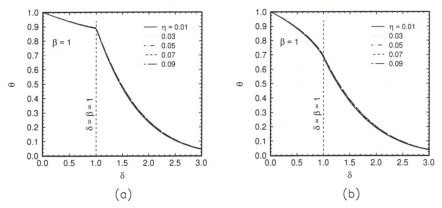

$$(a) \qquad\qquad (b)$$

Figure 10.9 Temperature profiles predicted by the single energy equation in a deformable conductor, equation (10.75a) at $\beta = 1$. Results from the boundary conditions of (a) $\partial^2 \theta / \partial \delta^2 = 0$ at $\delta = 0$ and (b) $\partial^3 \theta / \partial \delta^3 = 0$ at $\delta = 0$.

equation (10.65a) remains,

$$\nabla^2 R - \frac{1}{C_L^2}\frac{\partial^2 R}{\partial t^2} + C_p \eta \nabla^2 \dot{T} = 0, \quad \text{with} \quad R = \nabla \bullet \vec{q} + C_p \dot{T}, \quad (10.65a')$$

while Fourier's law in equation (10.67a) is replaced by the dual-phase-lag model shown by equation (10.1):

$$\vec{q}(\vec{x},t) + \tau_q \frac{\partial \vec{q}}{\partial t}(\vec{x},t) + \frac{\tau_q^2}{2}\frac{\partial^2 \vec{q}}{\partial t^2}(\vec{x},t) \cong -k\left\{ \nabla T(\vec{x},t) + \tau_T \frac{\partial}{\partial t}[\nabla T(\vec{x},t)] \right\}, \quad (10.1')$$

The τ_q^2 effect, i.e., the T-wave behavior, is included here for a broader coverage of the lagging response. In a one-dimensional situation, likewise, they become

$$\frac{\partial^3 q}{\partial x^3} - \frac{1}{C_L^2}\frac{\partial^3 q}{\partial x \partial t^2} = C_p \left[\frac{1}{C_L^2}\frac{\partial^3 T}{\partial t^3} - (1+\eta)\frac{\partial^3 T}{\partial x^2 \partial t} \right] \quad (10.80a)$$

$$q + \tau_q \frac{\partial q}{\partial t} + \frac{\tau_q^2}{2}\frac{\partial^2 q}{\partial t^2} \cong -k\left\{ \frac{\partial T}{\partial x} + \tau_T \frac{\partial^2 T}{\partial x \partial t} \right\}. \quad (10.80b)$$

Equations (10.80a) and (10.80b) provide two equations to be solved for two unknowns, the heat flux vector q and the temperature T. In contrast to the previous example, where the heat flux vector q was eliminated and the energy equation contained temperature only, we adopt the *mixed* formulation represented by equations (10.80a) and (10.80b) in this example and consider a flux-specified boundary condition:

$$q = \begin{cases} q_0, & \text{for} \quad 0 \le t \le t_0 \\ 0, & \text{for} \quad t > t_0 \end{cases} \quad \text{at} \quad x = 0. \quad (10.80c)$$

This is especially desirable for energy transport in a deformable conductor owing to the more complicated energy equation (10.80a) (than that in a rigid conductor) and the more general dual-phase-lag behavior described by equation (10.80b) (than Fourier's model shown by equation (10.67a)). In addition to the rectangular pulse of the heat flux, an additional boundary condition at $x = 0$ is needed owing to the high-order derivatives in space. As an example, let us consider a zero second-order derivative of temperature with respect to space,

$$\frac{\partial^2 T}{\partial x^2} = 0 \quad \text{at} \quad x = 0, \quad (10.80d)$$

which eliminates the temperature overshooting phenomenon (Figure 10.8). The initial conditions as $t = 0$ and the remote boundary condition as x approaches infinity remain the same, equations (10.72) and (10.73b), respectively.

Introducing the dimensionless variables,

$$\theta = \frac{T - T_0}{T_0}, \quad \delta = \frac{x}{(\alpha / C_L)}, \quad \beta = \frac{t}{(\alpha / C_L^2)}, \quad Q = \frac{q}{C_p T_0 C_L}, \tag{10.81}$$

which normalize space and time in the same fashion as those shown in equation (10.74), equations (10.80a) to (10.80d), the remote boundary condition (equation (10.73b)), and the initial condition (equation (10.72)) become

$$\frac{\partial^3 Q}{\partial \delta^3} - \frac{\partial^3 Q}{\partial \delta \partial \beta^2} = \frac{\partial^3 \theta}{\partial \beta^3} - (1 + \eta) \frac{\partial^3 \theta}{\partial \delta^2 \partial \beta} \tag{10.82a}$$

$$Q + z_q \frac{\partial Q}{\partial \beta} + \frac{z_q^2}{2} \frac{\partial^2 Q}{\partial \beta^2} = -\frac{\partial \theta}{\partial \delta} - z_T \frac{\partial^2 \theta}{\partial \delta \partial \beta}. \tag{10.82b}$$

$$\theta = 0, \quad \frac{\partial \theta}{\partial \beta} = 0, \quad \frac{\partial^2 \theta}{\partial \beta^2} = 0 \quad \text{as} \quad \beta = 0, \tag{10.82c}$$

$$Q = \begin{cases} Q_0, & 0 \le \beta \le \beta_0 \\ 0, & \beta > \beta_0 \end{cases} \quad \text{and} \quad \frac{\partial^2 \theta}{\partial \delta^2} = 0 \quad \text{at} \quad \delta = 0, \tag{10.82d}$$

$$\theta \to 0, \quad \frac{\partial \theta}{\partial \delta} \to 0 \quad \text{as} \quad \delta \to \infty. \tag{10.82e}$$

In addition to equation (10.81) for the dimensionless variables,

$$\beta_0 = \frac{t_0}{(\alpha / C_L^2)}, \quad Q_0 = \frac{q_0}{C_p T_0 C_L} \tag{10.82f}$$

are further introduced in equation (10.82d) for the dimensionless pulse duration and intensity of the boundary heat flux, respectively.

The method of Laplace transform is applied in the same manner, resulting in

$$\frac{d^3 \overline{Q}}{d\delta^3} - p^2 \frac{d\overline{Q}}{d\delta} = p^3 \overline{\theta} - (1 + \eta) p \frac{d^2 \overline{\theta}}{d\delta^2} \tag{10.83a}$$

$$\overline{Q} = G\frac{d\overline{\theta}}{d\delta}, \quad G(p) = \frac{1 + pz_T}{\left[1 + pz_q + \frac{1}{2}\left(pz_q\right)^2\right]}. \tag{10.83b}$$

$$\overline{Q} = \overline{Q}_b = Q_0\left(\frac{1 - e^{-p\beta_0}}{p}\right) \quad \text{and} \quad \frac{d^2\overline{\theta}}{d\delta^2} = 0 \quad \text{at} \quad \delta = 0, \tag{10.83c}$$

$$\overline{\theta} \to 0, \quad \frac{d\overline{\theta}}{d\delta} \to 0 \quad \text{as} \quad \delta \to \infty. \tag{10.83d}$$

The expression for $\overline{Q}_b$ results from the Laplace transform of the rectangular pulse shown in equation (10.82d). Solutions for the temperature and heat flux vector satisfying equations (10.83a) to (10.83d) are straightforward. They are

$$\overline{\theta}(\delta; p) = C_1 e^{-F_1\delta} + C_2 e^{-F_2\delta}, \quad \overline{Q}(\delta; p) = G\left[C_1 F_1 e^{-F_1\delta} + C_2 F_2 e^{-F_2\delta}\right], \text{ with}$$

$$C_1 = -\frac{\overline{Q}_b F_2}{GF_1\left(F_1 - F_2\right)}, \quad C_2 = \frac{\overline{Q}_b F_1}{GF_2\left(F_1 - F_2\right)},$$

$$F_{1,2} = \sqrt{\frac{b \mp \sqrt{b^2 - 4c}}{2}}, \quad b = \frac{p(1 + \eta + Gp)}{G}, \quad c = \frac{p^3}{G}, \tag{10.84}$$

In this formulation, note that the phase lags of the temperature gradient (z_T or τ_T) and the heat flux vector (z_q or τ_q) are absorbed in the parameter G defined in equation (10.83b).

The transformed solutions for the temperature and heat flux vector in equation (10.84) are used in the function subroutine, FUNC(P), in the Appendix for obtaining the Laplace inversion. For $z_T = z_q = 0$, the case of diffusion, Figure 10.10 shows the temperature profile in the deformable conductor. Again, the thermo-mechanical coupling factor ($\eta \sim 10^{-2}$) does not significantly affect the temperature distributions. Under the same boundary condition, $\partial^2\theta/\partial\delta^2 = 0$ at $\delta = 0$, the discontinuity of the temperature gradient across the wavefront (at $\delta = \beta = 1$) is much less pronounced than that in Figure 10.9(a) where a sudden rise of temperature occurs at the boundary rather than a suddenly applied heat flux.

Figure 10.11 shows the effect of the phase lag of the heat flux vector, z_q or τ_q, on the temperature profile at a constant value of $z_T = 10$. The temperature level increases with the value of z_q (τ_q). When the value of z_q (τ_q) deviates from zero, clearly, two wavefronts exist in the physical domain. One is in the vicinity of the mechanical wavefront,

$$x = C_L t \quad \text{or} \quad \delta = \beta, \tag{10.85a}$$

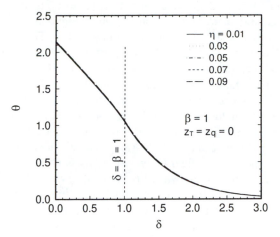

Figure 10.10 Temperature profiles resulting from the flux irradiation at the boundary of a half-space from equation (10.84) at $\beta = 1$, $Q_0 = 1$, $\beta_0 = 1$, $z_T = z_q = 0$ (the case of diffusion).

and the other is in the neighborhood of the T-wavefront,

$$x = C_T t = \left(\frac{\sqrt{2\alpha\tau_T}}{\tau_q} \right) t \quad \text{or} \quad \delta = \left(\frac{\sqrt{2z_T}}{z_q} \right) \beta . \tag{10.85b}$$

Sharp discontinuities exist at these locations, which become more distinct as the phase lag of the heat flux vector (z_q or τ_q) increases.

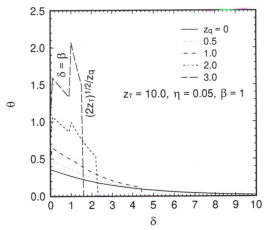

Figure 10.11 Effect of the phase lag of the heat flux vector, z_q or τ_q, on the temperature profiles from equation (10.84) at $\beta = 1$, $Q_0 = 1$, $\beta = 1$, and $z_T = 10$.

As shown in Table 10.2, the value of z_T to z_q (τ_T to τ_q) is of the order of 10^2 for metals. In the same domain of z_q as that shown in Figure 10.11, Figure 10.12 shows the corresponding effect at a larger value of $z_T = 100$. The physical domain of the heat-affected zone ($\delta \leq (\sqrt{2z_T}/z_q)\beta$) extends more into the solid, with a significantly lower temperature than that shown in Figure 10.11. The wavefront induced by the mechanical field persists in the neighborhood of $\delta = \beta = 1$ ($x = C_{Lt}$).

For $z_T \geq z_q^2/2$, the cases shown in Figures 10.11 and 10.12, the wavefront induced by the thermal field (T wave) leads the wavefront induced by the mechanical field (longitudinal stress wave). For the reversed case of $z_T \leq z_q^2/2$, the wavefront induced by the mechanical field (longitudinal stress wave) leads the wavefront induced by the thermal field (T wave). This is shown in Figure 10.13, where the value of z_T is taken as unity while the value of z_q varies from 2 to 10. The thermal wave speed decreases as the value of z_q increases, implying that the thermal wavefront is closer to the boundary of $\delta = 0$ at a larger value of z_q (τ_q).

Temperature levels in the heat-affected zone increase with the value of z_q. When approaching the mechanically induced wavefront (at $\delta = \beta$) from the thermally induced wavefront (at $\delta = (\sqrt{2z_T}/z_q)\beta$), contrary to the cases of $z_T \geq z_q^2/2$ shown in Figures 10.11 and 10.12, the temperature *increases* with the distance away from the heated boundary. This is a salient feature of T wave (accounting for the τ_q^2 effect) in transition from the case of flux precedence ($z_q < z_T$ or $\tau_q < \tau_T$ in Figures 10.11 and 10.12) to the case of gradient precedence ($z_T < z_q$ or $\tau_T < \tau_q$ in Figure 10.13) in transporting heat. Whether this special behavior exists in reality or not, however, needs experimental support.

In the presence of the τ_q^2 effect, namely, the T-wave behavior in heat transport, note that the case of diffusion is retrieved if and only if $z_T = z_q = 0$ ($\tau_T = \tau_q = 0$). An equal shift in the timescale from t to $t + \tau$, with $\tau_T = \tau_q = \tau$, is no longer trivial owing to the presence of the second-order term of τ_q^2, referring to equation

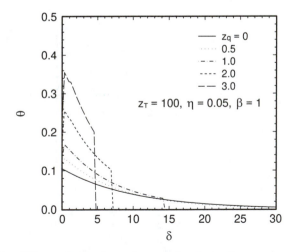

Figure 10.12   Effect of the phase lag of the heat flux vector, z_q or τ_q, on the temperature profiles from equation (10.84) at $\beta = 1$, $Q_0 = 1$, $\beta_0 = 1$, and $z_T = 100$.

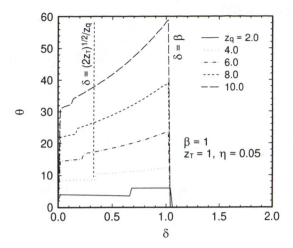

Figure 10.13 Temperature profiles for the case of $z_T \le z_q^2/2$ from equation (10.84) at $\beta = 1$, $Q_0 = 1$, $\beta_0 = 1$, and $z_T = 1$.

(10.1). Figure 10.14 demonstrates the difference by displaying the results of $z_T = z_q = 0$ (diffusion), 1, and 5 at $\beta = 1$ and $\eta = 0.05$. The mechanically induced wavefront is located at $\delta = \beta = 1$ for all cases, while the thermally induced wavefront is located at $\delta = (\sqrt{2}z_T/z_q)\beta$, having a value of $\sqrt{2}\beta \cong 1.414\beta$ for $z_T = z_q = 1$ and $(\sqrt{10}/5)\beta \cong 0.632\beta$ for $z_T = z_q = 5$. The temperature level in the heat-affected zone increases with the value of z_T (or z_q).

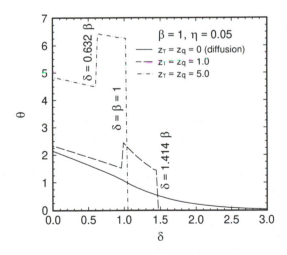

Figure 10.14 Temperature profiles under various values of $z_T = z_q$. Only the case of $z_T = z_q = 0$ reduces to the result of diffusion from equation (10.84) at $\beta = 1$, $Q_0 = 1$, $\beta_0 = 1$, and $\eta = 0.05$.

10.3.2 Momentum Equation

A *single* equation describing momentum transport in a nonisothermal solid can be derived based on the same concept. Since momentum transport is governed by the displacement field, which is a vector, the procedure rendering a single momentum equation would be greatly simplified by introducing the Lamé displacement potential:

$$\mu u_i = \phi_{,i} = \frac{\partial \phi}{\partial x_i} \quad \text{for} \quad i = 1, 2, 3. \tag{10.86}$$

Three components of the displacement vector, u_1, u_2, u_3, can thus be found by direct differentiation should the *scalar* potential ϕ be determined. Note that the Lamé displacement potential defined in equation (10.86) is identical to the heat flux potential defined previously in equation (2.17) in Section 2.4.

The energy and momentum equations in a deformable, nonisothermal conductor remain the same, equations (10.54) and (10.60),

$$- \nabla \bullet \vec{q} + S = C_p \dot{T} + \kappa_\sigma T_0 \dot{e}, \tag{10.87a}$$

$$\mu u_{i,jj} + (\lambda + \mu) u_{j,ji} - \kappa_\sigma T_{,i} = \rho \ddot{u}_i \quad \text{for} \quad i, j = 1, 2, 3,$$

$$\text{where} \quad e = \frac{\partial u_i}{\partial x_i}, \quad \kappa_\sigma = \kappa_\varepsilon (3\lambda + 2\mu). \tag{10.87b}$$

Substituting equation (10.86) into (10.87b) and noting that

$$\mu u_{i,jj} = \phi_{,ijj}; \quad (\lambda + \mu) u_{j,ji} = \left(\frac{\lambda + \mu}{\mu}\right) \phi_{,ijj}; \quad \ddot{u}_i = \ddot{\phi}_{,i}, \tag{10.88}$$

equation (10.87b) becomes

$$\frac{\partial}{\partial x_i} \left[\nabla^2 \phi - \left(\frac{\kappa_\sigma \mu}{\lambda + 2\mu}\right) T - \left(\frac{\rho}{\lambda + 2\mu}\right) \frac{\partial^2 \phi}{\partial t^2} \right] = 0. \tag{10.89}$$

A particular solution of equation (10.89), clearly, is a *wave* equation governing the displacement potential ϕ,

$$\nabla^2 \phi = \frac{1}{C_L^2} \frac{\partial^2 \phi}{\partial t^2} + \left(\frac{\kappa_\sigma \mu}{\lambda + 2\mu}\right) T, \quad C_L = \sqrt{\frac{\lambda + 2\mu}{\rho}} \tag{10.90}$$

with the temperature rise above the ambient, T, serving as the driving force.

Equations (10.87a) and (10.90) are the energy and momentum equations to be combined to give a single equation governing momentum transport in terms of ϕ. Fourier's law in heat conduction, equation (10.67a),

$$\bar{q} = -k\nabla T,$$ (10.67a')

provides a straightforward example illustrating this procedure. Combining equations (10.67a) and (10.87a) gives

$$k\nabla^2 T + S = C_p \dot{T} + \left(\frac{\kappa_\sigma T_0}{\mu} \right) \nabla^2 \dot{\phi}.$$ (10.91)

The temperature rise T above the ambient, alternatively, results from equation (10.90):

$$T = \left(\frac{\lambda + 2\mu}{\kappa_\sigma \mu} \right) \left[\nabla^2 \phi - \frac{1}{C_L^2} \frac{\partial^2 \phi}{\partial t^2} \right].$$ (10.92)

Substituting equation (10.92) into equation (10.91) yields

$$\nabla^4 \phi - \left(\frac{1 + \eta}{\alpha} \right) \nabla^2 \dot{\phi} = \frac{1}{C_L^2} \left[\nabla^2 \ddot{\phi} - \left(\frac{1}{\alpha} \right) \dddot{\phi} \right] - \left[\frac{\kappa_\sigma \mu}{k(\lambda + 2\mu)} \right] S.$$ (10.93)

Except for the heat source term, it has *exactly* the same form as equation (10.68) governing heat transport in a deformable conductor. Correspondence between temperature in transporting heat and displacement potential in transporting momentum is thus clear.

In the presence of the lagging behavior, exemplified by equation (10.1) containing the τ_q^2 effect, equation (10.67a) is replaced, while the other equations remain the same. Like the situation of heat transport discussed in Section 10.3, however, the single momentum equation incorporating the effect of nonisothermal deformation is too complicated to be instructive in revealing the fundamental behavior.

10.4 NONLOCAL BEHAVIOR

Equation (10.68) is the single energy equation describing heat transport in a deformable conductor. It absorbs Fourier's law in heat conduction and Hooke's law in elasticity in the same framework. Mathematically, it results from a direct combination of the momentum equation (10.60) and energy equation (10.65a) that accommodate the thermomechanical coupling effect. Equation (10.68) has a new appearance in the theory of heat conduction, but it is simply an alternate point of view describing the coupling between thermal and mechanical fields.

Recognizing that thermal expansion/contraction could be viewed as an *internal* mechanism in transporting heat, however, equation (10.68) can be *derived* on the basis of combined *lagging* and *nonlocal* responses between the heat flux vector and the temperature gradient without looping around the thermomechanical response. This is illustrated in Figure 10.15, where the lagging response in time, Figure 10.15(a), is interrelated to the nonlocal response in space, Figure 10.15(b). The lagging behavior is a special response in *time*. The heat flux flowing through a material volume located at a position $\bar{x}$ at time $(t + \tau_q)$ results in (or from) the temperature gradient established across the same material volume (at $\bar{x}$) at another instant of time $(t + \tau_T)$. For $\tau_T > \tau_q$, the heat flux vector leads the temperature gradient in the time history of heat transport. For $\tau_T < \tau_q$, on the other hand, the temperature gradient leads the heat flux vector. The concept of nonlocal behavior is extended from the lagging behavior at different times (Figure 10.15(a)) to the response at different locations (Figure 10.15(b)). The temperature gradient established across a material volume located at $(\bar{x} + \bar{\lambda}_T)$ results in (or from) the heat flux vector flowing through *another* material volume located at $(\bar{x} + \bar{\lambda}_q)$. Mathematically, this can be expressed as

$$\bar{q}(\bar{x} + \bar{\lambda}_q, t) = -k\nabla T(\bar{x} + \bar{\lambda}_T, t) .$$

(10.94)

In a one-dimensional case, equation (10.94) reduces to

$$q(x + \lambda_q, t) = -k\nabla T(x + \lambda_T, t) ,$$

(10.95)

which results in

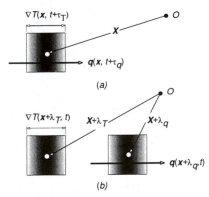

(a)

(b)

Figure 10.15 Constitutive relation between the heat flux vector and the temperature gradient. (a) Lagging behavior at the same location ($\bar{x}$) but different instants of time ($t+\tau_T$ and $t+\tau_q$) and (b) nonlocal behavior at the same instant of time (t) but at different locations ($\bar{x} + \bar{\lambda}_T$ and $\bar{x} + \bar{\lambda}_q$).

$$q(x,t) + \lambda_q \frac{\partial q}{\partial x}(x,t) + \frac{\lambda_q^2}{2!}\frac{\partial^2 q}{\partial x^2}(x,t) + \cdots =$$

$$-k\frac{\partial}{\partial x}\left[T(x,t) + \lambda_T \frac{\partial T}{\partial x}(x,t) + \frac{\lambda_T^2}{2!}\frac{\partial^2 T}{\partial x^2}(x,t) + \cdots\right] \qquad (10.96)$$

via Taylor's series expansion. Unlike the lagging behavior in time, the *spherical invariance* is the most important property in the nonlocal response. For isotropic and homogeneous heat transport in a three-dimensional body, as illustrated in Figure 10.16, influence of the material volumes located at $(x+\lambda)$ on the central volume located at x in transporting heat is identical, provided that the nonlocal length of influence, λ, is the same for all the neighboring material volumes. In terms of the one-dimensional constitutive equation describing the relation between the heat flux vector and the temperature gradient, equation (10.96), "the same influence" from the neighboring elements implies *the same* form of the constitutive equation under the same distance of λ. For the two neighboring elements located at the same distance λ from the central element located at x, one at $x - \lambda$ and the other at $x + \lambda$, an *invariant* form of equation (10.96) is possible if and only if all the *odd* power terms in λ_T and λ_q vanish:

$$q(x,t) + \frac{\lambda_q^2}{2!}\frac{\partial^2 q}{\partial x^2}(x,t) + \cdots = -k\frac{\partial}{\partial x}\left[T(x,t) + \frac{\lambda_T^2}{2!}\frac{\partial^2 T}{\partial x^2}(x,t) + \cdots\right]. \qquad (10.97)$$

Equation (10.97), containing only the *even* order terms in λ_T and λ_q, clearly, is independent of the signs of λ_T and λ_q. As long as the lengths of nonlocality, λ_T and λ_q, remain the same, regardless of the relative positions of the neighboring elements to the central element, equation (10.97) describes the nonlocal response between the heat flux vector and the temperature gradient.

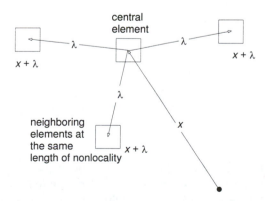

Figure 10.16 Spherical invariance in the nonlocal behavior in heat transport.

Nonlocal behavior in small scale may be present together with the lagging response in short times. A material volume undergoing a thermal expansion provides an excellent example illustrating this combined behavior. Driven by the heat flux (q) flowing through a material volume located at $(x+\lambda_q)$ at time $(t+\tau_q)$, as denoted by the dashed outline marked "old" in Figure 10.17, a temperature gradient (∇T) is established across the new configuration of the material volume located at $(x+\lambda_T)$ at a different instant of time $(t+\tau_T)$ (the solid outline marked "new"). The nonlocal response, q flowing through at $(x+\lambda_q)$ and ∇T established across at $(x+\lambda_T)$, is due to the motion of the center of the material volumes resulting from thermal expansion. The lagging behavior, q flowing through at $(t+\tau_q)$ and ∇T established across at $(t+\tau_T)$, on the other hand, results from the finite time required for the thermomechanical coupling to take place in heat transport. Mathematically, such dual behavior in both space and time can be described by

$$q(x + \lambda_q, t + \tau_q) = -k\nabla T(x + \lambda_T, t + \tau_T). \qquad (10.98)$$

Equation (10.98) is to be combined with the energy equation, established at a general position x and a general time t:

$$-\frac{\partial q}{\partial x}(x,t) = C_p \frac{\partial T}{\partial x}(x,t). \qquad (10.99)$$

In parallel to equation (2.7) for the dual-phase-lag model and equation (10.97) for the nonlocal behavior reflecting the spherical invariance, the Taylor series expansion is applied to equation (10.98) to convert the physical space and time to the same scale, x and t. Retaining up to the sixth-order terms in λ and τ, for an overview, results in

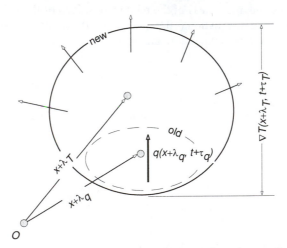

Figure 10.17   Heat flux vector (q) flowing through the old configuration of a material volume located at $(x+\lambda_q)$ at time $(t+\tau_q)$ and the temperature gradient (∇T) established across the new configuration of the material volume located at $(x+\lambda_T)$ at time $(t+\tau_T)$. Change of the configuration is caused by thermal expansion.

$$\frac{\partial q}{\partial x} + \tau_q \frac{\partial^2 q}{\partial x \partial t} + \frac{\tau_q^2}{2!} \frac{\partial^3 q}{\partial x \partial t^2} + \frac{\lambda_q^2}{2!} \frac{\partial^3 q}{\partial x^3} + \frac{\lambda_q^2 \tau_q}{2!} \frac{\partial^4 q}{\partial x^3 \partial t} + \frac{\lambda_q^2 \tau_q^2}{2!2!} \frac{\partial^5 q}{\partial x^3 \partial t^2}$$

$$+ \frac{\lambda_q^4}{4!} \frac{\partial^5 q}{\partial x^5} + \frac{\lambda_q^4 \tau_q}{4!} \frac{\partial^6 q}{\partial x^5 \partial t} + \frac{\lambda_q^4 \tau_q^2}{2!4!} \frac{\partial^7 q}{\partial x^5 \partial t^2}$$

$$= -k \left[\frac{\partial^2 T}{\partial x^2} + \tau_T \frac{\partial^3 T}{\partial x^2 \partial t} + \frac{\tau_T^2}{2!} \frac{\partial^4 T}{\partial x^2 \partial t^2} + \frac{\lambda_T^2}{2!} \frac{\partial^4 T}{\partial x^4} + \frac{\lambda_T^2 \tau_T}{2!} \frac{\partial^5 T}{\partial x^4 \partial t} \right.$$

$$\left. + \frac{\lambda_T^2 \tau_T^2}{2!2!} \frac{\partial^6 T}{\partial x^4 \partial t^2} + \frac{\lambda_T^4}{4!} \frac{\partial^6 T}{\partial x^6} + \frac{\lambda_T^4 \tau_T}{4!} \frac{\partial^7 T}{\partial x^6 \partial t} + \frac{\lambda_T^4 \tau_T^2}{2!4!} \frac{\partial^8 T}{\partial x^6 \partial t^2} \right]. \quad (10.100)$$

Equations (10.99) and (10.100) provide two equations for two unknowns, the heat flux vector q and the temperature T. Eliminating the heat flux vector from the two equations, as usual, should give the expected result of a single energy equation governing temperature. For this purpose, we first notice that equation (10.100) contains derivatives of the heat flux with respect to space (x) at least to the first order. The higher order derivatives, therefore, can all be expressed in terms of the time derivatives of temperature according to equation (10.99). The second terms on the left side of equation (10.100), for example, are

$$\frac{\tau_q^2}{2!} \frac{\partial^3 q}{\partial x \partial t^2} = \frac{\tau_q^2}{2!} \frac{\partial^2}{\partial t^2} \left(\frac{\partial q}{\partial x} \right) = -\frac{C_p \tau_q^2}{2!} \frac{\partial^3 T}{\partial t^3}. \quad (10.101)$$

Repeating this procedure for all terms on the left side of equation (10.100), the result is

$$\frac{\partial T}{\partial x} + \tau_q \frac{\partial^2 T}{\partial t^2} + \frac{\tau_q^2}{2!} \frac{\partial^3 T}{\partial t^3} + \frac{\lambda_q^2}{2!} \frac{\partial^3 T}{\partial x^2 \partial t} + \frac{\lambda_q^2 \tau_q}{2!} \frac{\partial^4 T}{\partial x^2 \partial t^2} + \frac{\lambda_q^2 \tau_q^2}{2!2!} \frac{\partial^5 T}{\partial x^2 \partial t^3}$$

$$+ \frac{\lambda_q^4}{4!} \frac{\partial^5 T}{\partial x^4 \partial t} + \frac{\lambda_q^4 \tau_q}{4!} \frac{\partial^6 T}{\partial x^4 \partial t^2} + \frac{\lambda_q^4 \tau_q^2}{2!4!} \frac{\partial^7 T}{\partial x^4 \partial t^3}$$

$$= \alpha \left[\frac{\partial^2 T}{\partial x^2} + \tau_T \frac{\partial^3 T}{\partial x^2 \partial t} + \frac{\tau_T^2}{2!} \frac{\partial^4 T}{\partial x^2 \partial t^2} + \frac{\lambda_T^2}{2!} \frac{\partial^4 T}{\partial x^4} + \frac{\lambda_T^2 \tau_T}{2!} \frac{\partial^5 T}{\partial x^4 \partial t} \right.$$

$$\left. + \frac{\lambda_T^2 \tau_T^2}{2!2!} \frac{\partial^6 T}{\partial x^4 \partial t^2} + \frac{\lambda_T^4}{4!} \frac{\partial^6 T}{\partial x^6} + \frac{\lambda_T^4 \tau_T}{4!} \frac{\partial^7 T}{\partial x^6 \partial t} + \frac{\lambda_T^4 \tau_T^2}{2!4!} \frac{\partial^8 T}{\partial x^6 \partial t^2} \right] \quad (10.102)$$

where $\alpha = k/C_p$. The energy equation (10.102) incorporating both the nonlocal and lagging behavior in heat transport introduces many high-order derivatives of temperature, rendering a capricious form of the energy equation in the theory of heat conduction.

Extracting the first- and the second-order terms in τ_T and τ_q, $O(\tau_T, \tau_q, \tau_T^2, \tau_q^2)$, and the second-order terms in λ_T and λ_q, $O(\lambda_T^2, \lambda_q^2)$, equation (10.102) becomes

$$\frac{\partial^4 T}{\partial x^4} - \frac{2}{\lambda_T^2}\left(\frac{\lambda_q^2}{2\alpha} - \tau_T\right)\frac{\partial^3 T}{\partial x^2 \partial t} - \frac{2}{\lambda_T^2}\left(\frac{\lambda_q^2 \tau_q}{2\alpha} - \frac{\tau_T^2}{2}\right)\frac{\partial^4 T}{\partial x^2 \partial t^2} - \frac{\tau_q^2}{\alpha\lambda_T^2}\frac{\partial^3 T}{\partial t^3} = 0. \quad (10.103)$$

Comparing to the one-dimensional form of equation (10.68),

$$\frac{\partial^4 T}{\partial x^4} - \left(\frac{1+\eta}{\alpha}\right)\frac{\partial^3 T}{\partial x^2 \partial t} - \frac{1}{C_L^2}\frac{\partial^4 T}{\partial x^2 \partial t^2} - \frac{1}{\alpha C_L^2}\frac{\partial^3 T}{\partial t^3} = 0, \quad (10.104)$$

which is the energy equation for a *deformable* conductor, equation (10.103) has an identical form, resulting in

$$\frac{2}{\lambda_T^2}\left(\frac{\lambda_q^2}{2\alpha} - \tau_T\right) = \frac{1+\eta}{\alpha}, \quad \frac{2}{\lambda_T^2}\left(\frac{\lambda_q^2 \tau_q}{2\alpha} - \frac{\tau_T^2}{2}\right) = \frac{1}{C_L^2}, \quad \frac{\tau_q}{\lambda_T} = \frac{1}{C_L}. \quad (10.105)$$

Solving for τ_T, τ_q, and λ_q in terms of λ_T yields

$$\tau_T = \left(\frac{\lambda_T}{C_L}\right)\left[1 + \sqrt{\frac{1 + [(1+\eta)/\alpha]\lambda_T C_L}{2}}\right], \quad \tau_q = \frac{\lambda_T}{C_L},$$

$$\lambda_q^2 = (1+\eta)\lambda_T^2 + \frac{2\alpha\lambda_T}{C_L}\left[1 + \sqrt{\frac{1 + [(1+\eta)/\alpha]\lambda_T C_L}{2}}\right]. \quad (10.106)$$

The value of λ_T can be arbitrarily chosen in these relations. Should the phase lags τ_T and τ_q and the lengths of nonlocality λ_T and λ_q be selected according to equation (10.106), obviously, the combined nonlocal and lagging responses capture the effect of thermoelastic deformation in a deformable conductor transporting heat.

According to equation (10.106), the longitudinal wave speed (C_L) is the only factor affecting the phase lag of the heat flux vector (τ_q). Thermal diffusivity α, the thermomechanical coupling factor η, and the longitudinal wave speed C_L, however, are all dominant parameters to the phase lag of the temperature gradient (τ_T) and the length of nonlocality of the heat flux vector (λ_q). Dependence of phase lags, τ_T and τ_q, on the length of nonlocality of the temperature gradient, λ_T, shows strong coupling between nonlocal and lagging behavior in the short-time transient. From equation (10.106), moreover,

$$\frac{\tau_T}{\tau_q} = 1 + \sqrt{\frac{1 + [(1 + \eta)/\alpha]\lambda_T C_L}{2}}.$$
(10.107)

For all physically admissible parameters, η, α, λ_T, and $C_L \geq 0$, equation (10.107) implies that $\tau_T > \tau_q$, corresponding to the case of *flux* precedence in the process of heat transport. In view of the lagging behavior, heat transport in deformable conductors is lead by the heat flux vector. The heat flux vector is the cause of the heat flow, while the temperature gradient is the effect in the short-time transient.

Correlation to the energy equation in a deformable conductor provides a possible example for the existence of the nonlocal and lagging responses in small scale and short time. Owing to the capricious form of equation (10.102), it should be capable of capturing other behavior should the high-order terms are further taken into account. The generalized energy equation (10.102) reflecting the combined nonlocal and lagging behavior seems to describe the most complete picture for heat transport in various scales of space and time. It reduces to the classical theory of diffusion ($\tau_T = \tau_q = \lambda_T = \lambda_q = 0$), CV wave ($\tau_T = \lambda_T = \lambda_q = 0$), dual-phase-lag model, phonon scattering model and parabolic two-step model ($\lambda_T = \lambda_q = 0$, the first-order effect in τ_q), and T-wave behavior as well as the hyperbolic two-step model ($\lambda_T = \lambda_q = 0$, including the second-order effect of τ_q^2). The high-order terms in equation (10.102) are certainly attractive from a mathematical point of view. Before a full extension into the high-order regimes is made, however, finding correlations to the existing model(s) and searching for the experimental phenomena that the lower-order model could explain are far more important. This bears the same merit as the early development of the linearized dual-phase-lag model.

FORTRAN 77 CODE FOR THE NUMERICAL INVERSION OF THE LAPLACE TRANSFORM

The FORTRAN 77 code that performs the Laplace inversion in terms of the Riemann-sum approximation, equation (2.49), is enclosed here. The sample program performs inversion of the transformed temperature, resulting from the solution of equation (2.55) for a semiinfinite solid subject to a suddenly raised temperature of unity at the boundary $z = 0$,

$$\overline{\theta}(z; p) = \frac{e^{-\sqrt{\frac{Ap + Dp^2}{1 + Bp}} \, z}}{p} \tag{A.1}$$

where z is the space variable (equivalent to δ used throughout the book), p is the Laplace transform variable to be inverted to the real time β (equivalent to S used in the program), and A, B, and D are constant parameters; see equation (2.55). The case of the dual-phase-lag model with $A = 2$, $B = 50$, and $D = 1$ is exemplified in the program. The temperature *distribution* is calculated at a fixed instant of time, $S = 2$ (the real time), in the physical domain from $Z = 0$ to 2.2. The "Do-loop" is thus placed on Z at a constant value of S. Should a time distribution of temperature be intended, the Do-loop should be placed on S instead. In all cases, the program will automatically adjust the optimal value of γ according to equation (2.50). Line numbers are generated at the beginning of every statement for later illustrations.

This program only shows a working example, with emphasis on illustrating the essence of the Laplace inversion technique via the Riemann-sum approximation. No intent has been made to tailor it for the best performance.

```
1          PROGRAM LAG
2          IMPLICIT DOUBLE PRECISION (A-H,O-Z)
```
***double precision is used due to possible intensive iterations (especially
***recommended for the use of personal computers (PC))
```
3          EXTERNAL FUNC
```
***FUNC defines the Laplace transformed solution
```
4          COMMON /DA/ Z,COEFA,COEFD,COEFB
5          OPEN(5,FILE='LAG',STATUS='UNKNOWN')
6          S=2.
```
***S is the real time. Its value can be the physical time with a dimension (t)
or ***the dimensionless time without a dimension (β).
***This sample program computes the distribution of lagging temperature in
***space (z) from 0 (z0) to 2.2 (zf) at s = 2.
```
7          NTERMS = 40000
```
***Maximum number of terms used in the Riemann-sum approximation,
***equation (2.55). Default value is set to 20000.
```
8          GMMA = 0.0
```
***GMMA is the value for the product of $\gamma\beta$ in equation (2.55). A value of 0.0
***intrigues the default value of $\gamma\beta$ = 4.7
```
9          COEFA=2.
10         COEFD=1.
11         COEFB=50.
```
***Set the values of coefficients in equation (2.61) with the following
***correspondence —
***Dual-phase-lag model: (A,D,B)=(2,1,NON-ZERO B). This is the case in
***this sample program.
***CV-Wave Model: (A,D,B)=(2,1,0);
***Diffusion Model: (A,D,B)=(2,0,0)
```
12         z0=0.
13         zf=2.2
```
***Define the initial (z0) and final (zf) positions for the temperature
***distribution
```
14         DO 10 I = 1,101
15         Z = Z0 + (Zf-Z0)*(I-1)/100.
```
***Z is the space variable. This sample program discretizes the physical
***domain from 0 to 2.2 into 100 intervals
```
16         CALL LAPINV(FUNC,GMMA,S,RESULT,NTERMS)
```
***LAPINV is the subroutine for the Riemann-sum approximation of the
***Laplace inversion. RESULT = temperature at (z,s)
```
17         WRITE(5,*) Z, RESULT
18   10    CONTINUE
19         CLOSE(5)
20         END
```
***The function subroutine FUNC(P) defines the solutions of temperature in
***the Laplace transform domain. This block needs to be modified for
***different problems with different solutions.

```
21      FUNCTION FUNC(P)
22      IMPLICIT DOUBLE PRECISION(A-H,O-Z)
23      COMMON /DA/ Z,COEFA,COEFD,COEFB
24      COMPLEX P,FACZ
```
***The Laplace transform variable P must be complex. Any other functions of ***P used in defining the transformed solution of temperature (FACZ in this ***program for example) must be declared to be complex accordingly.
```
25      FACZ = SQRT((COEFA*P+COEFD*P**2)/(1.+COEFB*P))
26      FUNC = EXP(-Z*FACZ)/P
27      RETURN
28      END
```
***Subroutine LAPINV performs the Riemann sum to approximate the ***Laplace inversion of the function specified in FUNC(P)
```
29      SUBROUTINE LAPINV(FUNC,GMMA,S,RESULT,NTERMS)
30      IMPLICIT DOUBLE PRECISION(A-H,O-Z)
31      EXTERNAL FUNC
32      COMMON /DA/ Z,TRASH1,TRASH2,TRASH3
33      COMPLEX GAM,B,CPR,PARTB,CHKCON
34      EPS = 1.0D-10
```
***EPS defines the convergence threshold for the ratio test of partial sums, ***EPS = (Temperature(N+1)-Temperature(N))/Temperature(N), ***with N denoting the partial sum of the first N terms in equation (2-55)
```
35      GAM = (0.0,0.0)
```
***Avoid the use of initial condition at S = 0 in this program. ***If needed, select a very small value of S, such as 0.001, to evaluate the ***initial temperature for validating purposes
```
36      IF (S.EQ.0.0) THEN
37      WRITE(*,*) 'LAPLACE VARIABLE CANNOT BE ZERO!'
38      RETURN
39      ENDIF
```
***The default value of GMMA*S = 4.7
```
40      IF (GMMA.EQ.0.0) GMMA = 4.7/S
41      GAM = GMMA
```
***Default number of terms used in the Riemann sum is 20000 terms
```
42      IF (NTERMS.EQ.0) NTERMS = 20000
43      PI = ACOS(-1.)
44      B = (0.0,1.0)
45      FIRST = (1./S)*EXP(GAM*S)
46      PARTA = 0.5*FUNC(GAM)
47      PARTB = (0.,0.)
48      I = 0
```
***Check convergence for the first NTERMS in the Riemann-sum ***Raise warning flags if EPS is larger than the specified value
```
49   5  IF (I.EQ.NTERMS) THEN
50      WRITE (*,*) 'NO CONVERGENCE FOR Z =',Z
51      GO TO 15
```

```
52        ENDIF
53        I = I + 1
54        CPR = GAM + B*(I*PI/S)
55        CHKODD = MOD(I,2)
56        CHKCON = PARTB
57        IF (CHKODD .EQ. 0) THEN
58        PARTB = PARTB+FUNC(CPR)
59        ELSE
60        PARTB = PARTB+FUNC(CPR)*(-1)
61        ENDIF
62        RESULT2 = FIRST*(PARTA+REAL(PARTB))
63        RESULT1 = FIRST*(PARTA+REAL(CHKCON))
```
***If summation is zero, apply different convergence check
```
64        IF(RESULT1 .EQ. 0.0)THEN
65        IF(ABS(RESULT2).LT.EPS)GO TO 15
66        GO TO 5
67        ENDIF
```
***The convergence check is then abs(abs(f(n) - f(n-1))/f(n)).
***First avoid divide by zero error
```
68        IF(RESULT2 .EQ. 0.0)THEN
69        IF(ABS(RESULT1).LT.EPS)GO TO 15
70        GO TO 5
71        ENDIF
72        CCON =ABS(ABS((RESULT2 - RESULT1))/RESULT2)
73        IF(CCON .LE. EPS)GO TO 15
74        GO TO 5
```
***If the Laplace variable is changed in successive calculations, such as the
***multiple-time calculations, make sure to reset gmma to zero for the next
***case
```
75   15   GMMA = 0.0
76        RESULT = RESULT1
77        RETURN
78        END
```

Explanations for the statements are placed in shaded areas, starting with asterisks. For the time history of the temperature at a fixed location in the solid, i.e., at $Z =$ constant for multiple values of S (physical instants of time), the constant value of Z is specified at line 6 instead and the do-loop from lines 12 to 18 is performed on the time variable (S). The program applies equally well in this case.

A.1 FLAG OF "NO CONVERGENCE FOR Z = "

A flag termed "NO CONVERGENCE FOR Z = " is implemented in the subroutine LAPINV, line 50, to warn users of dissatisfaction of the convergence criterion specified at line 34 after the use of NTERMS at line 7 (specified) or line 42 (default) in the Riemann-sum approximation. Appearance of this flag on the screen, however, does not necessarily imply a bad convergence in the numerical calculation. As a result of selecting a small threshold for convergence, EPS = 1.0D-10 at line 34, for example, the following warning messages appear on the screen:

```
NO CONVERGENCE FOR Z = 2.200000047683716E-002
NO CONVERGENCE FOR Z = 4.400000095367432E-002
NO CONVERGENCE FOR Z = 6.600000143051148E-002
NO CONVERGENCE FOR Z = 8.800000190734864E-002
NO CONVERGENCE FOR Z = 1.100000023841858E-001
```

The convergence criterion is satisfied at the rest of the locations from Z = 0 to 2.2. When enlarging the threshold value of EPS for convergence from 1.0D-10 to 1.0D-5, however, the distribution curve of temperature *remains* the same, implying that the threshold value of 1.0D-10 is in fact too small and the larger threshold, EPS = 1.0D-5, gives a satisfactory result as well. This situation is illustrated in Figure A.1, where the curves with EPS = 1.0D-5 and EPS = 1.0D-10 indeed merge onto each other. The temperature distribution starts to oscillate as the value of EPS further enlarges, evidenced by the gradually increasing oscillation of the curve as the value of EPS increases from 1.0D-3 to 1.0D-2. Like any other numerical algorithm, therefore, too large a value of EPS leads to inaccurate numerical results, while too small a value of EPS significantly wastes time in numerical computation. No optimal value for the convergence threshold of EPS has been intended yet.

In order to avoid the message of "NO CONVERGENCE FOR Z = ",

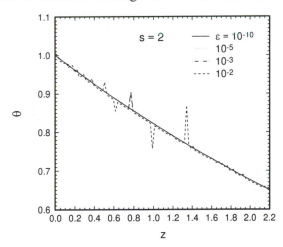

Figure A.1 Temperature distributions obtained under different thresholds for convergence. EPS = 1.0D-10, 1.0D-5, 1.0D-3, and 1.0D-2 from equation (A.1) at S = 2 and NTERMS = 40000.

another possibility is to increase the number of terms in the Riemann-sum approximation for the Laplace inversion, i.e., the value of NTERMS specified at line 7. Under the same convergent threshold, EPS = 1.0D-10, Figure A.2 shows the results obtained for NTERMS = 40000 (with the warning message of no convergence) and 1200000 (without the warning message of no convergence). The results are essentially the same, but the computation time in the case of NTERMS = 1200000 is significantly longer, reaching 1 minute and 15 seconds on an HP-715 workstation owing to intensive iterations.

In the absence of a quantitative guidance for the determination of an optimal value of EPS and an efficient algorithm to hasten the rate of convergence in the Riemann-sum approximation for the Laplace inversion, the results presented in this book are warranted in two ways: (1) convergent (close) results obtained under an increasing value of EPS at a constant value of NTERMS, and (2) convergent (close) results obtained under an increasing value of NTERMS at a constant value of EPS. With the assistance of digital computers, these two steps are adequate for most purposes in this book.

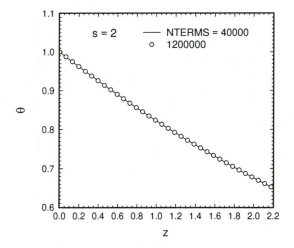

Figure A.2 Temperature distributions obtained under different numbers of terms, NTERMS, in the Riemann-sum approximation. NTERMS = 40000 (with warning messages) and 1200000 (without warning messages), S = 2, and EPS = 1.0D-10.

BIBLIOGRAPHY

The references are divided into two categories: The review articles summarize the research in microscale heat transport. They cover the subject matters from a broader point of view. The individual research papers, on the other hand, focus attention on more specific topics in microscale heat transfer.

B.1 MICROSCALE HEAT TRANSFER — REVIEW ARTICLES

Joseph, D. D. and Preziosi, L., 1989, "Heat Waves," *Reviews of Modern Physics*, Vol. 61, pp. 41-73.

Joseph, D. D. and Preziosi, L., 1990, "Addendum to the Paper on Heat Waves," *Reviews of Modern Physics*, Vol. 62, pp. 375-391.

These two articles give a complete survey of research in the wave theory of heat conduction. Unlike the microscale heat transfer models, which usually describe the small-scale effects in both space and time, the wave theory in heat conduction (sometimes called the hyperbolic heat conduction model or non-Fourier effect in heat conduction) describes the short-time response without considering the refined microscopic structure of the medium. In addition to an annotated bibliography of the literature on heat waves, the earlier article (1989) introduces the concepts of effective thermal conductivity, effective heat capacity, and relaxation functions along lines used to describe the elastic response of viscous liquids. From this point of view, the Cattaneo-type conductor (in the classical wave theory of heat conduction) appears as a special case of the Jeffreys-type conductor.

Tzou, D. Y., 1992, "Thermal Shock Phenomena Under High-Rate Response in Solids," in *Annual Review of Heat Transfer*, Edited by Chang-Lin Tien, Hemisphere Publishing Inc., Washington, D.C., Chapter 3, pp. 111-185.

This work also collected a large volume of literature on thermal wave theory. The major emphasis, however, is on the thermal shock phenomenon in a multi-dimensional body and its potential effects on the material failure in the short-time transient. Thermally induced mechanical deformation in the vicinity of thermo-mechanical wavefronts and applications of damage mechanics to the prediction of crack initiations/extensions and spreading of plastic zones around discontinuities are special features that distinguish this work from those assuming thermal wave propagation in rigid conductors.

Tien, C. L. and Chen, G., 1994, "Challenges in Microscale Conductive and Radiative Heat Transfer," *ASME Journal of Heat Transfer*, **Vol. 116, pp. 799-807.**

Tien et al. (1968) performed the pioneering research on the effect of film thickness on the thermal conductivity for metals at cryogenic temperatures. This article provides a detailed discussion of the theoretical bases for microscale heat conduction and radiation, including the size effect and characterization of microscale heat transport regimes in terms of the mean free path of energy carriers. The most valuable assets of this work include pinpointing the obstacles confronted in current research and recommendation of specific directions of future research (1994 and before). It also emphasizes applications of microscale heat transfer in cryogenic systems, material processing, and designs of microelectronic/optical/optoelectronic devices.

Özisik, M. N. and Tzou, D. Y., 1994, "On the Wave Theory in Heat Conduction," *ASME Journal of Heat Transfer*, **Vol. 116, pp. 526-535.**

In line with the review articles by Joseph and Preziosi (1989, 1990), this work categorizes research papers on the wave theory of heat conduction according to their individual emphases. It does not provide brief discussions for cited papers like those given in Joseph and Preziosis' works, but dividing the literature into eight representative categories assists readers in identifying the specific areas of thermal waves that they would be interested in. In addition, this work collects more updated advancement of the thermal wave model in the later stage, including thermal shock formation, the thermal resonance phenomenon, and reflection/refraction/transmission of thermal waves across material interfaces in layered media or free surfaces in an engineering system.

Duncan, A. B. and Peterson, G. P., 1994, "Review of Microscale Heat Transfer," *ASME Journal of Applied Mechanics Review*, **Vol. 47, pp. 397-428.**

This work provides a concise overview of the advances in the field of microscale heat transfer. It includes (1) the microscale effect in heat conduction, including the size effect on the effective thermal conductivity and the characteristics of laser heating in thin-flm structures, (2) the microscale mechanisms in forced convection, including development of micro heat pipes, and (3) the effect of small length scales on radiative heat transfer.

B.2 MICROSCALE HEAT TRANSFER — INDIVIDUAL PAPERS

Abramowitz, M. and Stegun, I. A., 1964, *Handbook of Mathematical Functions*, Dover Publications, New York.

Achenbach, J. D., 1968, "The Influence of Heat Conduction on Propagation of Stress Jumps," *Journal of Mechanics and the Physics of Solids*, Vol. 16, pp. 273-282.

Achenbach, J. D. and Bazant, Z. P., 1975, "Elastodynamic Near-Tip Stress and Displacement Fields for Rapidly Propagating Cracks in Orthotropic Materials," *ASME Journal of Applied Mechanics*, Vol. 42, pp. 183-189.

Alexander, S., Laermans, C., Orbach, R., and Rosenberg, H. M., 1983, "Fracton Interpretation of Vibrational Properties of Cross-Linked Polymers, Glasses, and Irradiated Quartz," *Physical Review (B)*, Vol. 28, pp. 4615-4619.

Alexander, S. and Orbach, R., 1982, "Density of States on Fractals: Fractons," *Journal of Physique Letters*, Vol. 43, pp. L625-L631.

Anderson, R. J., 1990, "The Thermal Conductivity of Rare-Earth-Transition-Metal Films as Determined by the Wiedemann-Franz Law," *Journal of Applied Physics*, Vol. 67, pp. 6914-6916.

Anisimov, S. I., Kapeliovich, B. L., and Perel'man, T. L., 1974, "Electron Emission from Metal Surfaces Exposed to Ultra-Short Laser Pulses," *Soviet Physics JETP*, Vol. 39, pp. 375-377.

Arfken, G., 1970, *Mathematical Methods for Physicists*, Academic Press, New York.

Aspnes, D. E., Kinsbron, E., and Bacon, D. D., 1980, "Optical Properties of Au: Sample Effects," *Physical Review B*, Vol. 21, pp. 3290-3299.

Bai, C. and Lavine, A. S., 1995, "On Hyperbolic Heat Conduction and the Second Law of Thermodynamics," *ASME Journal of Heat Transfer*, Vol. 117, pp. 256-263.

Baumeister, K. J. and Hamill, T. D., 1969, "Hyperbolic Heat Conduction Equation - A Solution for the Semi-Infinite Body Problem," *ASME Journal of Heat Transfer*, Vol. 91, pp. 543-548.

Baumeister, K. J. and Hamill, T. D., 1971, "Hyperbolic Heat Conduction Equation - A Solution for the Semi-Infinite Body Problem," *ASME Journal of Heat Transfer*, Vol. 93, pp. 126-128.

Beck, J. V. and Arnold, K. J., 1977, *Parameter Estimate in Engineering and Science*, John Wiley & Sons, New York.

Bernasconi, A., Sleator T., Posselt, D., Kjems, J. K., and Ott, H. R., 1992, "Dynamic Properties of Silica Aerogels as Deduced from Specific Heat and Thermal-Conductivity Measurements," *Physical Review (B)*, Vol. 35, pp. 4067-4073.

Bertman, B. and Sandiford, D. J., 1970, "Second Sound in Solid Helium," *Scientific American*, Vol. 222, pp. 92-101.

Bluhm, J. I., 1969, "Fracture Arrest," in *Fracture V*, Edited by Liebowitz, H., Academic Press, New York, pp. 1-63.

Boley, B. A. and Tolins, I. S., 1962, "Transient Coupled Thermoelastic Boundary Value Problems in Half-Space," *ASME Journal of Applied Mechanics*, Vol. 29, pp. 637-646.

Boley, B. A. and Weiner, J. H., 1960, *Theory of Thermal Stresses*, John Wiley & Sons, New York.

Brorson, S. D., Fujimoto, J. G., and Ippen, E. P., 1987, "Femtosecond Electron Heat-Transport Dynamics in Thin Gold Film," *Physical Review Letters*, Vol. 59, pp. 1962-1965.

Brorson, S. D., Kazeroonian, A., Moodera, J. S., Face, D. W., Cheng, T. K., Ippen, E. P., Dresselhaus, M. S., and Dresselhaus, G., 1990, "Femtosecond Room Temperature Measurement of the Electron-Phonon Coupling Constant in Metallic Superconductors," *Physical Review Letters*, Vol. 64, pp. 2172-2175.

Carslaw, H. C. and Jaeger, J. C., 1959, *Conduction of Heat in Solids*, 2nd edition, Clarendon Press, Oxford.

Casas-Vazquez, J., Jou, D., and Lebon, G. (editors), 1984, *Recent Developments in Nonequilibrium Thermodynamics*, Springer, Berlin.

Cattaneo, C., 1958, "A Form of Heat Conduction Equation Which Eliminates the Paradox of Instantaneous Propagation," *Compte Rendus*, Vol. 247, pp. 431-433.

Chester, M., 1963, "Second Sound in Solids," *Physical Review*, Vol. 131, pp. 2013-2015.

Chiffelle, R. J., 1994, "On the Wave Behavior and Rate Effect of Thermal and Thermomechanical Waves," M. S. Thesis, University of New Mexico, Albuquerque, NM.

Coleman, B. D., 1964, "Thermodynamics of Materials with Memory," *Archive for Rational Mechanics and Analysis*, Vol. 17, pp. 1-46.

Coleman, B. D., Fabrizio, M., and Owen, D. R., 1982, "On the Thermodynamics of Second Sound in Dielectric Crystals," *Archive for Rational Mechanics and Analysis*, Vol. 80, pp. 135-158.

Coleman, B. D., Hrusa, W. J., and Owen, D. R., 1986, "Stability of Equilibrium for a Nonlinear Hyperbolic System Describing Heat Propagation by Second Sound in Solids," *Archive for Rational Mechanics and Analysis*, Vol. 94, pp. 267-289.

de Oliveira, J. E., Page, J. N., and Rosenberg, H. M., 1989, "Heat Transport by Fracton Hopping in Amorphous Materials," *Physical Review Letters*, Vol. 62, pp. 780-783.

Eckert, E. R. G. and Drake, R. M., Jr., 1972, *Analysis of Heat and Mass Transfer*, McGraw-Hill, New York.

Elsayed-Ali, H. E., Norris, T. B., Pessot, M. A., and Mourou, G. A., 1987, "Time-Resolved Observation of Electron-Phonon Relaxation in Copper," *Physical Review Letters*, Vol. 58, pp. 1212-1215.

Elsayed-Ali, H. E., 1991, "Femtosecond Thermoreflectivity and Thermotransmissivity of Polycrystalline and Single-Crystalline Gold Films," *Physical Review B*, Vol. 43, pp. 4488-4491.

Fann, W. S., Storz, R., Tom, H. W. K., and Bokor, J., 1992a, "Direct Measurement of Nonequilibrium Electron-Energy Distributions in Sib-Picosecond Laser-Heated Gold Films," *Physical Review Letters*, Vol. 68, pp. 2834-2837.

Fann, W. S., Storz, R., Tom, H. W. K., and Bokor, J., 1992b, "Electron Thermalization in Gold, " *Physical Review (B)*, Vol. 46, pp. 13592-13595.

Flik, M. I., Choi, B. I., and Goodson, K. E., 1991, "Heat Transfer Regimes in Microstructures," ASME DSC-Vol. 32, pp. 31-47.

Flik, M. I. and Tien, C. L., 1990, "Size Effect on the Thermal Conductivity of High-T_C Thin-Film Superconductors," *ASME Journal of Heat Transfer*, Vol. 112, pp. 872-881.

Flügge, W., 1967, *Viscoelasticity*, Blaisdell Publishing, Boston, Massachusetts.

Fournier, D. and Boccara, A. C., 1989, "Heterogeneous Media and Rough Surfaces: A Fractal Approach for Heat Diffusion Studies," *Physica (A)*, Vol. 157, pp. 587-592.

Frankel, J. I., Vick, B., and Özisik, M. N., 1985, "Flux Formulation of Hyperbolic Heat Conduction," *Journal of Applied Physics*, Vol. 58, pp. 3340-3345.

Frankel, J. I., Vick, B., and Özisik, M. N., 1987, "General Formulation and Analysis of Hyperbolic Heat Conduction in Composite Media," *International Journal of Heat and Mass Transfer*, Vol. 30, pp. 1293-1305.

Freund, L. B., 1990, *Dynamic Fracture Mechanics*, Cambridge University Press, New York.

Fung, Y. C., 1965, *Foundation of Solid Mechanics*, Prentice-Hall, Englewood Cliffs, New Jersey.

Gefen, Y., Aharony, A., and Alexander, S., 1983, "Anomalous Diffusion on Percolating Clusters," *Physical Review Letters*, Vol. 50, pp. 77-80.

Goldman, C. H., Norris, P. M., and Tien, C. L., 1995, "Picosecond Energy Transport by Fractons in Amorphous Materials," 1995 National Heat Transfer Conference, Portland, Oregon.

Goodson, K. E., Flik, M. I., Su, L. T., and Antoniadis, D. A., 1993, "Annealing Temperature Dependence of the Thermal Conductivity of CVD Silicon Dioxide Layers," ASME HTD-Vol. 253, pp. 29-36.

Groeneveld, R. H. M., Sprik, R., Wittebrood, M., and Lagendijk, A., 1990, "Ultrafast Relaxation of Electrons Probed by Surface Plasmons at a Thin Silver Film," in *Ultrafast Phenomena VII*, Edited by Harris, C. B. et al., Springer, Berlin, pp. 368-370.

Guerrisi, M. and Rosei, R., 1975, "Splitting of the Interband Absorption Edge in Au," *Physical Review (B)*, Vol. 12, pp. 557-563.

Guyer, R. A. and Krumhansl, J. A., 1966, "Solution of the Linearized Boltzmann Equation," *Physical Review*, Vol. 148, pp. 766-778.

Havlin, S. and Bunde, A., 1991, "Percolation II," in *Fractals and Disordered Systems*, Edited by Bunde, A. and Havlin, S., Springer-Verlag, New York, pp. 97-150.

Hildebrand, F. B., 1976, *Advanced Calculus for Applications*, Prentice-Hall, Englewood Cliffs, New Jersey.

Jagannathan, A., Orbach, R., and Entin-Wohlman, O., 1989, "Thermal Conductivity of Amorphous Materials Above the Plateau," *Physical Review (B)*, Vol. 39, pp. 13465-13477.

Joshi, A. A. and Majumdar, A., 1993, "Transient Ballistic and Diffusive Phonon Heat Transport in Thin Films," *Journal of Applied Physics*, Vol. 74, pp. 31-39.

Jou, D., Casas-Vázquez, J., and Lebon, G., 1988, "Extended Irreversible Thermodynamics," *Report of the Progress in Physics*, Vol. 51, pp. 1105-1179.

Juhasz, T., Elsayed-Ali, H. E., Hu, X. H., and Bron, W. E., 1992, "Time-Resolved Thermoreflectivity of Thin Gold Films and Its Dependence on the Ambient Temperature," *Physical Review (B)*, Vol. 45, pp. 13819-13822.

Kaganov, M. I., Lifshitz, I. M., and Tanatarov, M. V., 1957, "Relaxation Between Electrons and Crystalline Lattices," *Soviet Physics JETP*, Vol. 4, pp. 173-178.

Kittel, C., 1986, *Introduction to Solid State Physics*, 6th edition, Wiley, New York.

Lambropoulos, J. C., Jolly, M. R., Amsden, C. A., Gilman, S. E., Sinicropi, M. J., Diakomihalis, D., and Jacobs, S. D., 1989, "Thermal Conductivity of Dielectric Thin Films," *Journal of Applied Physics*, Vol. 66, pp. 4230-4242.

Landau, L. D. and Lifshitz, E. M., 1959, *Fluid Mechanics*, Course of Theoretical Physics, Vol. 6, Pergamon Press, London, pp. 507-522.

Lord, H. W. and Shulman, Y., 1967, "A Generalized Dynamical Theory of Thermoelasticity," *Journal of Mechanics and the Physics of Solids*, Vol. 15, pp. 299-309.

Majumdar, A., 1992, "Role of Fractal Geometry in the Study of Thermal Phenomena," in *Annual Review of Heat Transfer*, Vol. IV, Edited by Tien, C. L., Hemisphere, Washington, D.C., pp. 51-110.

Majumdar, A., 1993, "Microscale Heat Conduction in Dielectric Thin Films," *ASME Journal of Heat Transfer*, Vol. 115, pp. 7-16.

Maxwell, J. C., 1867, "On the Dynamic Theory of Gases," *Philosophical Transactions, London*, Vol. 157, pp. 49-88.

Morse, P. M. and Feshbach, H., 1953, *Methods of Theoretical Physics*, Vol. 1, McGraw-Hill, New York.

Nath, P. and Chopra, K. L., 1974, "Thermal Conductivity of Copper Films," *Thin Solid Films*, Vol. 20, pp. 53-63.

Orlande, H. R. B., Özisik, M. N., and Tzou, D. Y., 1996, "Inverse Analysis for Estimating the Electron-Phonon Coupling Factor in Thin Metal Films," *Journal of Applied Physics*, accepted for publication.

Özisik, M. N., 1993, *Heat Conduction*, 2nd edition, John Wiley & Sons, New York.

Peshkov, V., 1944, "Second Sound in Helium II," *Journal of Physics, USSR*, Vol. VIII, 381 pp.

Popov, E. B., 1967, "Dynamic Coupled Problem of Thermoelasticity for a Half-Space Taking into Account the Finiteness of the Heat Propagation Velocity," *Journal of Applied Mathematics and Mechanics (PMM)*, Vol. 31, pp. 349-356.

Qiu, T. Q., Juhasz, T., Suarez, C., Bron, W. E., and Tien, C. L., 1994, "Femtosecond Laser Heating of Multi-Layered Metals - II. Experiments," *International Journal of Heat and Mass Transfer*, Vol. 37, pp. 2799-2808.

Qiu, T. Q. and Tien, C. L., 1992, "Short-Pulse Laser Heating on Metals," *International Journal of Heat and Mass Transfer*, Vol. 35, pp. 719-726.

Qiu, T. Q. and Tien, C. L., 1993, "Heat Transfer Mechanisms During Short-Pulse Laser Heating of Metals," *ASME Journal of Heat Transfer*, Vol. 115, pp. 835-841.

Qiu, T. Q. and Tien, C. L., 1994, "Femtosecond Laser Heating of Multi-Layered Metals - I. Analysis," *International Journal of Heat and Mass Transfer*, Vol. 37, pp. 2789-2797.

Rosei, R. and Lynch, D. W., 1972, "Thermomodulation Spectra of Al, Au, and Cu," *Physical Review (B)*, Vol. 5, pp. 3883-3893.

Rosei, R., 1974, "Temperature Modulation of the Optical Transitions Involving the Fermi Surface in Ag: Theory," *Physical Review (B)*, Vol. 10, pp. 474-483.

Savvides, N. and Goldsmid, H. J., 1972, "Boundary Scattering of Phonons in Silicon Crystals at Room Temperature, " *Physics Letters*, Vol. 41A, pp. 193-194.

Schlichting, H., 1960, *Boundary-Layer Theory*, McGraw-Hill, New York.

Sih, G. C., 1973, *Mechanics of Fracture 1: Methods of Analysis and Solutions of Crack Problems*, Noordhoff, Leyden, The Netherlands.

Sih, G. C. and Tzou, D. Y., 1985, "Discussions on Criteria for Brittle Fracture in Biaxial Tension," *Journal of Engineering Fracture Mechanics*, Vol. 21, pp. 977-981.

Sih, G. C. and Tzou, D. Y., 1986, "Heating Preceded by Cooling Ahead of a Crack - Macrodamage Free Zone," *Journal of Theoretical and Applied Fracture Mechanics*, Vol. 6, pp. 103-111.

Sih, G. C. and Tzou, D. Y., 1987, "Irreversibility and Damage of SAFC-40R Steel Specimen in Uniaxial Tension," *Journal of Theoretical and Applied Fracture Mechanics*, Vol. 7, pp. 23-30.

Sih, G. C., Tzou, D. Y., and Michopoulos, J. G., 1987, "Secondary Temperature Fluctuation in Cracked 1020 Steel Specimen Loaded Monotonically," *Journal of Theoretical and Applied Fracture Mechanics*, Vol. 7, pp. 79-87.

Taitel, Y., 1972, "On the Parabolic, Hyperbolic and Discrete Formulations of the Heat Conduction Equation," *International Journal of Heat and Mass Transfer*, Vol. 15, pp. 369-371.

Tavernier, J., 1962, "Sur l'equation de conduction de la chaleur," *Comptes Rendus*, Vol. 254, pp. 69-71.

Tien, C. L., Armaly, B. F., and Jagannathan, P. S., 1968, "Thermal Conductivity of Thin Metallic Films," Proceedings of the 8th Conference on Thermal Conductivity, October 7-10.

Tzou, D. Y., 1985, "Intensification of Externally Applied Magnetic Field Around a Crack in Layered Composite," *Journal of Theoretical and Applied Fracture Mechanics*, Vol. 4, pp. 191-199.

Tzou, D. Y., 1987, "Kinetics of Crack Growth: Thermomechanical Interaction," Ph. D. Dissertation, Lehigh University, Bethlehem, Pennsylvania.

Tzou, D. Y., 1988, "Stochastic Analysis of Temperature Distribution in the Solid with Random Thermal Conductivity," *ASME Journal of Heat Transfer*, Vol. 110, pp. 23-29.

Tzou, D. Y., 1989a, "On the Thermal Shock Wave Induced by a Moving Heat Source," *ASME Journal of Heat Transfer*, Vol. 111, pp. 232-238.

Tzou, D. Y., 1989b, "Shock Wave Formation Around a Moving Heat Source in a Solid with Finite Speed of Heat Propagation," *International Journal of Heat and Mass Transfer*, Vol. 32, pp. 1979-1987.

Tzou, D. Y., 1989c, "Thermoelastic Fracture Induced by the Thermal Shock Waves Around a Moving Heat Source," ASME HTD-Vol. 113, pp. 11-17.

Tzou, D. Y., 1989d, "The Effects of Thermal Shock Waves on the Crack Initiation Around a Moving Heat Source," *Journal of Engineering Fracture Mechanics*, Vol. 34, pp. 1109-1118.

Tzou, D. Y., 1989e, "Stochastic Modeling for Contact Problems in Heat Conduction," *International Journal of Heat and Mass Transfer*, Vol. 32, pp. 913-922.

Tzou, D. Y., 1990a, "Thermal Shock Waves Induced by a Moving Crack," *ASME Journal of Heat Transfer*, Vol. 112, pp. 21-27.

Tzou, D. Y., 1990b, "Thermal Shock Waves Induced by a Moving Crack - A Heat Flux Formulation," *International Journal of Heat and Mass Transfer*, Vol. 33, pp. 877-885.

Tzou, D. Y., 1990c, "Three-Dimensional Structures of the Thermal Shock Waves Around a Rapidly Moving Heat Source," *International Journal of Engineering Science*, Vol. 28, pp. 1003-1017.

Tzou, D. Y., 1991a, "Thermal Shock Formation in a Three Dimensional Solid Due to a Rapidly Moving Heat Source, *ASME Journal of Heat Transfer*, Vol. 113, pp. 242-244.

Tzou, D. Y., 1991b, "The Resonance Phenomenon in Thermal Waves," *International Journal of Engineering Science*, Vol. 29, pp. 1167-1177.

Tzou, D. Y., 1991c, "Resonance of Thermal Waves Under Frequency Excitations," ASME HTD-Vol. 173, pp. 11-27.

Tzou, D. Y., 1991d, "The Singular Behavior of the Temperature Gradient in the Vicinity of a Macrocrack Tip," *International Journal of Heat and Mass Transfer*, Vol. 33, pp. 2625-2630.

Tzou, D. Y., 1991e, "The Effect of Thermal Conductivity on the Singular Behavior of the Near-Tip Temperature Gradient," *ASME Journal of Heat Transfer*, Vol. 113, pp. 806-813.

Tzou, D. Y., 1991f, "On the Use of Node-Shifting Techniques for the Intensity Factor of Temperature Gradient at a Macrocrack Tip," *Numerical Heat Transfer*, Vol. 19(A), pp. 237-253.

Tzou, D. Y., 1991g, "The Effect of Internal Heat Transfer in Cavities on the Overall Thermal Conductivity," *International Journal of Heat and Mass Transfer*, Vol. 34, pp. 1839-1846.

Tzou, D. Y., 1991h, "A Universal Model for the Overall Thermal Conductivity of Porous Media," *Journal of Composite Materials*, Vol. 25, pp. 1064-1084.

Tzou, D. Y., 1992a, "Thermal Shock Phenomena Under High-Rate Response in Solids," in *Annual Review of Heat Transfer*, Edited by Chang-Lin Tien, Hemisphere Publishing Inc., Washington, D.C., Chapter 3, pp. 111-185.

Tzou, D. Y., 1992b, "Fracture Path Emanating from a Rapidly Moving Heat Source - The Effects of Thermal Shock Waves Under High Rate Response, *Journal of Engineering Fracture Mechanics*, Vol. 41, pp. 111-125.

Tzou, D. Y., 1992c, "Experimental Evidence for the Temperature Waves Around a Rapidly Propagating Crack Tip," *ASME Journal of Heat Transfer*, Vol. 114, pp. 1042-1045.

Tzou, D. Y., 1992d, "Thermal Resonance Under Frequency Excitations," *ASME Journal of Heat Transfer*, Vol. 114, pp. 310-316.

Tzou, D. Y., 1992e, "Damping and Resonance Phenomena of Thermal Waves," *ASME Journal of Applied Mechanics*, Vol. 59, pp. 862-867.

Tzou, D. Y., 1992f, "The Thermal Shock Phenomena Around a Rapidly Propagating Crack Tip: Experimental Evidence, *International Journal of Heat and Mass Transfer*, Vol. 35, pp. 2347-2356.

Tzou, D. Y., 1992g, "The Transonic Wave Solution in the Vicinity of a Rapidly Propagating Crack Tip in 4340 Steel, *International Journal of Engineering Science*, Vol. 30, pp. 757-769.

Tzou, D. Y., 1992h, "Characteristics of Thermal and Flow Behavior in the Vicinity of Discontinuities," *International Journal of Heat and Mass Transfer*, Vol. 35, pp. 481-491.

Tzou, D. Y., 1993a, "An Engineering Assessment to the Relaxation Time in Thermal Waves," *International Journal of Heat and Mass Transfer*, Vol. 36, pp. 1845-1851.

Tzou, D. Y., 1993b, "Reflection and Refraction of Thermal Waves from a Surface or an Interface Between Dissimilar Materials," *International Journal of Heat and Mass Transfer*, Vol. 36, pp. 401-410.

Tzou, D. Y., 1994, "Deformation Induced Degradation of Thermal Conductivity in Cracked Solids," *Journal of Composite Materials*, Vol. 28, pp. 886-901.

Tzou, D. Y., 1995a, "A Unified Field Approach for Heat Conduction from Micro- to Macro-Scales," *ASME Journal of Heat Transfer*, Vol. 117, pp. 8-16.

Tzou, D. Y., 1995b, "The Generalized Lagging Response in Small-Scale and High-Rate Heating," *International Journal of Heat and Mass Transfer*, Vol. 38, pp. 3231-3240.

Tzou, D. Y., 1995c, "Experimental Support for the Lagging Response in Heat Propagation," *AIAA Journal of Thermophysics and Heat Transfer*, Vol. 9, 686-693.

Tzou, D. Y., 1995d, "Anisotropic Overall Thermal Conductivity in Porous Materials due to Preferentially Oriented Pores," *International Journal of Heat and Mass Transfer*, Vol. 38, pp. 23-30.

Tzou, D. Y. and Chen, E. P., 1990, "Overall Degradation of Conductive Solids with Mesocracks," *International Journal of Heat and Mass Transfer*, Vol. 33, pp. 2173-2182.

Tzou, D. Y. and Li, J., 1993a, "Thermal Waves Emanating from a Fast-Moving Heat Source with a Finite Dimension," *ASME Journal of Heat Transfer*, Vol. 115, pp. 526-532.

Tzou, D. Y. and Li, J., 1993b, "Local Heating Induced by a Nonhomogeneously Distributed Heat Source," *International Journal of Heat and Mass Transfer*, Vol. 36, pp. 3487-3496.

Tzou, D. Y. and Li, J., 1994a, "The Overall Thermal Conductivity in a Straining Body with Microcrack Evolution," *International Journal of Heat and Mass Transfer*, Vol. 36, pp. 3887-3895.

Tzou, D. Y. and Li, J., 1994b, "Thermal Wave Behavior in High-Speed Penetration," *International Journal of Engineering Science*, Vol. 32, pp. 1195-1205.

Tzou, D. Y. and Li, J., 1995, "Some Scaling Rules for the Overall Thermal Conductivity in Porous Materials," *Journal of Composite Materials*, Vol. 29, pp. 634-652.

Tzou, D. Y., Özisik, M. N., and Chiffelle, R. J., 1994, "The Lattice Temperature in the Microscopic Two-Step Model," *ASME Journal of Heat Transfer*, Vol. 116, pp. 1034-1038.

Tzou, D. Y. and Sih, G. C., 1985, "Crack Growth Prediction of Subsurface Crack in Yielded Material," *ASME Journal of Applied Mechanics*, Vol. 52, pp. 237-240.

Tzou, D. Y. and Zhang, Y. S., 1995, "An Analytical Study on the Fast-Transient Process in Small Scales, " *International Journal of Engineering Science*, Vol. 33, pp. 1449-1463.

Vadavarz, A., Kumar, S., and Moallemi, M. K., 1991, "Significance on Non-Fourier Heat Waves in Microscale Conduction," ASME DSC-Vol. 32, pp. 109-121.

Vernotte, P., 1958, "Les Paradoxes de la Théorie Continue de l'équation De La Chaleur," *Compte Rendus*, Vol. 246, pp. 3154-3155.

Vernotte, P., 1961, "Some Possible Complications in the Phenomena of Thermal Conduction," *Compte Rendus*, Vol. 252, pp. 2190-2191.

Volklein, F. and Kessler, E., 1986, "Analysis of the Lattice Thermal Conductivity of Thin Films by Means of a Modified Mayadas-Shatzkes Model, The Case of Bismuth Films," *Thin Solid Films*, Vol. 142, pp. 169-181.

Williams, M. L., 1952, "Stress Singularities Resulting From Various Boundary Conditions in Angular Corners in Extension," *ASME Journal of Applied Mechanics*, Vol. 19, pp. 526-536.

Zallen, R., 1983, *The Physics of Amorphous Materials*, Wiley, New York.

Index

Alternating behavior, 58, 271, 284
Amorphous materials
    anomalous diffusion, 171
    experiment, 168
    heated mass concept, 171
    lagging behavior, 174
    rate of heat transport, 169, 172,
       173, 178, 179
Amplification, 185, 215, 222, 285
    (*see also* Localization)
Anomalous diffusion, 167, 168, 171-
    173, 178, 179, 181, 182
Bromwich contour integration, 39-41,
    43, 45, 167, 250
Characteristic times, 3, 27, 30, 57, 64
    defects, 219, 226
    metal films, 120, 122, 133 (*see*
      also phase lags)
Conductor, rigid, 231, 232, 237, 243,
    255, 267, 269
    deformable, 243, 244, 271, 294-
      298, 300-303, 305-308, 310,
      315
Constitutive behavior (*see*
    Constitutive equation)
Constitutive equation, 65, 71, 75, 76,
    78, 79, 82, 84, 85, 192, 241,
    242, 274, 286, 295, 297, 310,
    311
Correlation length, 167, 171-174
Cracks, moving,
    material coordinates, 191, 192,
    194, 208
    subsonic, 197-198, 200, 201, 203,
      205, 207
    supersonic, 201-203, 205-208,
      212, 213, 215
    trajectory, 191, 209
    transonic, 205-207
Curvature, geometric, 183, 186, 199,
    215
CV wave (*see* Thermal waves)
Defects
    accumulation, thermal energy,
      183, 185, 189, 199, 200, 201,
      207, 208, 213, 215, 230
    asymptotic behavior, 188, 194-
      196, 200, 208 (*see also*
      Cracks)
    cracks, 183, 186, 187, 189,
      190-201, 203-210, 212, 213,
      215
    microvoids, 215-217, 221, 222,
      226, 228-230
    near-tip behavior, 187-192, 194-
      196, 198-201, 203, 206-209,
      212-215 (*see also* asymptotic
      behavior)
    shock formation, around, 192,
      201, 202, 205-208
    voids, 183, 189
Delayed response, 8, 12, 20, 21, 22,
    24, 26, 32, 45, 55
    amorphous materials, 168, 174